Tributes
Volume 52

Waves of Trust:
Science, Technology and Society
Essays in Honor of Rino Falcone

Tributes Series Editor
Dov Gabbay

dov.gabbay@kcl.ac.uk

Waves of Trust:
Science, Technology and Society
Essays in Honor of Rino Falcone

Edited by

Alessandro Sapienza

Filippo Cantucci

Fabio Paglieri

Luca Tummolini

ISBN 978-1-84890-475-0

College Publications
Scientific Director: Dov Gabbay
Managing Director: Jane Spurr

http://www.collegepublications.co.uk

Cover design by Laraine Welch

CONTENTS

I   Scientific contribution

Andrea Omicini and Alessandro Ricci

II   Socio-political contribution

Francesco Profumo

III   Afterwords

Rino, Una Vita di Impegno e Dedizione
Rino, a Life of Commitment and Dedication

Caro Rino, uomo e scienziato di grande talento,
Il tempo corre, in giro c'è aria di... cambiamento!

Dear Rino, a man and scientist so bright,
Time flies, bringing winds of change in sight!

Non ti auguriamo certo di riposarti,
Perché sappiamo bene che nella tua missione continuerai ad impegnarti.

We won't wish you rest, not even a bit,
For we know in your mission, you'll never quit.

Sei per noi fonte di ispirazione,

Con il tuo esempio e la tua dedizione.

You're our inspiration, a true motivation,

With your example and your dedication.

Per celebrare questa tua carriera brillante,
Ti dedichiamo un libro, omaggio toccante.

To celebrate your brilliant career's might,
We've made you this book, a touching delight.

Gli amici tutti, con grande partecipazione,
Han contribuito per questa bella occasione.

Friends all around, with hearts so true,
Have gathered together just for you.

Racconti, aneddoti, e storie a profusione,
Ci hanno fatto sudare, ma con passione.

Pieno di amici che ti voglion bene, con grande affezione
E colleghi che ti stimano con ammirazione.

Rino caro, questo è solo un piccolo segno,
Di un affetto grande, un ringraziamento degno.

Stories, tales, and memories so dear,
Brought us to toil, yet filled with passion sincere.

Full of friends who hold you near,
And colleagues who value you with warmth sincere.

Dear Rino, this is just a small token,
Of great affection, a thank you unspoken.

PREFACE

After months spent chasing down authors and hiding from Rino, we have finally managed to complete what could be described as a "book operation". In fact, with the exception of the final stages, this volume was created in total secrecy while the unsuspecting recipient was busy with the arduous management of our Institute. Now, anyone familiar with his personality can imagine how challenging it was to continue working side by side with Rino while concealing such an initiative.

Consider also that the situation was exponentially complicated by the involvement of a significant number of our closest colleagues, with whom we periodically met for updates on the project. Many times, we were caught red-handed organizing last-minute meetings in the Institute hallways or in some of our colleagues' offices. To his classic question "Aho, che state a fa?" (roughly translatable as "Ehi, what the heck are you doing?"), we came up with the most creative excuses: "We're developing a new cognitive model for the coffee machine"; "It's a preliminary test on how to improve the synergy between lunch breaks and brainstorming"; "We've found an innovative approach to making researchers' interactions more trustworthy". Although we didn't investigate further, we are reasonably confident that there is a state-of-the-art reference for each of these valid justifications.

Over time, our dissimulation techniques became so refined that we developed an elaborate predictive model of Rino's movements (we're considering submitting the results to *Nature*), meticulously planning the moments to devote to the book. In short, now that the endeavor is complete, we believe we officially deserve the title of *masters of clutching at straws*, a sport even more thrilling than parkour.

And when it comes to sports, Rino is undoubtedly a great expert. Readers will notice, in several contributions, numerous references to football, both to Roma, his favorite team, and to epic football matches that have gone down in history—perhaps not in the history of football but certainly in that of the CNR. The choice of the title, *Waves of Trust*, was also inspired by another of Rino's sporting passions: swimming. "Waves" refers to one of his true feats: crossing the Strait of Messina (as evidenced by the image on the back cover of this book). It was a dream he had worked on for years but had to postpone due to the pandemic. On August 28, 2023, in 1 hour, 8 minutes, and 8 seconds, Rino swam nearly 4 kilometers of open sea, braving strong currents and stormy winds. An ambitious undertaking, the result of meticulous planning and immense dedication, it perfectly embodies his spirit: challenging himself and the sea, with a healthy dose of courage and determination.

Yet *Waves of Trust* is not prominently focused on the athletic side of Rino. Like waves, trust propagates dynamically in many different directions and affects many different domains, from science to technology to society. The title reflects not only Rino's brilliant achievements in the understanding of this challenging and crucial topic, but also his deep engagement with political and social issues and the institutional roles he has held over time, which have significantly impacted both the academic community and society.

In light of Rino's dual identity, we decided to structure the book into two separate sections: one devoted to scientific results, the other focused on political and societal engagement. True to Rino's legacy, this distinction is only meant as a coarse-grained organizational criterion, not as a well-defined divide: in reality, science, politics, and society are deeply intertwined, a fact that will often be apparent in the contributions included in this volume.

The first section, *"Scientific contribution"*, highlights the broad value of Rino's scientific and intellectual activity, offering insights for future developments in cognitive science, encompassing theory, practice, and technology. The topics addressed span from the concept of trust in its various aspects and application fields to social norms, autonomy in artificial systems, human-machine collaboration, artificial intelligence, and so on. Each scientific contribution in this volume provides a unique perspective on key topics, connected by a common thread that weaves through theory, simulation, and practical application.

Trust, norms, and values

The initial contributions focus on the relationship between trust, norms, and values. Munindar P. Singh, in *Social Artificial Intelligence and the Interplay of Norms, Values, and Trustworthiness*, explores how norms and values shape trust in agent interactions, highlighting scenarios where even norm violations can enhance trust. This discussion connects to the work of Mario Paolucci and Giulia Andrighetto, *Trust, Reputation and Norms: A Simulative Approach to Cognitive Artifacts*, which adopts a simulation-based approach to analyze how trust and reputation emerge in complex systems, integrating social norms as guides for agent behavior.

Delegation, coordination, and cooperation

The theme of delegation emerges in Anna Borghi and Luca Tummolini's *Semantic Delegation: Understanding Concepts With A Little Help From Our Friends*, where the authors explore the role of semantic delegation in concept representation, emphasizing how trust in experts shapes our understanding. Similarly, Francesca Camilli, Tarek el Sehity, and Raffaella Pocobello, in *Trust the Process: Exploring Trust as a Foundational Element in Open Dialogue*, examine how trust can facilitate open dialogic processes in collaborative contexts. On the theme of cooperation, particularly between humans and artificial agents, Amedeo Cesta, Gabriella Cortellessa, and Angelo Oddi, in *Adjustable Autonomy in AI and Robotics: Lessons Learned and Experiences*, analyze how trust influences the autonomy of robotic agents, emphasizing the importance of dynamically adapting decision-making capabilities to user needs and context. In *Coordination and Trust in MAS Towards Intelligent Socio-technical Systems*, Andrea Omicini and Alessandro Ricci focus on trust in multi-agent systems, exploring the role of trustworthy coordination in the modeling and engineering of such systems.

Trust, explainability, and transparency

Valeria Seidita and Antonio Chella, in *From Trust Theory to Explain-*

able Agents: Enhancing Agent-Based Systems, expand the discussion by proposing trust-based agent systems capable of explaining their behavior to enhance comprehensibility and trustworthiness. Aron Szekely and Luca Tummolini, in *When Transparency Shapes Trust: A Signalling Approach*, challenge the intuitive link between transparency and trust as a policy intervention.

Social implications of artificial intelligence

Pierluigi Zoccolotti and Fabio Lucidi, in *How Will Artificial Intelligence Affect Our Lives?*, reflect on the social impact of AI, with a particular focus on work and health. Fabio Paglieri, in *Trust, Delegation, and Cognitive Diminishment: The Case of Generative AI*, explores the topic of trust in generative AI, discussing how cognitive delegation can reduce users' critical awareness. Vieri Giuliano Santucci and Giovanni Pezzulo, with *Harmonizing Artificial Intelligence, Minds, and Society: A Cognitive Science Perspective*, adopt a cognitive perspective to integrate AI and society, emphasizing trust as a bridge between technology and human needs. Robin Cohen, in *From Trust Modeling to Trusted AI: Enlightened by the work of Falcone*, celebrates Rino's work, offering an integrated perspective on trust theory and its applications in AI.

Epistemic logics and trust in information

Stefania Costantini and Andrea Formisano, in *Epistemic Logics Programs: A Survey of Motivations, Recent Contributions and New Developments*, provide an overview of developments in epistemic logics, highlighting their importance in modeling trust in information. Robert Demolombe, in *Reasoning About Trust in Information Transmitted by a Sequence of Agents*, discusses trust as a mediator in the transmission of information between agents, showing the reasoning process that the receiving agent can do to conclude on the information he has received in different types of situations. Moreover, Jaime Simão Sichman, in *Reputation Interoperability in Multi-Agent Systems*, introduces and investigates the concept of reputation interoperability, which involves agents and systems interpreting and utilizing reputation evaluations from heterogeneous sources.

Cognitive modeling of trust

The work *Should I Trust? On the Epistemic Foundation of Arguments* by Emiliano Lorini analyzes the concept of trust through an epistemic formalization of argumentation, proposing a semantics for epistemic logic based on belief bases to represent and justify an agent's beliefs and to develop a logical theory of trust support. Elisa Colì, in her *Trusting in the Online World: From Theoretical Studies to the Application of the Socio-Cognitive Model*, analyzes the important role trust plays in the digital environment, reviewing key studies on the topic and different perspectives, while emphasizing the potential of the socio-cognitive theoretical model of trust in understanding the determinants of online trust. Filippo Cantucci and Alessandro Sapienza, in *Trust and Cognition in Human and Hybrid Societies*, explore, through a series of significant contributions developed alongside Rino, the role of trust in interactions between humans and artificial agents. This theme connects to that addressed by Antonio Lieto in *Trust in the Rational and Social Bands of Cognitive Architectures*, which introduces a dual model where trust operates at both rational and social levels within cognitive architectures. This section concludes with Cristiano Castelfranchi, Rino's longtime mentor, whose chapter *T3: A Multidimensional, Integrated and Dynamic Theory of Trust* addresses various aspects of the complexity, multidimensionality, and dynamics of trust, reaffirming the fundamental role of theory and models in defining new technologies. The overall objective of the chapter is to provide a vision of trust that is a tribute to the past but also a starting point for future developments, integrating theory and practice in human and technological contexts.

The second part of the book, titled *Socio-political contribution*, collects testimonials from prominent institutional figures, both in the research community and in Italian politics. Personal reflections, critical evaluations of his work in research policy, and analyses of the relationship between science and politics help clarify Rino's role not only as a scientist but also as a key figure for institutions and society.

Personal tributes and testimonials

Lettera a Rino Falcone in occasione del suo pensionamento by Lucio Bianco is an open and intimate letter reflecting on the personal relationship between Rino and the former CNR president, a bond founded on mutual respect and recognition of both human and scientific value. Similarly, *Omaggio a Rino Falcone* by Gaetano Manfredi celebrates Rino's professional and personal roles as a promoter of reflections on research dynamics and the need to raise societal awareness about the value of science. Francesco Lenci, in *Il mio cammino con Rino*, and Settimo Termini, in *A commitment on all fronts*, recount their relationships with Rino, highlighting his central role in promoting scientific research in Italy and defending its autonomy. The authors recall his prominent role concerning the "Osservatorio sulla Ricerca", the "Manifesto Per un'Europa di Progresso", his work as an advisor to Minister Mussi, and the audience with President Giorgio Napolitano. Corrado Bonifazi, in *Life as Directors*, recounts his experience as director of the IRPPS and his collaborative and mutually respectful relationship with Rino, then director of the ISTC. Maurizio Franzini, in *Nel dedalo della fiducia, seguendo Rino Falcone*, analyzes Rino's approach to the concept of trust, emphasizing how his work has provided theoretical and empirical tools to understand a complex and multidimensional phenomenon. Oliviero Stock, with *Rino Falcone in quattro dimensioni*, offers a personal and professional reflection exploring Rino's distinctive qualities across four "dimensions" of his life and career: human, scientific, organizational, and research politics. *Dedica a Rino Falcone* by Nando Dalla Chiesa highlights the depth of relationships built over the years, describing the professional and personal connection between the author and Rino, in particular their collaboration at the Ministry of Universities and Research during the second Prodi government. Finally, *Io e Rino* by Marco Pagani reflects on significant moments of their relationship, both within the CNR and in private life. The contribution also considers the organizational and political challenges of the CNR, emphasizing Rino's role as a leader and visionary in this context.

Science and society

The role of science as a foundation for democracy and social participation is a theme to which Rino dedicated considerable effort throughout his career. Lucia Votano, in *Scienza, società e democrazia*, explores the critical role of science as a tool for societal progress, advocating for a balance between scientific knowledge, social values, and democracy. Salvatore Capasso's chapter, *Scienze sociali e ICT: innovazione al servizio dell'umanità*, examines the role of humanities and social sciences in the development and implementation of new technologies based on digital innovation, highlighting the importance of trust in this context. Paolo De Nardis, in *Dalla ricerca scientifica alla politica della ricerca: la struttura di un impegno pubblico*, celebrates Rino Falcone as a figure embodying the intersection of science and public engagement, uniting scientific dedication with a fight for a more just society. Luigi Nicolais, with *The integration of humanistic and scientific cultures: a critical approach to AI development*, explores the need to overcome the dichotomy between humanistic and scientific disciplines, emphasizing their joint necessity for human development and technological innovation. Daniele Archibugi, in *Rino Falcone, champion of the academic community*, celebrates Rino as a central figure in defending the Italian academic community. Rino is appreciated for his ability to unite the scientific community and his contributions to positioning research as a pillar of social and economic progress.

The governance of science

Maria Chiara Carrozza, in *Un percorso nel Consiglio Nazionale delle Ricerche a servizio della Scienza e del Paese*, recognizes the value of Rino's respect for the scientific approach. Rino has always regarded science as serving the collective good, with particular attention to research autonomy and the responsibility of scientists toward society. *Consigliere Rino Falcone e il ruolo chiave del Ministero dell'Università e della Ricerca (2006-2008)* by Francesco Profumo reflects on Rino's experience as advisor to the Ministry of Universities and Research from 2006 to 2008, within the context of the Lisbon Strategy and the 7th Framework Programme of the EU. Profumo recalls Rino's contributions to improving national policies for research

9

and higher education to enhance Italy's competitiveness and position in Europe. *Missione Ricerca, per lo sviluppo del Paese*, by Daniela Palma, examines the state of scientific research in Italy, highlighting structural problems and proposing strategies to revitalize the research system.

Visions for the future of science

Pierpaolo Campostrini, in *The courage to choose science-based solutions*, discusses the importance of science, recalling how crucial Rino's intervention was in the realization of the MOSE project in Venice. Alberto Silvani, with *Uno zibaldone di pensieri, parole e (potenziali) azioni*, discusses on various aspects of Rino's work and legacy, offering an open reflection on ideas and proposals for science and innovation. Massimo Cocco, in *Shared experiences for a participatory approach to scientific research*, illustrates the value of public scientific research and the importance of participation and transparency in scientific governance.

In the daunting task of putting together such a rich and diverse collection of high-profile contributions, we were not alone. We had the help of several accomplices, and we are now eager to thank them. First and foremost, we need to acknowledge the key role played by Ivana Della Portella, Rino's wife, who not only helped us identifying some of the authors who contributed to this initiative but was also the mastermind in organizing the surprise event that made Rino aware of the book. Moreover, readers can thank Ivana for the amazing front cover, which depicts Rino contemplating the "robot/computer boy" artwork by Banksy: an invitation to reflect on the themes of digitization and human-machine relationships.

Equally essential was Daniele Archibugi's contribution, for which he deserves credit as "shadow co-editor" of this collection: Daniele, like a true secret agent of the scientific community (as in a Sidney Pollack spy story), was in fact our special infiltrator and helped us contact some of the important authors of the book.

Additionally, thanks go to Francesco Stella, who joined our research group in the last year. Our thanks to him for assisting in the formatting of the various chapters, a task that required precision and attention. Skills that Francesco certainly possesses.

While working on this book, one thing impressed us profoundly. Since the onset we were quite confident of the great participation from colleagues and friends, but what surprised us was the enthusiasm with which each of them decided to contribute. It seemed that participating in this tribute was not just a gesture of respect for a dear friend or colleague, but also a precious opportunity to express the affection and gratitude we all feel toward him. At times, it was as if the invited contributors were thanking us for giving them the chance to be part of this collection. In many cases, we interacted with long-time colleagues, friends of Rino whom we did not know directly, and we were moved by the sincere affection that transpired in every exchange, every email, every conversation. It was not just about offering a scientific contribution in his honor, but about bringing to light the special and personal bond that each of them had built with Rino over the years. The involvement and affection that the authors put into this project is something that cannot be fully captured by written words. This book is not just an academic recognition: it is a gesture of deep respect and great affection for the extraordinary figure of Rino Falcone, a person whose worth transcends any role or career.

PART

SCIENTIFIC CONTRIBUTION

Social Artificial Intelligence and the Interplay of Norms, Values, and Trustworthiness

Munindar P. Singh
North Carolina State University
mpsingh@ncsu.edu

Abstract

In his distinguished career in artificial intelligence, Rino Falcone has bridged the worlds of psychology and computer science and made groundbreaking contributions to a sociocognitive foundation for trust. This note shares some thoughts on that broad theme. Informally, we would expect there to be strong relationships between norms, values, trust, and trustworthiness. This short note highlights some of the ways in which those relationships arise. Clearly, respecting norms should enhance trust and trustworthiness. Additionally and more interestingly, this note brings out counterintuitive cases where the violations of norms can enhance trust and trustworthiness.

1 Introduction

Rino Falcone is a foremost scholar of trust in the computational era. Right from the inception of our field of autonomous agents and multiagent systems, he has led the direction of research into trust in agents. More importantly, taking advantage of his expertise not only in technology but also in social psychology, he was one of the proponents of the sociocognitive treatment of trust when most of the Artificial

Thanks to the NSF for partial support under grant IIS-2116751.

Intelligence trust field was stuck on conceptually incomplete models of trust.

Research in AI has conventionally focused on single-agent systems and tackles problems such as learning, knowledge representation, and planning. Whereas these are important topics, the social themes where we consider not one but many intelligent agents are far more interesting to many of us in the Multiagent Systems community. Trust between agents is, of course, a concern only when more than one agent is involved. Despite this unavoidable fact, a lot of the AI approaches to trust tackle the multiagent aspects of it somewhat reluctantly with a focus on aspects such as, for example, reliability that do not involve too strong of a social component. But, achieving not just AI but Social Artificial Intelligence requires us to dig deeper into how trust arises as a sociocognitive phenomenon and how it interplays with other sociocognitive phenomena.

So, I think it would be fitting for this brief article in a festschrift volume honoring Rino Falcone to take a closer look at how two constructs originating in social psychology, namely norms and values, relate to trust and trustworthiness. Understanding these concepts together would give us a new way of thinking about Social Artificial Intelligence.

2 Norms

By norms, I mean *social* norms, meaning that they apply to two or more parties. And, I further direct our attention to regulative as opposed to constitutive norms.

Norms, in a broad sense, characterize legitimate expectations team members have of each other. What distinguishes norms from other kinds of expectations is the point about *legitimacy*: a norm establishes the standard by which a party's behavior and interaction are judged.

Here are some example norms. Anyone driving a motor vehicle should drive on the right of the road (in the US and Italy, for instance). Other drivers and pedestrians can thus expect that a vehicle approaching them will keep to its own side. Moreover, this expectation on their part is legitimate—to be contrasted with an expectation that they may have that the incoming vehicle will bring them gold coins.

The legitimacy accorded by norms is more fundamental than the idea that a norm by propped up by sanctions, as some definitions of norms require. Sanctions are indeed a valuable mechanism for the emergence of norms as well as for the enforcement of established norms [1]. However, the idea that sanctions are essential to a norm sounds far-fetched. This is because a sanction is up to the potential sanctioning party, and that party may have other reasons to sanction someone (without any norm having been violated by the latter) or to not sanction someone (when they did violate a norm).

In support of this argument, I would draw the distinction between illocution and perlocution introduced by Austin [2] as part of his speech act theory. An illocution describes the core intuition about a communication (i.e., a speech act), whereas a perlocution describes how the communicating parties may further interact. For example, a receiver of a request may act on it, ignore it, or lash back at the requester. Those are potential perlocutions that ought not to affect the core meaning of the request to begin with, which is what the illocution captures.

A similar situation arises with the notion of belief. An agent ought to form beliefs that are justified by perception or reasoning and consistent, but we surely would not deny that there are people with beliefs that are not adequately justified or with beliefs that are inconsistent with their other beliefs.

Back to norms. Thus, norms may or may be combined with sanctions and other social signals, and usually are. Norms may be formally or informally expressed. Norms may have greater or lesser support from institutions and may or may not carry legal import.

Norms may arise by being explicitly designed or engineered as when an institution legislates new norms (let's call these *de jure*) or by emerging through the social practices of a community and the mutual sanctioning of its members (let's call these *de facto*). But most often in a long-lived society, norms would arise through a combination or even alternation of the engineered and emergent paradigms, blending the de jure and the de facto elements.

Norms are prevalent because they serve important societal functions. Norms can guide the decision-making of an agent. For example, an agent may decide to comply with a norm by choosing an action that complies with the applicable norms.

Norms, through their purported legitimacy, provide a basis for *accountability* [3]. Accountability is not sanctioning, for an argument similar to why sanctioning is not an intrinsic part of the definition of a norm; accountability is not the same as recording who performed what action [4].

3 Values

Values in the sense of social psychology correspond to deeply held self-characterizations of the considerations an agent prioritizes [5, 6]. Values can provide rationales for how an agent feels or behaves. There is some controversy about the motivational power of values: that is, whether values can lead an agent to act or whether values merely lead to post hoc rationalizations of behaviors [7]. Similarly, there is debate about how context sensitive or fixed values might be [8]. These important aspects won't directly concern us here.

The main point about values relevant to the present discussion is that values provide, from the personal perspective of an agent, a more foundational basis for evaluating options than social norms. That is, we can judge the suitability of a member of a community or sociotechnical system of the norms in the community or sociotechnical system based on how well the norms promote behaviors and outcomes aligned with that member's values [9].

4 Trust and Trustworthiness

Rino, in his long record of research on trust, has identified important features that trust possesses from the sociocognitive perspective. These include a belief by the trustor of some dependence on the trustee and the voluntary nature of the trustor placing themself in a position of vulnerability with respect to the trustee [10, 11].

In passing, I do want to take issue with Rino on his recent view on trustworthiness. Rino was one of the early advocates of the relational understanding of trust, e.g., [12]. So I was surprised to find trustworthiness described as "an intrinsic property of the trustee" [11, Section 2.1]. I would argue that trustworthiness is very much relational and reflects how well the trustee goals, interests, and values of the trustor [13]—this work adapts the ability-benevolence-integrity model [14] to understand trustworthiness in AI.

5   Interplay

I want to build on the crucial connection between trust and dependence that Rino and his colleagues [11, 15] have observed. Specifically, the dependence—whether subjective or objective—is based not only with respect to the resources needed by the dependent party (or equivalently, the trustor, in this setting) to achieve its goals but also with respect to its interests being advanced and values being promoted (even if they conflict with its goals).

Or, another way to put it, we can think of norms as capturing social dependence relations—social not only because they involve more than one party but also because they are expressed in high-level terms. This point relates intimately to an architectural view of trust as dependence that I developed some years ago [16], but which I will largely sidestep in this note.

Let's now consider some of the main connections between norms and values, and trust. First, the most obvious connection is that compliance with norms promotes trust and trustworthiness. In the social world where the trustor and trustee live, the trustee must have the resources and capabilities to satisfy the norms (when satisfaction demands additional resources) and the planning ability and moral fortitude to satisfy the norms (when their satisfaction demands a careful allocation of resources and restraint to forestall violations). Suppose the trustee satisfies the applicable norms but does not meet the trustor's expectations regarding resources. That is, the trustee's behavior may not serve to satisfy the trustor's goals. Presumably, the trustee remains trustworthy because of their compliance with the norms. Thus, the trustor could rationally still continue to trust the trustee.

Surprising, perhaps, is the converse connection. The violation of a norm [17] may promote trust and trustworthiness as well. The more natural and uncontroversial setting where this would be so is when the violation leads to satisfying the trustor's goals and advancing their interests.

However, suppose the trustee violates a norm in a way that does not satisfy the trustor's goals or advance their interests. Even in such cases, the trustor's trust and the trustee's trustworthiness may improve when the norm violation is appropriately justified. A norm demands accountability, and its violation thus demands an accounting from the violator. Importantly, the provisioning of that account can lead to the violator being seen as trustworthy. This is because even if the trustee violates the specific norm under consideration, they satisfy the *metanorm* corresponding to providing an account. Thus, the trustee provides evidence of a high moral character as a society member by respecting the metanorm of accountability.

Yet another factor in the previous case is the nature of the explanation offered by the trustor as part of its accounting of its decision. If that explanation includes the values that led to the trustor violating the specified norm, that provides evidence of their trustworthiness with respect to those values. When the trustor shares those values, they may find the trustee meritorious and thus not reduce and possibly improve their trust in the trustee.

We have talked about the influence of norm satisfaction and violation on trust. What about the influence of trust on the norms? The satisfaction of a norm in general should strengthen the norm, although the satisfaction of a norm by an untrustworthy party may diminish it—or be used as an argument to diminish it (e.g., in the style of "Hitler was a vegetarian"—a claim that may not be accurate).

The violation of a norm by an untrustworthy party should not generally weaken the norm because their untrustworthiness would raise questions about their motives rather than the suitability of the norm. The violation of a norm by a trustworthy party should generally weaken the norm because it indicates that the norm must have some limitations, e.g., to the circumstances in which it applies, for a trustworthy party to disregard it. However, this would not be the only outcome. When a norm violation by a trustee is accompanied by a strong explanation, it helps refine and strengthen the norm and enhances the trustworthiness of the trustee. The explanation outlines the specific circumstances in which the norm does not naturally employ, and the presentation of the explanation serves as evidence that the trustee reasoned carefully about the violation before engaging in it. In such cases, the norm is strengthened for cases other than the specified circumstances where it is weakened (and potentially eventually removed).

When, over the course of multiple interactions and norm violations, multiple explanations are shared by different parties in the role of trustees, we have grounds for a social reconsideration of the norm, and such reconsideration guides the emergence of new, refined norms. The more trusted these trustees are the faster the learning of refined norms.

As we have all seen play out in the media time and again, politicians frequently do violate the norms that they hold up, but their supporters neither lower their trust in the violator nor in the strength of the norm. I posit that this is because of the apology offered by the violator. The apology is apparently believable to their supporters, who see it as evidence of the trustworthiness of the politician because of the metanorm of apologizing having been respected (even if the politician doesn't fully take accountability).

6 Discussion

Pulling these ideas together, we can see that the above-mentioned constructs are essential to understanding social intelligence and developing a foundation for social artificial intelligence. Moreover, capturing formal computational models of the above illustrations of the interplay of norms and trust can provide a basis for a theory of *sociotechnical systems* [18]. Such sociotechnical systems combine people and organizations with computational entities such as agents, devices, and data. Current approaches largely emphasize norms to regulate the interactions in such systems but largely do not relate it to trust.

Understanding trust in an integrated manner and relating it to the norms in the system can help us understand how the norms may change. In this way, we can think of a *self-renewing* sociotechnical system whose norms may change in light of changes to the environment or to the goals of the people and organizations involved.

7 Personal Note

Our first joint work was a workshop on Delegation and Trust that Rino was kind enough to invite me to participate in co-organizing. Rino was a great collaborator—he did nearly all of the work himself for this workshop, and it produced one of the earliest interdisciplinary volumes on trust [19]. We had several more collaborations, some leading to volumes [20, 21] and some to a journal special issue [22].

Rino's diligence in running these events is legendary. After he stepped back from the organization, as his administrative duties at work increased, it was difficult for others to pick up the work—despite their continued interest in the topic of trust in societies, they could not carry out the tasks that Rino had made look easy.

Besides these formal organizational efforts, I have greatly enjoyed our many conversations and they have deeply shaped my thinking toward the sociocognitive aspects of trust (and related concepts such as dependence, delegation, and deception).

References

[1] L. G. Nardin, T. Balke-Visser, N. Ajmeri, A. K. Kalia, J. S. Sichman, and M. P. Singh, "Classifying sanctions and designing a conceptual sanctioning process model for socio-technical

systems," *The Knowledge Engineering Review (KER)*, vol. 31, pp. 142–166, Mar. 2016.

[2] J. L. Austin, *How to Do Things with Words*. Oxford: Clarendon Press, 1962.

[3] A. K. Chopra and M. P. Singh, "Sociotechnical systems and ethics in the large," in *Proceedings of the AAAI/ACM Conference on Artificial Intelligence, Ethics, and Society (AIES)*, (New Orleans), pp. 48–53, ACM, Feb. 2018.

[4] A. K. Chopra and M. P. Singh, "Accountability as a foundation for requirements in sociotechnical systems," *IEEE Internet Computing (IC)*, vol. 25, pp. 33–41, Sept. 2021.

[5] M. Rokeach, *The Nature of Human Values*. New York: Free Press, 1973.

[6] S. H. Schwartz, "An overview of the Schwartz theory of basic values," *Online Readings in Psychology and Culture*, vol. 2, pp. 3–20, Dec. 2012.

[7] S. Hitlin and J. A. Piliavin, "Values: Reviving a dormant concept," *Annual Review of Sociology*, vol. 30, pp. 359–393, Aug. 2004.

[8] E. Liscio, M. van der Meer, L. C. Siebert, C. M. Jonker, and P. K. Murukannaiah, "What values should an agent align with? An empirical comparison of general and context-specific values," *Journal of Autonomous Agents and Multi-Agent Systems (JAAMAS)*, vol. 36, no. 1, p. 23, 2022.

[9] P. K. Murukannaiah and M. P. Singh, "From machine ethics to Internet ethics: Broadening the horizon," *IEEE Internet Computing (IC)*, vol. 24, pp. 51–57, May 2020.

[10] C. Castelfranchi and R. Falcone, *Trust Theory: A Socio-Cognitive and Computational Model*. Agent Technology, Chichester, United Kingdom: John Wiley & Sons, 2010.

[11] R. Falcone and A. Sapienza, "The role of trust in dependence networks: A case study," *Information*, vol. 14, p. 652, Dec. 2023.

[12] C. Castelfranchi, R. Falcone, and F. Marzo, "Being trusted in a social network: Trust as relational capital," in *Trust Management: Proceedings of the iTrust Workshop*, vol. 3986 of *Lecture Notes in Computer Science*, (Berlin), pp. 19–32, Springer, 2006.

[13] A. M. Singh and M. P. Singh, "Wasabi: A conceptual model for trustworthy artificial intelligence," *IEEE Computer*, vol. 56, pp. 20–28, Feb. 2023.

[14] R. C. Mayer, J. H. Davis, and F. D. Schoorman, "An integrative model of organizational trust," *The Academy of Management Review*, vol. 20, pp. 709–734, July 1995.

[15] R. Falcone and C. Castelfranchi, "From dependence networks to trust networks," in *Proceedings of the 11th AAMAS Workshop on Trust in Agent Societies (Trust)*, pp. 13–26, 2009.

[16] M. P. Singh, "Trust as dependence: A logical approach," in *Proceedings of the 10th International Conference on Autonomous Agents and MultiAgent Systems (AAMAS)*, (Taipei), pp. 863–870, IFAAMAS, May 2011.

[17] A. M. Singh and M. P. Singh, "Norm deviation in multiagent systems: A foundation for responsible autonomy," in *Proceedings of the 32nd International Joint Conference on Artificial Intelligence (IJCAI)*, (Macau), pp. 289–297, IJCAI, Aug. 2023.

[18] M. P. Singh, "Norms as a basis for governing sociotechnical systems," *ACM Transactions on Intelligent Systems and Technology (TIST)*, vol. 5, pp. 21:1–21:23, Dec. 2013.

[19] R. Falcone, M. P. Singh, and Y.-H. Tan, eds., *Trust in Cyber-Societies: Integrating the Human and Artificial Perspectives*, no. 2246 in Lecture Notes in Computer Science, (Berlin), Springer, 2001.

[20] R. Falcone, S. Barber, L. Korba, and M. P. Singh, eds., *Trust, Reputation, and Security: Theories and Practice*, no. 2631 in Lecture Notes in Computer Science, (Berlin), Springer, 2003.

[21] R. Falcone, S. Barber, J. Sabater-Mir, and M. P. Singh, eds., *Proceedings of the 11th International TRUST Workshop*, no. 5396 in Lecture Notes in Computer Science, (Heidelberg), Springer, 2008.

[22] R. Falcone and M. P. Singh, "Introduction to special section on trust in multiagent systems," *ACM Transactions on Intelligent Systems and Technology (TIST)*, vol. 4, pp. 23:1–23:2, Mar. 2013. Guest editorial.

TRUST, REPUTATION, AND NORMS: A SIMULATIVE APPROACH TO COGNITIVE ARTIFACTS

MARIO PAOLUCCI

Institute of Cognitive Sciences and Technologies (ISTC) and Institute for Research on Population and Social Policies (IRPPS), CNR, Rome

mario.paolucci@istc.cnr.it

GIULIA MISSIKOV-ANDRIGHETTO

Institute of Cognitive Sciences and Technologies (ISTC), CNR, Rome

giulia.andrighetto@istc.cnr.it

Abstract

In this paper, we examine the interplay between trust – a key concept for human and non-human social interactions extensively explored by Rino Falcone – and two other social cognitive artifacts: reputation and social norms. Trust and reputation formation, as well as social norm emergence, involve interactions among individuals over repeated periods. These elements are part of an interdependent, continuously changing, complex, and multilevel system, so that the overall effect turns out as more than the simple combination of individual elements, and it cannot be inferred from the response of each single component. We discuss how agent-based modeling and social simulation contribute to understanding the micro and macro dynamics of social interaction and the feedback loops involved. These contributions also reflect the original approach of Rosaria Conte.

1 Norms, Reputation and Trust

The intellectual history of reputation and trust, as modeled in the Institute of Psychology – later known by its acronym ISTC (Institute for Cognitive Science and Technology) – traces back to its role as a foundational center for cognitive studies.

The cognitive approach, focusing on the description of cognitive artifacts such as norms and reputation, originated from the collaboration between Cristiano Castelfranchi and a network of researchers worldwide (for reference, see [1]).

This approach, summarized in the book *Cognitive and Social Action* [2], presents social and mental artifacts as entities with their own reality, capable of being manipulated as independent constructs within a logical framework.

But what are these social and mental artifacts, and how can a mind be built from them? The best analogy is that of a machine-specifically, a cybernetic machine [3] – or, as Cristiano himself described it[1], a form of *theoretical psychology*. Combining these perspectives, we propose calling this approach "Theoretical Cyber-Psychology" (TCP) for the purposes of this paper, where reasoning occurs through symbolic manipulation. In essence, minds are represented as machines that manipulate social artifacts following well-defined rules.

The manipulation can be guided by artifacts, or have them as objects of manipulation, or both. One of the most important artifacts is the concept of a *goal*, as a state of the world that the agent wants to reach. Thus, a goal is a guide for action, and part of the architecture that connects the mind and the world where this mind is located. This idea inspired various architectures, including the widely-used BDI (Belief-Desire-Intention) architecture for artificial agents. Interestingly, the BDI architecture minimizes the role of goals, emphasizing instead the dynamic interplay between intentions (I) and desires (D) in selecting a plan based on beliefs (B).

Goals, a cornerstone of Castelfranchi's theorization, offer a prototypical example of a representative mental artifact. However, in *Cog-*

---

[1]Castelfranchi, ISTC presentation for his 80th birthday

nitive and Social Action, three other artifacts are introduced. These are not tools for connecting mind and action, do not represent a state of the world but an idea that is found to be central to social interaction, easily categorized under an overarching label that guides its formalization and application: examples are Norms, Trust, and Reputation. While we will address norms later, here we focus on trust and reputation and the evolution of their study after the publication of *Cognitive and Social Action.*

Within the ISTC, two of Castelfranchi's colleagues – Rosaria Conte, the mentor of the authors of this article, and Rino Falcone – took charge of the development of these artifacts.

Science is built not only on ideas, theories, and models but also by people and the incentives driving them. In this case, the need for academic recognition encouraged the two researchers to pursue separate lines of inquiry, despite the evident overlap between trust and reputation. How did this divergence play out, and what lessons can be learned?

In Castelfranchi and Falcone's recent work *Trust Theory* [4], the theme is revisited through the lens of Theoretical Cyber-Psychology. The book begins with a precise delineation of the key concepts – trust, trustor, and trustee – aiming to provide "an explicit anatomy of trust, an integrated and justified model of its components, their interaction, and how trust functions" (p. 359). Their approach stands in opposition to reductionism, which, as the authors argue, often results in oversimplification or "quantificatio precox". However, some degree of simplification is inevitable when transitioning from theory to experiments or simulations.

This transition is urgent; the amount of theoretical reasoning in the works of the TCP is massive, the reputation of its authors outstanding, but the lack of consistent application cases is disheartening. When turning the theory into experiments, we are always met with a quantification problem, or a lack of proper instruments. For reputation, an attempt to create an actual model has been made with the Repage approach, but for trust not even that has been realized; in trust-based simulations, trust is too often reduced to a single dimension. For what concerns the key players of the new online, platformed world, instead, we don't actually have a hint if some or these ideas have found application, for example, on your trust on an Uber to arrive, on your elaborated post not to be shadow banned by the algorithm, although there's a reasonable guess that platform overlords have their own representation of this (see also all the literature on data enclaves and their enshittification, for example [5])

2 Norms and Trust

Also the relationships between the concepts of trust and norms are manifold and by no means obvious. Rino and Cristiano [6] examine the relationships between norms and trusts, analyzing in particular how one can be considered the basis of the other and how this leads to multiple loops providing the grounds for more advanced forms of social coordination and cooperation.

Social norms are informal rules that prescribe what one should and should not do and unlike legal norms they are neither codified nor enforced through formal sanctions, but complied with because of shared expectations and peer punishment [2, 7, 8, 9, 10, 11, 12]. Compliance with norms thus depends on the expectations one has of the behavior of others (known as empirical expectations) and the belief that a sufficiently large number of people in their community think that they ought to conform to the rule and may be willing to sanction transgressions (the so called normative expectations). These social expectations that others will behave in a certain way and think one should behave in a certain way on which compliance with norms is based is none other than trust. Social norms, conventions, coordination between agents cannot exist without trust. Trust – as Rino and Cristiano argue – is a prerequisite for the emergence and maintenance of social norms. Moreover, once social norms emerge and get established in a community, they create a new basis for expecting others to behave in a certain way and new reasons for trusting others. What is interesting is that this dynamic of mutual reinforcement between norms and trust allows for the development of new forms of social coordination and cooperation that would not be possible just relying on interpersonal forms of trust.

Unraveling this complex loop between norms and trust is key to understanding the smooth and effective functioning and regulation of societies. Social simulations might help to model this loop and how it is possible to move from forms of coordination based on specific mutual expectations to more advanced forms based on social norms and institutions. This requires having sophisticated agents able to form expectations based on the behavior of the agents they interact with, derive the trustworthiness of an agent regarding a particular situation, take decisions based on this information, update their expectations, recognise the emergence of social norms, and change their beliefs and expectations due to the macro phenomena that emerged.

3 A call for simulation of Trust, Reputation and Norms

Despite its significance, the simulation of trust remains underexplored. This paper advocates for an increased focus on simulating trust, highlighting its necessity in understanding complex social dynamics and improving systems that rely on social norms, reputation and trust-based interactions. If trust is inherently complex, comprising various dimensions that can influence individual and collective behavior and not merely a binary state of trust or distrust, what is the proper way to represent it computationally?

Simulations can help unravel these complexities by allowing researchers to manipulate variables and observe outcomes in controlled environments. By simulating different contexts and agent behaviors, we can better understand how trust evolves, how it can be fostered or eroded, and the mechanisms by which it influences decision-making processes.

By introducing elements such as malicious agents or varying levels of information availability, researchers can observe how trust is affected, not only to aid in refining existing models but to help planning for effective policy decisions.

31

While simulations provide valuable insights, real-world testing remains crucial. Developing hybrid approaches that combine simulations with field experiments could validate findings and enhance their applicability.

In conclusion, enhancing the simulation of trust is vital for advancing our understanding of this complex construct. Lacking some ordering principle like the one that a comprehensive simulation could provide, TCP will stay as it is, constrained in the realm of theory and overshadowed by the advance of the moment, be it big data or generative AI. The question is, why has a comprehensive simulation approach not been attempted already? Perhaps because it is just too complex to be implemented; but we think that here, once more we are faced with an issue of diverging incentives. The only serious chance of an expanding collaboration, the original FuturICT initiative, has never been allowed to take over; while the single researcher can profit by producing advances along a specific dimension, models of "simpler minds" are the only that a single researcher, or a small group, can put forward. We have asked above what was the lesson to be learned. Perhaps, in this newly divided world, it's time to put away with at least some of the walls, and work together for robust, TCP-enabled, multiple-artifact simulations; perhaps it's time to try again for large-scale simulations including norms, reputation, and trust.

References

[1] F. Paglieri, M. Tummolini, R. Falcone, and M. Miceli, eds., *The Goals of Cognition: Essays in Honor of Cristiano Castelfranchi.* College Publications, 2012.

[2] R. Conte, C. Castelfranchi, *et al.*, *Cognitive and social action.* Garland Science, 1995.

[3] G. Pezzulo, "Re-founding cognitivism based on the cybernetic idea of goal-directed action," in *The Goals of Cognition: Essays in Honor of Cristiano Castelfranchi* (F. Paglieri, M. Tummolini, R. Falcone, and M. Miceli, eds.), pp. 13–23, College Publications, 2012.

[4] C. Castelfranchi and R. Falcone, *Trust theory: A socio-cognitive and computational model.* John Wiley & Sons, 2010.

[5] K. Birch, "Data paradoxes," in *Data Enclaves*, pp. 107–124, Springer, 2023.

[6] R. Falcone, C. Castelfranchi, H. L. Cardoso, A. Jones, and E. Oliveira, "Norms and trust," *Agreement technologies*, pp. 221–231, 2013.

[7] R. Conte, G. Andrighetto, and M. Campennl, *Minding norms: Mechanisms and dynamics of social order in agent societies*. Oxford University Press, USA, 2014.

[8] C. Bicchieri, *The grammar of society: The nature and dynamics of social norms*. Cambridge University Press, 2005.

[9] J. Elster, "The cement of society: A study of social order," 1989.

[10] E. Ostrom, "Collective action and the evolution of social norms," *Journal of economic perspectives*, vol. 14, no. 3, pp. 137–158, 2000.

[11] C. Horne and S. Mollborn, "Norms: An integrated framework," *Annual Review of Sociology*, vol. 46, no. 1, pp. 467–487, 2020.

[12] G. Andrighetto, S. Gavrilets, M. Gelfand, R. Mace, and E. Vriens, "Social norm change: drivers and consequences," 2024.

Semantic Delegation: Understanding Concepts with a Little Help from our Friends

Anna M. Borghi

Department of Dynamic, Clinical Psychology and Health
Sapienza University of Rome
Via degli Apuli 1, Rome, 00185, Italy
anna.borghi@uniroma1.it

Luca Tummolini

Institute of Cognitive Sciences and Technologies (ISTC)
National Research Council of Italy
Via Giandomenico Romagnosi 18A, 00196, Rome, Italy
luca.tummolini@istc.cnr.it

Abstract

Even if concepts are the basic building blocks of private thoughts, not only do most of our mental representations are acquired from others, but we also routinely rely on experts in many knowledge domains. Evidence suggests that in many of these situations people believe they know what in fact is known by someone else instead. In this contribution, we explore how trust and delegation of semantic expertise influence the way in which concepts are represented and ask whether it is possible to actually ground one's understanding via the understanding of others. We conclude by deriving some implications for the debate over understanding in contemporary AI systems.

AMB acknowledges the support by the project PRIN 2022 Prot 2022YJA4TB and by the Sapienza Excellence Project grant n. RG123188B09BDF14. LT acknowledges the support by project PRIN WHIM (Prot. 2022LYRT8E) and PRIN 2022 PNRR NOJA (Prot. P2022YYRK3, CUP B53D23030380001).

1   Introduction

Although we tend to think of concepts as the basic building blocks of our private thoughts, much of our everyday conversation with others routinely employs concepts and words for whose understanding we are fundamentally dependent on more expert thinkers or speakers. The standard example is the meaning of scientific concepts like QUARKS or ELECTRONS for which we defer to scientific experts who are assumed to have the competence required to fix what these terms are about and on whom we implicitly rely when we use these terms. For a less standard example, consider the concept of TRUST, which, even if it is often claimed to be the glue that holds society together, has evaded any agreed-upon definition so far. Still, if there are semantic experts on this topic out there - who, if not Rino Falcone and Cristiano Castelfranchi might be up to the task? - we might defer to *them*, and start using the concept with a little help from our friends, i.e. trusting and relying on their cognitive expertise to fix our concepts.

This phenomenon of "division of linguistic labor" was originally proposed in philosophy to account for the semantics of terms that refer to natural kinds [1]. In the case of natural kinds, these terms are supposed to refer to classes of entities whose structure is accessible only to a few experts in society like electrons are for physicists. Even if one is ignorant in matters of physics, there might still be a reason to acquire the concept of ELECTRON and use the word ELECTRON in thinking and public conversation. By being disposed to consult physicists about its use, the concept can become part of the stock of one's conceptual repertoire. In this view, some terms exist in every community, the use of which depends on some kind of cooperation with a subset of other speakers who originally acquired the terms. Reliance on this special subclass of speakers - deference to experts - is considered a way to acquire and master concepts alternative to direct perceptual experience with their referents.

This paper aims to extend the scope of the sociocognitive approach to trust, reliance, and delegation [2] to this semantic domain, to which it has not been applied so far. With this aim in mind, we will identify what knowledge domains are the best candidates for this socially mediated representation; we will then discuss whether the underlying cooperative processes constitute a distinct form of knowledge and conclude by asking whether grounding knowledge in the experiences of others can indeed be possible.

2 Others are more important to learn and master complex knowledge domains

Trusting others is crucial for learning. According to Vygotsky, in order to learn, children need social scaffolding [3]. Initially, adults help them to develop their cognition, and then progressively, children become autonomous. Studies show that, over time, behavior regulation occurs through inner speech, an internal mechanism that is initially acquired in a social context [4]. The voice of the adult becomes progressively an inner voice. To help promote children's autonomy, adults - parents, teachers, etc. - develop a tutelary relationship with children [5], which might involve conflict, but not necessarily.

The role of others might be particularly critical when children and adults have to manage complex situations and deal with complex concepts. Studies show that children from age 5-6 and adults rely more on others in domains where their knowledge is lacking. Hence, trust in others might be particularly crucial in some areas. For example, Kominsky et al. [6] found that the more complex a causal mechanism was, the more people tended to seek experts. Notably, to be able to recognize the lackings of their knowledge and decide to seek experts' support, children need to develop metacognitive abilities [7], together with epistemic curiosity [8].

2.1 Others are more important to learn and use abstract concepts and words

When are concepts particularly difficult to learn? Here, we contend that a source of difficulty might be the abstractness of concepts. Recent research has shown that learning abstract concepts (e.g., JUS-

TICE) might be more challenging than learning concrete ones (e.g., BOOK) because the first do not have a clear referent and their meaning might be vague or indeterminate. Indeed, abstract concepts are typically acquired later, and linguistically rather than perceptually, than concrete concepts. The same difficulty characterizes conceptual processing: the well-known concreteness effect reveals that abstract concepts are processed slower and recalled worse than concrete ones. We hypothesize that, when processing abstract and complex concepts, people use more metacognitive and monitoring activities [9]: the outcome of these processes might reveal that their knowledge has limitations, and that to proficiently process and use abstract concepts, they have insufficient resources in themselves and need others. Besides requiring more resources during processing, abstract concepts are perceived by people as more difficult. Across various rating studies in adults, we found a consistent difference between perceived knowledge of abstract and concrete concepts and the perceived need for others to learn them. Ratings reveal that, with abstract concepts, people feel more uncertain, less confident in knowing the word's meaning, and feel the need to rely more on others (social metacognition) [10, 11, 12]. Once it is determined that social scaffolding is particularly crucial when people learn and use abstract concepts, it has to be determined how and when they rely on others and how they choose the others on which to rely to enhance their knowledge.

3 Reliance on others might improve knowledge or improve the illusion of knowing

What effects might have reliance on others to learn and process complex and abstract concepts? Reliance on others might solve several needs. First, and more trivially, it can help better understand a concept's meaning. Other people, particularly experts in a domain, can be an important source of information, and transmit their knowledge. Second, reliance on others' knowledge might help people strengthen their sense of knowing. Importantly, this does not necessarily need to be accompanied by a real deepening of one's own knowledge. Consider some principles according to which people typically revolve to others. A classical one is the principle of authority. E.g., "I propose this statement, in line with what Plato said". The authority of the source allows people to make their statements stronger and legitimates them. Importantly, however, sometimes using this principle is a rhetorical expedient, useful to strengthen one's own argument, and does not generate changes in one's own perceived knowledge or in its perception. This is typically the case when the source is a classical one, such as an important writer or thinker of the past. A different case is suggested by the literature on communities of knowledge [13]. Knowing that within a given domain, there are experts and that they are active and eventually reachable might make people use concepts and words that they would not use and the meaning of which they do not know in depth. Clearly this phenomenon is intimately connected with Putnam's notion of division of linguistic labor we have introduced above and according to which people recurrently delegate specific aspects of knowledge to others. What is the impact of the existence of a community of knowledge on one's own knowledge?

When people use concepts benefiting from the advantages a community of knowledge might offer, does their knowledge change? In the most trivial and probably more frequent cases, knowledge augments when people interact with authoritative others. However, this might not always be the case. Some studies suggest that when people rely on sources such as the Internet, they tend to know less in terms of knowledge content - what is important for them is how to reach relevant information [14]. The phenomenon might be similar when people rely on experts - we might not need to deepen our knowledge but simply memorize how and through whom to reach it. Let us consider the case in which people's knowledge does not change or increase - or, at least, it does so minimally - but the feeling of knowing and the confidence in knowing change. Two contrasting cases might occur. First, people might experience a form of an illusion of knowledge, feeling more certain and confident because there are potentially available experts. Second, and alternatively, they might feel humbler and less confident in their knowledge because they know less than experts and experience the gap between others' knowledge and their own knowledge.

Some experimental results suggest that knowing that experts are present might induce in people the illusion of knowing more than they effectively do. These studies make use of paradigms typically employed to investigate the illusion of explanatory depth (IoED) [15]: people rate their knowledge of a mechanism - for example, how a bike might work -, then they have to explain it, and then rate it once again. Typically, people tend to overestimate their knowledge. Kominsky and Keil [16] showed that a similar phenomenon, which they called the "misplaced meaning effect", also occurs with word meanings: participants had pairs of words and had to estimate how many differences they would be able to produce. Both children and adults tended to overestimate their knowledge; the effect was particularly pronounced in kindergarten.

In a further experiment, the authors compare two different situations, one in which participants have to rate how many differences exist that an expert would know. If people feel intimidated because the experts know more, then the effect of overestimation should be smaller. If, instead, they tend to illusionary believe they know more because of the presence of experts, then the overestimation should be greater. The results indicate that the greater the knowledge of experts participants do not possess, the greater the effect of overestimation. In sum: participants believe they know more because of the presence of experts. (Notably, the tendency to overestimate their knowledge might lead participants not to rely on others and experts when they would need to). We have claimed that people tend to rely on experts more when they deal with complex concepts, and we have made the case for abstract ones. However, their recognition that relying on experts might be crucial does not necessarily mean that they trust that the knowledge of experts will be decisive. An important role is played by the domain of knowledge. When concepts are particularly complex or abstract, people might feel that their meaning is intrinsically indeterminate, both for naive and expert people. Falcinelli et al. [17] found that people rate that they need to rely on others more and that they need experts more with ecological concepts, which are perceived as more abstract and more scientific, i.e. likely more difficult to know in-depth, compared to technological concepts. Interestingly, however, the more people felt the need to rely on others and on experts, the less they trusted the experts' knowledge.

This coexistence of the perceived need for experts to know more and the scarce trust in experts in matters that need further scientific research - for example, ecological concepts - is intriguing. Apparently, it contrasts with evidence showing that, the more difficult the concept and people know there are some experts, the more people tend to overestimate their knowledge. At the same time, this result does not imply that participants feel a strong gap between themselves and the experts. They simply evaluate the concepts so difficult to understand that, despite they need to rely on experts, even experts cannot be very helpful. This might be a characteristic of scientific concepts, compared, for example, to technological ones, because people perceive scientific knowledge as an ongoing process rather than a finite one (for an extensive discussion see [18]).

4 How do we estimate that someone is worth of being trusted

A further critical point concerns what is necessary to outsource our knowledge. In the previously mentioned experiment by Kominsky and Keil, the simple mention of the fact that experts know something exerts an effect of knowledge overestimation (see also [19]). But in real life, which criteria do people adopt to select experts? According to the sociocognitive model of trust [2], the trustor needs to estimate and infer how reliable its potential partners would be in executing the task of interest. These learning mechanisms rely on direct experience, reputation/recommendation, and categories/stereotypes/inferential reasoning as channels of information to evaluate agents' trustworthiness. Notably, the problem of selecting the right experts is particularly critical in abstract and complex domains, where the possibility of one's own empirical verification of the truth of some statements is limited. In this respect, literature on testimony and its development provides interesting insights (see [20], for an overview related to abstract concepts; see [21], for a review). People might trust experts because they know them personally. For example, children trust more people they know. They start trusting their parents, then people they know, independent of their effective knowledge within a domain. However, already 4-5-year-olds start to lose their trust in familiar people if the information they provide is not accurate. In addition, they tend to trust people who flaunt their own knowledge and do not have a clear verification criterion.

Alternatively, people might trust them because they know they are experts within a given domain. Older children and adolescents start relying on experts who do not boost their knowledge but prefer to say they do not know about matters that are not knowledgeable. Children also prefer informants who are popular among others and who favor information on which many people consent. Interestingly, there might be cultural differences related, for example, to the role collective plays in a given society: for example, consensus is more important for Chinese Americans than for European American children [22]. To summarize, the capability to select the right informant develops with age and metacognitive abilities and is modulated by the culture.

5 Beyond trust: exploring the role of delegation for grounding concepts socially

The possibility that some of our concepts - and the words that we use to express them - may be connected to their content via some social mediation has also been accepted in the framework of embodied and grounded approaches to cognition (e.g. [23]), although so far it has received only limited attention (but see [24, 25, 26]).

According to the standard model of grounded cognition, knowledge domains for which we can have direct experience - and especially those who are more concrete - are represented by reusing the perception and action systems that were originally responsible for online interaction. In this view, representing via re-enactment *grounds* our concepts in the environment because it preserves the reliable causal connection that is characteristic of perceptual experience. Since, however, the possibility to perceive with our senses is limited, consistently with the division of linguistic labor hypothesis, it has been assumed that some of our concepts might instead reach out in the environment mostly thanks to our interaction with others, especially through language. Jesse Prinz, in particular, has been the first to propose that some of our concepts, while not grounded via perceptual experience, can instead be grounded by "word tracking" [23], that is, by tracking how experts who have direct access to the relevant states use the words competently. The hypothesis is that deference to such expert use establishes a social chain that would provide the required reliable causal connection between the relevant mental representations (concepts, words) and their referent. Essentially, meaning in these cases is grounded because "we know a guy who knows a guy who knows a guy who has directly observed the referent" [27].

An important drawback of the standard grounded approach, which is still not fully appreciated, lies in its reliance on a fundamentally passive spectator-observation model that is unable to explain how conceptual representations may be at the same time properly connected with the environment but also interpretable from the point of view of the system itself. Spectator-observation models of cognition are unable to explain how an organism can detect representational errors without relying on the mediation of a third-party who has access to both the representation and what it is supposed to be about (see for instance the "encodingist" argument in [20]). This limitation can be overcome by adopting an action-based view in which the grounding process becomes supported by a perception-action cycle (closed loop interaction) in which output (action) changes sensory input via the environment and, therefore, influences subsequent internal states, enabling the agent to (implicitly) verify his or her own representations directly (see [28] for more detailed argument and computational model). What is important in this context is that the socially mediated strategy of grounding by deference to experts suffers from the same fundamental problem of the passive spectator model. One way to see this point is to consider that the relation of deference between speakers has been mostly conceived as a mere "mental attitude" but not as an action proper which, according to the sociocognitive model, is necessary to account for trust [2]. In this model, the result of the decision to trust and rely on someone else is the action of "delegation" which implies an active role for the trustor too and further creates a social relation between the agents [29]. We conjecture that this active component which is missing in standard passive accounts of deference to experts, may instead be important for a proper social grounding process. To begin sketching this possibility, we propose the following thought experiment.

Consider a congenitally blind person who cannot perceive colors. Being unable to perceive colors, the blind person's understanding that the "traffic light is green" cannot be grounded via visual perception and, in this sense, the person cannot connect this term with their outside environment. Classical empirical findings support this view by demonstrating that understanding color words like GREEN activates brain areas associated with color perception in sighted individuals [30, 31]. However, recent studies also show that blind individuals can still form a conceptual understanding of colors through linguistic experiences, allowing them to grasp the concept GREEN despite the absence of visual input [32]. Learning the semantic space of color terms from the statistics of language use alone may not be itself grounded but may be sufficiently aligned with the semantic space of the sighted companion to support effective communication with those who have instead the necessary visual representations to link the word GREEN to the actual presence of a green traffic light in the environment. In this case, we may consider the companion as the "expert" to whom the blind person "delegates" the cognitive task of sensory discrimination in the environment. We suggest that when the sighted companion crosses the street arm-in-arm with the blind person, the informational link between them creates a contextual relationship that might help ground the linguistic concept of GREEN of the blind person through the expert's visual representations. In this partnership, the blind person can influence the companion's behavior through language while simultaneously verifying their knowledge through their joint actions. This reflects the idea of "coupled dynamics" as explored, for instance, by Hasson and Frith [33], where information exchange occurs across different channels: the blind person communicates verbally, the companion engages in visually-guided action, and the blind person experiences the results through other available sensory modalities like touch and proprioception. It's not just trust in the expert companion as a mere attitude but a full-fledged delegation of the cognitive activity of sensing and categorizing the environment, which is crucial in this process. Delegation requires the blind person to adopt a complementary course of action, thanks to which the appropriate connection with the environment becomes possible. But is this more active social interaction sufficient to establish a grounding process? Unfortunately the interaction between the blind person and their companion is not sufficiently systematic but only limited to a restricted domain of actions. High-level actions like being able to safely cross the street do not obviously exhaust the space of possible contingencies that characterize interaction with colored objects. Indeed a color term grounded through this very limited range of socially-mediated actions would be a concept very different from that shared by the rest of sighted individuals. But imagine now that the blind person has at their disposal a pair of smart glasses incorporating a very efficient vision-language model, a future release of OpenAI CLIP, for instance [34] . A model of this kind might be able to take as input any sensory image sampled from the smart glasses and generate the proper color term as output. Hooked up in this way, the blind person would be able learn over time the meaning of color terms not only by learning the statistics of how color terms are used by sighted speakers (the so called experts in their speaking community) but also by exploring the contingencies between one's own own actions and these linguistically-mediated outcomes. We conjecture that deference to an expert agent that establishes a systematic closed-loop interaction of this kind is indeed sufficient to vindicate the possibility of a proper form of social grounding.

6 Conclusions

Under what conditions does trust and delegation to semantic experts constitute the conceptual knowledge of the individual? When does it generate a more volatile illusion of understanding instead? Given the limits of our possibilities for direct experience, individuals often need to delegate the understanding of concepts to those who are more knowledgeable. Here we have argued that this delegation can operate as a social mechanism for acquiring and mastering complex knowledge. Building on the philosophical notion of division of linguistic labor, we have discussed how communities depend on experts to fix the meanings of terms, allowing individuals to use complex language without fully comprehending the underlying concepts. We have distinguished between the actual acquisition of knowledge and the illusion of knowledge and discussed evidence suggesting that individuals may feel more knowledgeable simply because they know experts exist, even if their actual understanding remains superficial. We have shown how individuals assess the reliability of experts and decide whom to trust, revealing that familiarity and perceived authority play significant roles in this process. Finally we have explored how concepts can be socially grounded through delegation in a collaborative interaction, imagining a possible interaction between a congenitally blind person and an advanced AI model. We conclude by noting that the same process might also illuminate current debates over understanding in AI models.

To see this suppose that the congenitally blind agent is not a human but a standard Large Language Model (LLM). Contemporary LLM like GPT-4 have been shown to have a semantic space for color terms that is strongly aligned with that of humans without the need of representations derived from sensory modalities [35]. It has been suggested that even if these kinds of models lack direct access to the physical environment, they might nevertheless be grounded through us in such a way that we humans end up playing the role of their sighted expert companions [36, 27]. However, if our argument in this paper is on the right track, this conclusion might be premature. Grounding the linguistic knowledge of LLMs - enabling a form of understanding in AI systems that is similar to what humans have - requires that these machines become autonomous agents capable of trusting and delegating the relevant cognitive tasks to us. But building machines that are able to enter this kind of systematic social relationships with us will thus require us to complete the project that Rino and Cristiano have started. Now that we have all the help we need, let's get to work.

References

[1] Putnam, H. (1973). Meaning and reference. The Journal of Philosophy, 70(19), 699-711.

[2] Castelfranchi, C., & Falcone, R. (2010). Trust theory: A socio-cognitive and computational model. John Wiley & Sons.

[3] Vygotsky, L. S. (2012). The collected works of LS Vygotsky: Springer Science & Business Media.

[4] Fernyhough, C., & Borghi, A. M. Inner speech as language process and cognitive tool. Trends in Cognitive Sciences, S1364-6613.

[5] Castelfranchi, C. (2022). A theory of tutelary relationships. Cham: Springer.

[6] Kominsky, J. F., Zamm, A. P., & Keil, F. C. (2018). Knowing when help is needed: A developing sense of causal complexity. Cognitive Science, 42(2), 491-523.

[7] Kuhn, D., & Dean, Jr, D. (2004). Metacognition: A bridge between cognitive psychology and educational practice. Theory into practice, 43(4), 268-273.

[8] Piotrowski, J. T., Litman, J. A., & Valkenburg, P. (2014). Measuring epistemic curiosity in young children. Infant and Child Development, 23(5), 542-553.

[9] Borghi, A. M., Fini, C., & Tummolini, L. (2021). Abstract concepts and metacognition: searching for meaning in self and others. Handbook of embodied psychology: thinking, feeling, and acting, 197-220.

[10] Villani, C., Lugli, L., Liuzza, M. T., & Borghi, A. M. (2019). Varieties of abstract concepts and their multiple dimensions. Language and Cognition, 11(3), 403-430.

[11] Mazzuca, C., Falcinelli, I., Michalland, A. H., Tummolini, L., & Borghi, A. M. (2022). Bodily, emotional, and public sphere at the time of COVID-19. An investigation on concrete and abstract concepts. Psychological research, 86(7), 2266-2277.

[12] Fini, C., Falcinelli, I., Cuomo, G., Era, V., Candidi, M., Tummolini, L., Mazzuca, C. & Borghi, A. M. (2023). Breaking the ice in a conversation: abstract words prompt dialogs more easily than concrete ones. Language and Cognition, 15(4), 629-650.

[13] Rabb, N., Fernbach, P. M., & Sloman, S. A. (2019). Individual representation in a community of knowledge. Trends in Cognitive Sciences, 23(10), 891-902.

[14] Sparrow, B., Liu, J., & Wegner, D. M. (2011). Google effects on memory: Cognitive consequences of having information at our fingertips. science, 333(6043), 776-778.

[15] Rozenblit, L., & Keil, F. (2002). The misunderstood limits of folk science: An illusion of explanatory depth. Cognitive science, 26(5), 521-562.

[16] Kominsky, J. F., & Keil, F. C. (2014). Overestimation of knowledge about word meanings: The "misplaced meaning" effect. Cognitive science, 38(8), 1604-1633.

[17] Falcinelli, I., Fini, C., Mazzuca, C., & Borghi, A. M. (2024). The TECo Database: Technological and ecological concepts at the interface between abstractness and concreteness. Collabra: Psychology, 10(1).

[18] Fini, C., Falcinelli, I., & Borghi, A.M. (in press). Technological concepts: object and tool of knowledge outsourcing at different ages. Topoi.

[19] Sloman, S. A., & Rabb, N. (2016). Your understanding is my understanding: Evidence for a community of knowledge. Psychological Science, 27(11), 1451-1460.

[20] Bickhard, M. H. (2009). The interactivist model. Synthese, 166, 547-591.

[21] Harris, P. L., Koenig, M. A., Corriveau, K. H., & Jaswal, V. K. (2018). Cognitive foundations of learning from testimony. Annual Review of Psychology, 69(1), 251-273.

[22] DiYanni, C. J., Corriveau, K. H., Kurkul, K., Nasrini, J., & Nini, D. (2015). The role of consensus and culture in children's imitation of inefficient actions. Journal of Experimental Child Psychology, 137, 99-110.

[23] Prinz, J.J. (2002) Furnishing the Mind: Concepts and their Perceptual Basis. Cambridge.: MIT Press.

[24] Shea, N. (2018). Metacognition and abstract concepts. Philosophical Transactions of the Royal Society B: Biological Sciences, 373(1752), 20170133.

[25] Borghi, Mazzuca & Tummolini (submitted). A social route to abstractness.

[26] Borghi, A. M. (2023). The freedom of words: abstractness and the power of language. Cambridge University Press.

[27] Pavlick, E. (2023). Symbols and grounding in large language models. Philosophical Transactions of the Royal Society A, 381(2251), 20220041.

[28] Mannella, F., & Tummolini, L. (2023). Kick-starting concept formation with intrinsically motivated learning: the grounding by competence acquisition hypothesis. Philosophical Transactions of the Royal Society B, 378(1870), 20210370.

[29] Castelfranchi, C., & Falcone. R. (1998). Towards a theory of delegation for agent-based systems. Robotics and Autonomous Systems, 24(83-4), 141-157.

[30] Chao, L. L., & Martin, A. (1999). Cortical regions associated with perceiving, naming, and knowing about colors. Journal of Cognitive Neuroscience, 11(1), 25-35.

[31] Simmons, W. K., Ramjee, V., Beauchamp, M. S., McRae, K., Martin, A., & Barsalou, L. W. (2007). A common neural substrate for perceiving and knowing about color. Neuropsychologia, 45(12), 2802-2810.

[32] Lewis, M., Zettersten, M. & Lupyan, G. (2019) Distributional semantics as a source of visual knowledge. Proceedings of the National Academy of Sciences, 116(39), 19237-19238.

[33] Hasson, U., & Frith, C. D. (2016). Mirroring and beyond: coupled dynamics as a generalized framework for modelling social interactions. Philosophical Transactions of the Royal Society B: Biological Sciences, 371(1693), 20150366.

[34] Radford, A., Kim, J. W., Hallacy, C., Ramesh, A., Goh, G., Agarwal, S., Sastry, G., Askell, A., Mishkin, P., Clark, J., Krueger, G., & Sutskever, I. (2021). Learning transferable visual models from natural language supervision. CoRR, abs/2103.00020

[35] Marjieh, R., Sucholutsky, I., van Rijn, P., Jacoby, N., & Griffiths, T. L. (2024). Large language models predict human sensory judgments across six modalities. Scientific Reports, 14(1), 21445.

[36] Mollo, D. C., & Millière, R. (2023). The vector grounding problem. arXiv preprint arXiv:2304.01481.

49

From Trust Theory to Explainable Agents: Enhancing Agent-Based Systems

Valeria Seidita

Department of Engineering, University of Palermo, Italy
ICAR, National Research Council (CNR), Italy
valeria.seidita@unipa.it

Antonio Chella

Department of Engineering, University of Palermo, Italy
National Research Council (CNR), Italy
antonio.chella@unipa.it

Abstract

This paper examines the development of agent systems on the basis of Falcone and Castelfranchi's trust theory towards the realization of agent models capable of self-explication and self-disclosure. Starting from the trust model, the work integrates the Belief-Desire-Intention (BDI) paradigm to extend the deliberation and belief representation capabilities of agents. The practical implementation uses the NAO robot to validate these ideas by endowing it with self-modeling and justification capabilities to enhance human-robot interaction and collaboration. Next, the concept of self-disclosure is introduced, enabling agents to explain their actions and increase the transparency and explainability of their decisions. This approach combines advanced theoretical models with practical implementations, laying the foundation for further developments in the integration of emotions, mental states and ethical values into the reasoning of agents, making them more reliable and understandable when interacting with humans.

1 Introduction

For several years, the main goal of robotics researchers has been to design and program robots that can perform specific tasks, now we are moving towards create true collaborative partners that can make autonomous decisions, adapt and work in harmony with humans. With advances in artificial intelligence technologies and autonomous robots, the ability of robots to be part of mixed teams is rapidly improving, opening up new possibilities in many areas.

The term Human-Robot Teaming Interaction (HRTI) [1, 2, 3] refers to the discipline that aims to develop collaboration between humans and robots in a team setting, where both work together to achieve common goals. In a typical HRTI configuration, the robot is an active partner that collaborates with the human, contributing to decision making, executing tasks and communicating effectively for mission success, rather than a passive tool. A classic example is a team game, e.g. soccer, in which all team members pursue the same goal and have the same knowledge of the world in which they are operating: the playing field with all its objects, the rules of the game and behavior, the characteristics of the other team members and of themselves (the latter information, for example, contributes greatly to the decision on the actions to be taken).

This is a very complex scenario, it is not just about team interaction, but there are other complex aspects to consider. For example, in a team consisting of humans and robots, the robots must be able to make autonomous decisions. If the robot has a goal in mind, it can decide for itself how to achieve it without having to constantly wait for human instructions.

Proactivity means that the robot is able to anticipate the needs of the team or, if necessary, take the initiative to react to changes in the environment or the requirements of the interaction. In such a scenario, the robot must be able to communicate clearly and effectively with humans. This includes both verbal communication, e.g. explanations of the actions the robot is performing, and non-verbal communication, e.g. gestures or visual cues that help the human to understand the robot's intentions. In this context, transparency means that the robot must make its goals, decisions and limitations understandable in order to reduce the perception of uncertainty in humans and at the same time increase their trust in the robot itself. Through interaction and communication, robots must be able to learn and change their behavior according to the needs of the team. This is especially important in dynamic environments where circumstances can change quickly. In addition, the process of controlling actions can be shared with the human team member - the robot can perform some parts of a task autonomously, while others can be delegated to the human or require their direct intervention.

Trust plays a crucial role in social interactions, both between humans and between artificial agents. In society, it is trust that enables people to delegate tasks and work together to achieve common goals. The concept of trust, as outlined by Falcone and Castelfranchi, can be described as a mental state to predict and evaluate the behavior of other agents. This concept applies not only to human interactions, but also to multi-agent systems [4, 5, 6], where the degree of trust between agents is crucial for deciding which actions to delegate and how to cooperate in dynamic and partially known environments. A major challenge in HRTI interactions is to establish an appropriate level of trust between humans and robots. Humans must be able to rely on the robots to perform their tasks safely and effectively. Trust is built over time and depends on various factors, such as transparency, consistency of actions and the robot's ability to recover from errors.

According to Castelfranchi and Falcone [7, 8, 9], trust in the context of agent systems is a social factor and also has a computational value. It underlies the delegation of tasks and decisions between agents, and its computation depends on an agent's knowledge of its environment and the capabilities of other agents. In our work, we aim to develop an agent system that integrates the trust model with the Belief-Desire-Intention (BDI) paradigm [10, 11, 12] to improve the deliberative and representational capabilities of agents.

In this chapter, we describe how Calstelfranchi and Falcone's trust model was a cornerstone and an important starting point for endowing the robot with self-modeling capabilities that form the basis for decision making and also allow the robot to justify itself and provide an explanation for its actions. The next sections describe the trust model and the basic concepts of the BDI agent paradigm that guided us in the concrete implementation of working scenarios.

2 The Trust Theory

The concept of trust is a very general concept that serves to explain how and why people manage to rely on each other in daily life to achieve their goals. In the literature, from philosophy to computer science, there is no single accepted definition of trust. They all revolve around the same concept, which we find perfectly reflected in one of Gambetta's most quoted definitions [13]: *Trust is the subjective probability by which an individual, A, expects another individual, B, to perform a certain action on which his welfare depends.*

In this definition, reference is made to the welfare of the individual. This is a very strong concept that one should be careful with when talking about the welfare of an individual who is dependent on a robot (or intelligent agent).

Trust is highly dependent on how well one knows oneself and the surrounding world in which other individuals, artificial or not, also reside. Falcone and Castelfranchi define trust as *a complex state of mind and a social relationship.* Trust is seen as a combination of beliefs, goals and evaluations that influence the behavior of one actor towards another. They describe trust as a form of delegation in which one actor (the trustor) entrusts another actor (the trustee) with a task or goal because they believe that the latter can perform the task competently and predictably [14].

One of the key elements of this theory is that trust requires a delegation component: The trustor believes that the trustee can do something and in addition it decides to entrust him with the responsibility for carrying it out. However, trust cannot be reduced to mere subjective probability, as it also includes beliefs about certain aspects such as the trustee's competence and disposition.

To summarize:

- the *trustor* - is an "intentional entity" like a cognitive agent based on the BDI agent model that has to pursues a specific goal.

- the *trustee* - is an agent that can operate into the environment.

- the *context* - is a context where the trustee performs actions.

- τ - is a "causal process". It is performed by the trustee and is composed of a couple of act α and result p, g_X is surely included in p and sometimes it coincides with p.

- the *goal* g_X - is defined as $Goal_X(\mathbf{g})$.

Falcone and Castelfranchi tie all these elements together through the following function:

$$TRUST(\text{X Y C } \tau \ g_X) \tag{1}$$

From this we can see that, according to the two authors, trust is a probabilistic assessment of success and also a real state of mind, which, as already mentioned, encompasses various cognitive and social aspects. The trustor assesses the abilities of the trustee and also his willingness to fulfill the task. This process always involves a certain degree of uncertainty and risk, because trust means accepting the risk that the trustee might fail or disappoint the trustor's expectations.

Falcone and Castelfranchi establish a close relationship between trust and delegation, stating that trust is the mental counterpart of delegation. In other words, trust is the cognitive process by which an actor (the trustor) decides to entrust a task to another actor (the trustee) based on a set of beliefs and evaluations. Delegation thus represents a concrete act that follows the trust process in which the trustor transfers responsibility for achieving a goal to the trustee.

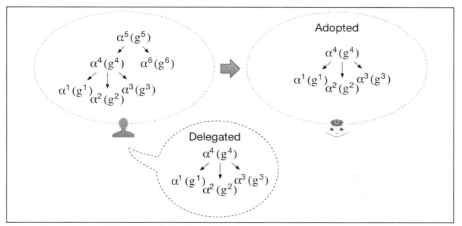

Figure 1: The model of adoption and delegation of actions.

According to their theory, at least two fundamental beliefs must be formed in order for the trustor to delegate an action:

- Competence: the trustor must believe that the trustee is capable of performing the delegated act. This belief relates to the trustee's ability or competence to perform the task.

- Predictability: In addition to competence, the trustor must be convinced that the trustee will actually perform the desired action, i.e. that his behavior is reliable. These two beliefs form the cognitive core of trust and support the trustor's decision to delegate the task. Essentially, trust sets the stage for delegation by providing a positive assessment of the trustee's capabilities and availability. Delegation is then a next step, a concrete action involving a decision based on the trust earned.

Falcone and Castelfranchi's theory defines different levels of delegation, ranging from the case in which the trustor delegates the trustee directly to the case in which the trustee acts completely independently in place of the trustor [15, 14]. In our work, we first looked at the simplest level of delegation, the so-called *literal help*. The Figure 1 shows the case where a human deselects a subgoal with its list of plans to assign them to a robot. This is the case of literal help, where there is an interaction between two people, the trustor, who is the principal, and the trustee, who is the agent, and where the trustee literally takes over the subgoals assigned to it by the trustor [14]. We have extended this concept by enabling the robot (through an implementation that uses intelligent agents) to commit to a specific goal while justifying the outcome. To this end, we have integrated the model (Equation 1) of trust into the reasoning cycle of a BDI agent, as explained below.

The core idea is to utilize an agent's (or human's) practical reasoning, which consists of two main steps, namely reasoning that leads to the choice of intentions, and reasoning about the means to an end (means-end reasoning), i.e. about the state of the world one wants to achieve and how to achieve it [16]. Practical thinking is the basis of the BDI agent paradigm.

The BDI paradigm is a cognitive model that enables agents to reason and make decisions based on their beliefs, desires and intentions. It represents a practical reasoning cycle that starts from the representation of a goal and leads to its realization through the execution of planned actions. In a multi-agent system, the BDI paradigm is particularly useful for managing trust between agents, as it enables the modeling of deliberation, i.e. the process by which an agent decides whether to perform or delegate a task based on its trust in other agents.

In this work, the integration of the trust model into the BDI cycle is done by extending the deliberation process and the belief representation. Agents are able to reason about how to achieve their goals and to develop a model of themselves that allows them to justify their actions. This ability to model themselves, or "self-modeling", is also crucial for improving transparency and explainability in human-agent and agent-agent interactions.

3 From Justification to Explainability

Starting from the trust model set out in the previous section and based on the properties and characteristics of the BDI agent paradigm, the first goal achieved was to extend the traditional BDI (Belief-Desire-

Intention) reasoning cycle to improve the explainability and transparency of decisions made by autonomous agents by including a system for justifying actions. This extension allows agents to communicate not only the actions they perform, but also the motivations and beliefs that lead them to make certain decisions, improving interaction with humans and other agents. First, we extended the reasoning cycle to justify one's actions.

```
1.  B ← B₀;         /* B₀ are initial beliefs */
2.  I ← I₀;         /* I₀ are initial intentions */
3.  while true do
4.      get next percept ρ via sensors;
5.      B ← brf(B,ρ);
6.      D ← options(B,I);
7.      I ← filter(B,D,I);
8.      π ← plan(B,I,Ac); /* Ac is the set of actions */
9.      while not (empty(π) or succeeded(I,B) or impossible(I,B)) do
10.         α ← first element of π;
11.         execute(α);
12.         π ← tail of π;
13.         observe environment to get next percept ρ;
14.         B ← brf(B,ρ);
15.         if reconsider(I,B) then
16.             D ← options(B,I);
17.             I ← filter(B,D,I);
18.         end-if
19.         if not sound(π,I,B) then
20.             π ← plan(B,I,Ac)
21.         end-if
22.      end-while
23. end-while
```

Figure 2: The BDI theretical reasoning cycle

In the traditional BDI cycle, the agent gathers information from the environment (beliefs), establishes its goals (desires), and plans actions to achieve those goals (intentions). This cycle repeats itself continuously as the environment and available information evolve, making agent behavior flexible and adaptive. In a dynamic environment and to implement the ability to justify one's actions, thus realizing the described trust model, an agent must be able to develop a self-model. This model must include the agent's own goals, beliefs and even motivations and body. That is, everything external and internal to the agent that drives a decision.

In short, the self-modeling model we implemented for BDI agents is based on the integration of the BDI reasoning cycle with trust theory. The central concept is that the agent (in this case the robot) builds a model of itself that includes knowledge of its capabilities and the actions it can perform in the environment. This model allows the robot to justify its actions, especially in situations of failure, by providing an explanation of why an action was not successful [16].

$$A_c \leftarrow action(B_{\alpha_i}, Cap) \tag{2}$$

where B_{α_i} and Cap are respectively portions of belief base related to the action α_i and the set of agent's capability for that action.

To do this, the robot uses a representation of beliefs related to actions. Each action in the plan is associated with a portion of the belief base, which provides the necessary context for performing that action. If an action fails, the robot is able to justify the failure by analyzing the beliefs involved, making explicit why the goal could not be achieved. This justification process is implemented through a module that links actions to relevant beliefs, allowing the agent to know, at any time, if and why an action is going wrong. During the execution

foreach α_i **do**
  $evaluate(\alpha_i)$;
  $J \leftarrow justify(\alpha_i, B_{\alpha_i})$;
end

Algorithm 1: Practical reasoning with justification.

of actions, the process can be explained using the BDI reasoning cycle. Once the robotic system has been analyzed, designed and put into operation, all agents involved in the system acquire knowledge.

They explore the belief base and any initial goals for which they are responsible (points 1, 2, 3, 4 - Fig. 2. Next, the module that implements deliberation and means-end reasoning (points 5, 6, 7 - Fig. 2) is enriched with a new function. At this point, during the execution of the BDI cycle, the sequence of actions described in the algorithm for each plan is processed so that the agent can decide which action to perform.

Self-modeling is enabled by the agent's ability to reflect on its beliefs and relate these beliefs to the planned actions. The robot thus becomes autonomous in managing the plan and, by continuously monitoring its actions, can request additional information or create a new schedule if necessary, which strengthens trust in the human-robot interaction. Self-modeling underlies the other aspect of our work that leads to the introduction of the concept of self-disclosure in agents. Self-disclosure, i.e. the ability of an agent to justify his actions and decisions, is crucial for increasing transparency and trust in the interaction between agents and between agents and humans. In the context of human-robot interaction, this ability enables humans to better understand the robot's decisions and trust its operational capabilities.

Following the same principle that led us to determine the justification function and its placement in the BDI reasoning cycle, in order to implement explainability, agents are provided with an "inner speech" function and then with a so-called self-disclosure, which allows them to articulate the reasons for their decisions verbally or through other communication channels. This capability not only improves collaboration, but also makes it easier for agents to learn and adapt to new contexts. In our experiment, the NAO robot was equipped with these capabilities to test the validity of the model and measure the effect of explainability in building trust between humans and robots.

The concept of *inner speech* is complex and difficult to define clearly. In the literature, it is often associated with subjective experiences related to language, actions, feelings and personal experiences. It is often used as a synonym for thoughts or mental processes. In psychology, however, inner speech is closely linked to self-perception and self-reflection.

Morin [17, 18] first associated this concept with the ability to think about oneself. In the cognitive domain, self-talk is an important tool for cognitive and behavioral regulation as well as problem solving and planning. This was our starting point for adding another function to the reasoning cycle, *rehearsal*. This is a mechanism by which the agent mentally simulates the execution of an action before actually performing it in order to evaluate the consequences and see if it can be successfully completed. In practice, the *rehearsal* function allows the agent to imagine the results of an action, comparing it with the current conditions and beliefs and decide whether or not to perform it. Before executing an action, the agent uses the *rehearsal* function to evaluate the selected action α_i, compare it with the beliefs about this action α_i, the available capabilities and the final goal D to be achieved. In this phase, the agent analyzes the preconditions and postconditions associated with the action and tries to anticipate whether the action will be successful or not. The algorithm 2 shows what has just been

foreach α_i **do**
  | $evaluate(\alpha_i)$;
  | $R \leftarrow rehearsal(\alpha_i, B_{\alpha_i}, D)$;
  | $update(B, D)$;
  | $J \leftarrow justify(\alpha_i, B_{\alpha_i})$;
end

Algorithm 2: Practical reasoning with the rehearsal function and inner speech

said in analogy to what has been described for the algorithm 1.

4 Practical implementation using NAO

The practical validation of the proposed ideas was performed with the NAO robot, a humanoid robot equipped with a programming platform to implement BDI agents. The robot was equipped with self-modeling capabilities which allow it to maintain an internal model of its actions and capabilities and to justify the results obtained when interacting with humans.

During the human-robot interaction, the NAO was programmed to perform a series of autonomous actions in cooperation with humans, such as recognizing and carrying objects. Each robot action was associated with a set of beliefs that determined success or failure. For example, to accomplish the task of grasping an object, the robot had to constantly update its beliefs about the position and properties of the object itself and motivate any failures by communicating the reasons for failure to the human.

The concept of inner speech was implemented using the paradigm of speech acts [19, 20], in which the agent communicates with itself to reflect on its actions. This "inner dialog" allows the agent to constantly revise his beliefs and decisions, improve decision making and make his actions transparent to any observer. With the help of speech acts, the agent can verbalize its reasoning and thus make the process that led to the action performed explicit. Each time the agent performs an action, a speech act is triggered that explains the reason for this action based on the agent's current beliefs and capabilities.

To realize this implementation, we used the Jason platform [21, 22] to program BDI agents and CArtAgO [23, 24] to represent the environment. The agents use Jason for reasoning based on the BDI cycle and CArtAgO for interacting with the environment by performing actions with artifacts in the virtual environment. Each agent can perceive the environment through the extended reasoning cycle, update its belief base and finally explain its actions using a built-in justification mechanism.

The NAO robot was used to investigate how the ability to explain itself affects people's trust in the robot. Preliminary results show that people's trust increases significantly when the robot is able to justify its actions and provide clear explanations, making the interaction smoother and more effective. This study demonstrates the importance of self-disclosure in robotic systems, especially in dynamic environments where mutual trust is essential [25, 26].

5 Conclusions and future perspectives

In this work, we illustrated different solutions to endow agents and robots with advanced cognitive capabilities, focusing in particular on the ability to self-model and justify actions. By using Falcone and Castelfranchi's trust model and integrating it into the BDI cycle, we have created agents that are able to understand and explain their actions and adapt in unknown and dynamic environments. This approach allows the robots not only to act autonomously, but also to justify their mistakes and request help in the future if needed.

The use of Jason and CArtAgO has facilitated the implementation of this model and allows a tight integration between the theoretical cognitive model and its practical implementation. The ability of agents to manage their internal language (inner speech), justify their actions and make decisions autonomously is an important step towards creating reliable, explainable and credible agents.

The work presented paves the way for further developments in the design of reliable and explainable agent systems. One of the future goals will be to extend the model of trust to include not only the ability of agents to justify their actions, but also to incorporate emotional elements and ethical values into their decisions. This approach could enable agents to interact with greater sensitivity to human needs, making them more understandable and reliable.

Another research direction will be the integration of these techniques into more complex environments, such as healthcare and social care, where trust and transparency in agent decisions will become even more important. Using the BDI paradigm with self-modeling and explainability could significantly improve human-agent interaction in critical contexts and enable robots and software agents to play a more significant and responsible role.

References

[1] G. Hoffman and C. Breazeal, "Collaboration in human-robot teams," in *AIAA 1st Intelligent Systems Technical Conference*, p. 6434, 2004.

[2] J. R. Hollenbeck, D. S. DeRue, and R. Guzzo, "Bridging the gap between I/O research and HR practice: Improving team composition, team training, and team task design," *Human Resource Management: Published in Cooperation with the School of Business Administration, The University of Michigan and in alliance with the Society of Human Resources Management*, vol. 43, no. 4, pp. 353–366, 2004.

[3] E. Salas, N. J. Cooke, and M. A. Rosen, "On teams, teamwork, and team performance: Discoveries and developments," *Human factors*, vol. 50, no. 3, pp. 540–547, 2008.

[4] M. J. Wooldridge and N. R. Jennings, "Intelligent agents: Theory and practice," *The knowledge engineering review*, vol. 10, no. 2, pp. 115–152, 1995.

[5] M. Wooldridge and P. Ciancarini, "Agent-Oriented Software Engineering: The State of the Art," *Agent-Oriented Software Engineering: First International Workshop, AOSE 2000, Limerick, Ireland, June 10, 2000: Revised Papers*, 2001.

[6] M. Wooldridge, *An Introduction to MultiAgent Systems*. John Wiley & Sons, 2009.

[7] C. Castelfranchi and R. Falcone, *Trust theory: A socio-cognitive and computational model*, vol. 18. John Wiley & Sons, 2010.

[8] R. Falcone and C. Castelfranchi, "Socio-cognitive model of trust," in *Human Computer Interaction: Concepts, Methodologies, Tools, and Applications*, pp. 2316–2323, IGI Global, 2009.

[9] Y. Guo, C. Zhang, and X. J. Yang, "Modeling trust dynamics in human-robot teaming: A bayesian inference approach," in *Extended Abstracts of the 2020 CHI Conference on Human Factors in Computing Systems*, pp. 1–7, 2020.

[10] M. Georgeff and A. Rao, "Rational software agents: from theory to practice," in *Agent technology*, pp. 139–160, Springer, 1998.

[11] A. S. Rao, M. P. Georgeff, *et al.*, "BDI agents: from theory to practice.," in *ICMAS*, vol. 95, pp. 312–319, 1995.

[12] A. S. Rao, "Agentspeak (l): BDI agents speak out in a logical computable language," in *European Workshop on Modelling Autonomous Agents in a Multi-Agent World*, pp. 42–55, Springer, 1996.

[13] D. Gambetta, "Can we trust trust? trust: Making and breaking cooperative relations, department of sociology, university of oxford," 2000.

[14] C. Castelfranchi and R. Falcone, "Towards a theory of delegation for agent-based systems," *Robotics and Autonomous Systems*, vol. 24, no. 3-

4, pp. 141–157, 1998.

[15] R. Falcone and C. Castelfranchi, "Levels of delegation and levels of adoption as the basis for adjustable autonomy," in *Congress of the Italian Association for Artificial Intelligence*, pp. 273–284, Springer, 1999.

[16] C. Castelfranchi, A. Chella, R. Falcone, F. Lanza, and V. Seidita, "Endowing robots with self-modeling abilities for trustful human-robot interactions," in *20th Workshop" From Objects to Agents", WOA 2019*, vol. 2404, pp. 22–28, CEUR-WS, 2019.

[17] A. Morin, "Possible links between self-awareness and inner speech theoretical background, underlying mechanisms, and empirical evidence," *Journal of Consciousness Studies*, vol. 12, no. 4-5, pp. 115–134, 2005.

[18] A. Morin, "A neurocognitive and socioecological model of self-awareness," *Genetic, social, and general psychology monographs*, vol. 130, no. 3, pp. 197–224, 2004.

[19] J. Searle and J. R. Searle, *Speech acts: An essay in the philosophy of language*, vol. 626. Cambridge university press, 1969.

[20] G. Soon, C. K. On, P. Anthony, and A. Hamdan, "A review on agent communication language," *Computational Science and Technology*, pp. 481–491, 2019.

[21] R. Bordini, J. Hübner, and M. Wooldridge, *Programming multi-agent systems in AgentSpeak using Jason, Volume = 8*. John Wiley & Sons, 2007.

[22] O. Boissier, R. Bordini, J. Hübner, A. Ricci, and A. Santi, "Multi-agent oriented programming with jacamo," *Science of Computer Programming*, vol. 78, no. 6, pp. 747–761, 2013.

[23] A. Ricci, M. Viroli, and A. Omicini, "Cartago: A framework for prototyping artifact-based environments in mas," in *International Workshop on Environments for Multi-Agent Systems*, pp. 67–86, Springer, 2006.

[24] A. Omicini, A. Ricci, and M. Viroli, "Artifacts in the a&a meta-model for multi-agent systems," *Autonomous agents and multi-agent systems*, vol. 17, no. 3, pp. 432–456, 2008.

[25] V. Seidita, A. M. P. Sabella, F. Lanza, and A. Chella, "Agent talks about itself: an implementation using jason, cartago and speech acts," *Intelligenza Artificiale*, vol. 17, no. 1, p. 7 – 18, 2023.

[26] A. Pipitone, A. Geraci, A. D'Amico, V. Seidita, and A. Chella, "Robot's inner speech effects on trust and anthropomorphic cues in human-robot cooperation," *arXiv preprint arXiv:2109.09388*, 2021.

Trusting in the Online World: From Theoretical Studies to the Application of the Socio-Cognitive Model

Elisa Colì

Institute of Cognitive Sciences and Technologies (ISTC), CNR, Rome

elisa.coli@istc.cnr.it

Abstract

The increasing proliferation of digital media has led more and more scholars to question, among other things, the relationships that take shape and develop in the online world. Literature has widely explored both the reasons that drive people to interact online (e.g., the need for connection) and the ways in which these relationships occur (e.g., aggressive or prosocial behaviors). Less attention, however, has been directed toward the role that trust plays in such relationships. In studying the complexity of digital dynamics and the bonds formed through screen mediation, a spontaneous question arises: "Is trusting in the online world possible?" Trust, a fundamental pillar in human relationships, takes on new facets in the virtual environment, where physicality dissolves, giving way to digital identities and distant connections. This chapter aims to highlight the important role that trust plays in the digital environment by providing a review of the main studies that have focused on this topic and the different perspectives from which it has been approached. It also aims to shed light on the potential of the socio-cognitive theoretical model of trust for understanding the determinants of trust in online relationships, presenting a series of studies - ranging from remote psychotherapy to online dating - in which the model has been applied.

1 Introduction

What a challenging task it is to write about trust for a scholar and expert in the field! But how could I possibly decline the request to pay tribute to such an esteemed individual with whom I have had the pleasure and honor of working? This chapter, written after an initial period of "writer's block" triggered by the request - leading me to question my ability, worry about making a poor impression, and doubt my own competence - is dedicated to you, Rino! Primarily, to the person you are: in you, I have found a father figure, a compassionate and understanding individual with whom I could speak openly, always certain of receiving the right advice and reassurance.

In you, I have found someone who believed in me and recognized and valued my skills. I have found a brilliant researcher who ignited my passion for trust, directing my research over the past few years towards this topic. A researcher not only adept at conducting research but also at building a team: I remember as if it were yesterday the onset of the COVID-19 pandemic, about which little was known at the time, and how, just days before we were confined at home due to the lockdown, you called me into your office. It was you and Cristiano, with your whiteboard full of diagrams and plans illustrating your research project. After presenting it to me, you said, "Could you draft a questionnaire to help develop this research?" Honored by the request, I set to work, and shortly after, we began the administration. Without fixed hours, sometimes without weekends (since every day at home felt the same), but with a wonderful spirit of collaboration, driven by curiosity for knowledge. From this wonderful experience together, I learned the importance of teamwork, discussion, sharing ideas, and the speed at which sometimes action is required because certain dynamics evolve rapidly. And of the pleasure that can come from a challenge aimed at expanding knowledge and understanding. In short, I owe you so much, dear Rino. For someone like me, who isn't particularly skilled with words, I'd say I've already written too much. So, I will simply close by expressing all my gratitude and wishing you a good/new life! With love.

This chapter stems from the desire to delve deeper into the topic of online trust. With the advent of social media, online relationships have gained significant momentum, and concurrently, studies aimed at understanding human behavior in relation to these new technologies have increased.

While numerous studies aim to investigate the reasons that drive people to interact online (e.g., the need for connection) and the ways in which these relationships occur (e.g., aggressive or prosocial behaviors), research focusing on trust, a fundamental pillar of human relationships both online and offline, remains largely concentrated in the fields of marketing and e-commerce. This focus often overlooks the dynamics that characterize interpersonal relationships. Starting with a definition of online trust and reviewing the main studies that have focused on this topic, this chapter aims to highlight the significant role trust plays in the digital environment, attempting to answer the question, "Is it possible to trust in the online world?".

In the second part of the chapter, the focus will shift to the socio-cognitive theoretical model of trust [1, 2], highlighting its potential for understanding the determinants of trust in online interpersonal relationships. A series of studies will be presented—ranging from remote psychotherapy to online dating—where this model has been applied.

2 Offline trust and online trust – the same definition?

Trust, trustworthy, and entrust are highly recurrent terms in our vocabulary, and trust is central to our daily lives. Every day, we perform countless acts of trust: we rely on doctors to take care of us, on websites for making purchases, on information we find on the web, and on unknown individuals with whom we interact online. Trust permeates all areas of an individual's associative life, from interpersonal relationships, involving one or more people, to institutional relationships, involving entities that are not identifiable as individuals, such as public and private institutions or economic organizations [3]. Given the importance of trust in everyday life, it has become the subject of extensive research and has been studied from various disciplinary perspectives. However, despite agreement on the complexity and multidimensional nature of trust, there is no consensus on its definition [4]. This lack of consensus is partly due to the various definitions of trust found in the literature, which reflect the paradigms of different disciplines. For example, sociologists view trust as a social structure, psychologists as a personality trait, and economists as a rational choice mechanism [5]. Additionally, as McKnight and Chervany[6] pointed out, empirical research has shaped most definitions of trust, leading each researcher to develop a narrow conceptualization of trust that fits their specific area of study. Broadly, the main definitions of trust can be grouped into two categories: the first focuses on trust as an expectation regarding the behavior of an interaction partner [7], and the second sees trust as the acceptance of and exposure to vulnerability [8, 9]. Castelfranchi and Falcone [2] emphasized the need to identify a strong common conceptual core to characterize the general notion of trust. According to these authors [1, 2], trust is a complex mental attitude based on two essential mental components: the goals and beliefs about both the trustor and trustee, and the trustee's behavior. It reflects the outcome of internal mental processes [2] that encompass both cognitive and social aspects, such as perceptions of the other involved in the interaction in terms of competence, dependence [10], and common membership in and affiliation with a salient social group [11].

Regarding interpersonal trust specifically, researchers have demonstrated that it is an important component of social capital [12] and, in particular, a key element in the development and maintenance of positive social interactions [13, 14, 15]. Furthermore, interpersonal trust has been positively associated with life satisfaction [16], although some aspects of this relationship still need clarification. Some studies have highlighted the positive consequences of high levels of interpersonal trust in social life, particularly in interactions with others. These studies found that people with high levels of trust are less likely to lie, cheat, or steal than those with low levels of interpersonal trust; additionally, they are more likely to respect the rights of others. Moreover, these individuals are less likely to be unhappy, conflicted, or maladjusted, and are more likely to have many friends [17].

Turning to the examination of online trust, it is evident that the main definitions developed in this context often refer to e-commerce. Specifically, Bart et al. [18], building on the offline definition of trust, developed a definition based on Rousseau's work, emphasizyng that "online trust includes consumer perceptions of how the site would deliver on expectations, how believable the site's information is, and how much confidence the site commands". In essence, trust is developed when consumers form positive impressions of an electronic merchant's site, and are willing to accept vulnerability. Another definition of online trust also originates from the field of e-commerce. Specifically, online trust is defined as "a confident expectation in an online situation of risk that one's vulnerabilities will not be exploited" [19]. Based on this definition, several researchers argue that the nature and meaning of online trust are not fundamentally different from the concept of face-to-face trust [20, 21, 22, 23], even though a human must trust an object created by another human rather than another human directly [22]. Consequently, online trust has been identified as a critical component of business strategy because it reduces perceived risk and generates positive word-of-mouth, which in turn influences customer purchasing decisions [24, 25, 26]. In particular, consumer trust and satisfaction with the transaction form the foundation for a long-term business relationship between a company and its customer [27]. Referring to trust in online economic exchanges, Shankar et al. [20] view online trust as the reliance by a firm's stakeholders on the firm's business activities in the electronic medium, particularly its website. Moving away from the context of e-commerce and considering the definition of online trust more generally, Tagliaferri [28] emphasizes that the definition itself depends on the conception of trust being assumed and the characteristics of the environment. According to the author, the main problem in the debate on online trust is the focus on specific definitions of trust or specific online environments, which leads to identifying the necessary environmental conditions for trust

71

based on the adopted definition. This approach not only diminishes the richness of the trust construct but also risks ignoring the rapid evolution of online environments and the need for constantly updated design features. Other authors question whether there is a difference between how people trust others in the physical world and how they trust others in an online environment. Specifically, Corritore et al. [22] argue that understanding online trust should draw on existing work on offline trust, which indicates that a considerable number of studies on offline trust are applicable to trust in online settings. They add that the common denominator between the two is their grounding in exchange, which, in both contexts, is impeded by risks, fears, costs, and complexities. Therefore, the notions of trust as acceptance and exposure to vulnerability, and as an expectation regarding the behavior of an interaction partner, are valid when applied to online relationships and exchanges. Regarding differences, Shankar et al. [20] contend that offline and online trust differ in terms of trust objects. In offline trust, the object of trust is typically a person or an entity (organization), while in an online context, the technology (primarily the Internet) and the organization implementing the technology are the true objects of trust. Similarly, Koranteng et al. [29] underline that although offline trust definitions are applicable in online settings, the situational elements influencing the formation of trust differ. For instance, exchange appears to be a common factor in both settings, but exchanges in offline environments differ from those in online settings [30]. Issues regarding physical distance also vary between these environments, with social media in particular eliminating geographical boundaries [31]. Moreover, human network attributes such as non-verbal language, which underpin trust in traditional environments, are absent on social media. Thus, the lack of elements related to face-to-face relationships reduces the richness of communication among members [32].

Regarding the role of trust in online environments, Turilli et al. [33] highlight how trust facilitates the emergence of social behavior by providing individuals with opportunities for beneficial social interactions. Although this effect characterizes both offline and online trust, the authors note that episodes of online trust have a peculiar impact on the development of the social context. Specifically, online trust makes interactions advantageous for the trustor, thus incentivizing them to engage with others and increasing their social interactions and networks. Additionally, as Yamagishi and Kikuchi [34] note, individuals refine their social intelligence, namely their ability to assess the trustworthiness of others and avoid risky interactions. This initiates a virtuous circle leading to a selection process, whereby trustworthy individuals engage in an increasing number of interactions, while untrustworthy individuals become progressively marginalized and excluded from the social system. These dynamics, while common to both offline and online interactions, are particularly evident in online communities. Trust not only occurs in online interactions but also provides the means for significant evolution of social behaviors in these environments. In particular, some authors [35, 36] emphasize that trust is one of the most important elements in promoting successful online interactions, while Kim & Ahmad [37] specify that trust plays a key role in the success of social interaction by fostering the sharing and dissemination of content in social media sharing communities. According to other scholars [38, 20], trust in online communities acts as a moderator that facilitates mutual communication and leads to better relationships, while Hoff [39] considers it the precursor to active engagement in online environments. Other authors [40] emphasize that trust, particularly online interpersonal trust, is influenced by internet usage behaviors, individual factors, and cultural factors, while it can also influence online behaviors such as self-disclosure, knowledge sharing, behavioral decisions, and group cooperation.

3 Trends in Online Trust Research

A literature review reveals that much of the research on online trust focuses primarily on trust in e-commerce transactions [41]. This body of work explores aspects such as building customer trust in online communications [42, 43] and maintaining trust relationships through online communication channels [44, 45]. Typically, online communications in e-commerce involve two parties: the buyer and the seller. Sellers provide electronic services to buyers through online communication, aimed at facilitating smooth online transactions. While interpersonal trust between these parties has been considered [46, 42], the emphasis tends to be on business levels that assist with marketing issues and customer loyalty.

Another group of studies identifiable in the literature relates to online communication within the context of virtual collaboration. For example, Jarvenpaa and Leidner [47] examined the relationship between communication and interpersonal trust in virtual teams. Other studies have focused on trust development in remote work groups, particularly emphasizing conflict management and negotiation. For instance, Drolet and Morris [48] demonstrated that people are more collaborative and inclined towards conflict resolution in face-to-face rather than online interactions. Some authors [49, 50], studying trust in cooperative games, acknowledged that trust development is closely tied to proximity, as remote exchanges make it more challenging to detect potentially useful interaction signals for goal achievement. However, the authors suggest that trust can develop remotely, albeit at a slower pace due to the time required for sharing social information. In this regard, other authors have explained that trust formation is delayed in relationships mediated by computer systems because interaction requires nearly four times the number of messages transmitted in face-to-face interactions [51, 52]. Bos et al. [53], analyzing trust development across face-to-face, online, telephone, and chat interactions, confirmed that online trust development is feasible but inevitably characterized by greater delays and fragility compared to in-person meetings, despite reports of high collaboration levels among respondents. Therefore, in the online mode, trust appears to be present but rather vulnerable. However, it has also been demonstrated that with technical support, there is no substantial difference in terms of trust between face-to-face and online encounters [54].

Regarding the specific context of social media, Koidl and Kapanova[40] highlight the challenge in identifying publications specifically focused on online interpersonal trust, as much of the research tends to address security-related topics. A brief review of studies concerning interpersonal trust in social media reveals a focus on the following themes. 1. Fake news, filter bubbles, and echo chambers: recent discussions indicate that users tend to trust content from known peers more than from unknown sources. Moreover, users prioritize the engagement garnered by their own content rather than their engagement with others' content [55]. 2. Online participation [56] and the influence of social media use on trust development: most studies indicate a positive effect of social media on trust (e.g., [57, 58, 59]). 3. Factors influencing trust in social media communication, including information quality, shared preferences, familiarity, privacy, chat topic, convenience, and time savings [60].

4 Is it possible to trust online?

Some scholars have questioned whether trust can be established in an online environment and have attempted to identify the conditions under which this can occur. Specifically, they have pondered whom or what one can trust, given that online interactions involve not only individuals but also a mix of human and technological actors. Friedman et al. [61] argue that trust can only be placed in people, and that the technological environment within which relationships occur, due to its characteristics (e.g., anonymity, reliability, and security of technology), may either support or hinder trust between individuals.

Other authors [33] have highlighted a key issue in this research area concerning the definition of trust, particularly whether trust - a relationship where a trustor relies on a trustee's behaviors to meet their expectations - can adequately describe relationships between parties that interact exclusively online. They have questioned whether online social interactions meet the minimum requirements for trust to emerge, noting specifically that two conditions are often lacking online: a shared institutional and cultural background and certainty about the identity of the trustee. These authors underscore that the debate on online trust has led to two predominant viewpoints in the literature: some scholars (e.g., [62, 63]) argue that the aforementioned conditions cannot be fulfilled online, thus positing that trust cannot exist in this environment; while others (e.g., [64, 65]) contend that trust can indeed exist online, emphasizing that the conditions mentioned are not necessary for trust to emerge and that both conditions can actually be met in online interactions.

According to Tagliaferri[28], to understand whether trust can emerge in the online environment, two elements must be evaluated. Firstly, it is crucial to determine if the online environment meets the necessary conditions for trust development, including identifying the specific characteristics that an online environment should have to either foster or hinder trust. Secondly, it is essential to specify the definition of trust being referenced. He focuses specifically on interpersonal trust, highlighting its defining elements: trust is an attitude that agent A1 (the trustor) holds towards agent A2 (the trustee) for a specific purpose. For trust to qualify as such, at least two elements are necessary: 1. The trustor must rely on the trustee to achieve the purpose. 2. The circumstances under which trust is elicited must involve at least two elements of freedom: the trustor must be free to choose whether or not to rely on the trustee, and the trustee must be free to choose whether or not to betray the trust by not contributing to achieving the purpose within the trust relationship. Tagliaferri further defines an online environment as a virtual space where two or more agents can interact through an interface following the protocols of the OSI application layer model. This definition underscores how an online environment enables interactions among various types of agents, each potentially requiring different interfaces and protocols. For instance, two human agents might communicate via a social media news feed, which manages user communication using the HTTPS protocol. Returning to the initial question of whether online trust is possible, Tagliaferri's answer is nuanced: it depends. The first determinant is the theory of trust being employed. Doxastic theories of trust, for example, may better facilitate the emergence of online trust, whereas affective theories of trust might pose challenges, especially when the trustee is an AI agent lacking relevant characteristics such as affective capabilities and authentic moral values that enable certain types of trust to develop. Nonetheless, other forms of trust appear compatible with the ability of AI agents to trust each other, particularly basic forms of trust based on dependency. The second determinant influencing the answer concerns the specific nature of online environments. A blanket affirmation for online environments in general is unlikely (and similarly challenging for physical environments). How specific online environments are constructed can profoundly influence the emergence of trust within them. Notably, it is feasible to implement features in online environments that could facilitate the emergence of trust.

In light of this brief review, assuming that online trust is indeed possible under certain conditions, the chapter proceeds by presenting some theoretical insights into the socio-cognitive model of trust [1, 2], accompanied by concrete examples related to interactions on social media, and concludes with a description of some studies on online trust in which the theoretical model of trust has been applied practically.

5 The socio-cognitive model of trust – theoretical mentions supported by examples

The socio-cognitive model of trust ([1, 2] was developed to identify a common and significant conceptual core of trust, which takes into account the most pertinent and insightful definitions. In particular, according to the model's authors, trust encompasses three interrelated concepts. First, trust is defined as a pure mental attitude - an opinion, judgment, or prior evaluation regarding a potential trustee and their requisite characteristics. Second, trust involves a decision - a more intricate assessment where various potential trustees are compared before choosing whom to trust. Lastly, trust is characterized as an action - the fiduciary behavior that defines the relationship between trustor and trustee.

For example, consider a group of first-class high school students who create a WhatsApp group to coordinate their work on an essay assigned by their Italian teacher for their oral exam. This collaborative task not only facilitates coordination but also fosters mutual acquaintance among students. Through initial interactions on social media, students begin to form impressions of their peers' reliability (mental attitude). As the exchanges progress, each student can make a comparative evaluation based on their peers' attitudes (decision). Only when a student begins collaborating with a chosen peer on the essay (after making a decision) does genuine trust manifest, potentially exposing them to risk if the peer fails to fulfill their responsibilities. This example illustrates trust as a relational construct involving a trustor (X), who must trust someone else to carry out a specific task, and a trustee (Y), who is the subject of that trust. When a trust relationship is established, the trustor delegates a task to the trustee, which entails actions by the trustee that lead to changes and eventual outcomes. The trustor's intention in fostering this trust relationship is to achieve a specific purpose, which defines the nature of the trust relationship itself. The trustee's actions occur within a specific context or environment that significantly influences those actions.

According to the socio-cognitive model, trust is founded on two fundamental mental components: goals (the mental representation that identifies the desired state) and beliefs, with the most crucial being competence and willingness attributed to the trustee. When the trustor assigns a task to the trustee, they must believe that the trustee possesses the necessary capabilities to perform the required actions (competence belief). Additionally, the trustor must believe that the trustee intends to carry out the task and will execute it as agreed (willingness belief). Thus, the first belief concerns the possession of skills, competencies, and resources needed to achieve a goal, while the second belief pertains to the possession of motivational attitudes towards the task, including intentionality, disposition, and motivation. In addition to these beliefs, there are other relevant factors. The belief in unharmfulness refers to the potential that, whether voluntarily or involuntarily, the trustee may hinder the accomplishment of the goal. The dependence belief relates to the extent to which the trustor can achieve the goal independently or requires the trustee's assistance. Lastly, the contextual trust belief assesses how the surrounding environment can positively or negatively impact the attainment of the goal.

For example, returning to the student scenario, Mario trusts Francesca to help him excel in the Italian exam. Therefore, Mario must believe that Francesca, the classmate he relies on, possesses various competencies. These include expertise in conducting bibliographic research, proficiency in using the internet effectively, the ability to discern reliable online information sources from unreliable ones, and familiarity with the library system in their country, among others. Additionally, Mario must beliieve that Francesca is genuinely motivated and willing to collaborate effectively with him to produce a high-quality essay for their Italian teacher. He gauges Francesca's motivation by observing factors such as the time she dedicates to their collaborative work, her availability on WhatsApp for discussions, her collaborative attitude, and the promptness of her responses to his messages. Furthermore, Mario must believe in Francesca's unharmfulness, trusting that she will not engage in behaviors like sharing their work with other classmates without consent, which could compromise the essay and affect their exam results. Lastly, Mario needs to believe that the virtual environment where he and Francesca interact supports their relationship of trust. For instance, while communication on WhatsApp might occasionally lead to misunderstandings, it generally facilitates rapid information exchange and effective coordination for their thesis development. Any act of trust involves a gamble and thus entails risk. By trusting, the trustor becomes dependent on the trustee and exposes themselves, becoming vulnerable to the trustee's actions. There is always a risk that the intended goal may not be achieved, leading to failure and potentially wasting both material and immaterial resources.

In the student example, specifically, failing to submit the essay or producing a substandard one could jeopardize their exam performance. Furthermore, receiving a poor grade could lower their overall exam average or, when coupled with other negative evaluations, necessitate credit recovery during summer break. This outcome would result in wasted time, energy, and potentially financial resources, as students might require private tutoring to catch up. The socio-cognitive model [1, 2] views trust as inherently dynamic, evolving over time in response to changes in its underlying components. This dynamic nature is influenced by direct experiences, where the trustor gains trust from interactions with the same or different agents, on specific or general tasks, and in various environments. Additionally, trust changes when other influencing factors evolve, such as shifts in people's attitudes, the testimonies to which individuals are exposed, or alterations in the reputation of the trustee with whom they engage.

In the earlier example, Mario's trust in Francesca could evolve over time because the type of help offered by Francesca may not have led to the realization of a good job for the Italian teacher, or because in the meantime she may have been discredited by her classmates due to some event that happened at school.

This model offers the advantage of pinpointing a foundational core that comprehensively addresses the key processes underlying interpersonal trust. Moreover, it operationalizes the trust construct by delineating its constituent components and the sequential process - from assessment to decision-making - that culminates in the act of trusting someone.

6 Applications of the socio-cognitive model to online interactions

Below, will be presented some of the main studies conducted in collaboration with Prof. Falcone, including some already published and others unpublished, in which the socio-cognitive model of trust [1, 2] has been applied to online interactions (with specific reference to interpersonal trust and social media). While these are complex studies that extend beyond the scope of trust alone, I will focus here exclusively on the aspects related to the application of the model itself.

I will first refer to two qualitative exploratory studies: the first aimed at investigating the elements on which trust among adolescents interacting on social media is based, and the second aimed at exploring the elements underlying trust among mothers interacting within

Facebook support groups, as well as between mothers and professionals. In both studies, the socio-cognitive model of trust [1, 2] served as the foundation to identify the determinants of trust through its beliefs.

Next, I will consider two quantitative studies: the first focusing on remote psychological support and aimed at analyzing trust within the therapeutic relationship, and the second centered on online dating and trust in long-distance intimate relationships.

6.1 Study 1 - Adolescents and Trust in Online Social Interactions: A Qualitative Exploratory Study

In this study [66] the socio-cognitive model [1, 2] was used to explore the trust that adolescents place in others with whom they interact on social media. Specifically, the authors analyzed the socio-cognitive determinants of trust, with particular reference to the beliefs (competence, willingness, unharmfulness, dependence, and context) on which adolescents' online trust is based. In particular, the authors aimed to answer the following questions: From adolescents' point of view, what skills should the others to whom they rely online have? On which basis should an adolescent evaluate someone online as available and as non-dangerous? What are the factors that make adolescents feel dependent on others online? Finally, what are the context-related factors that foster or hinder the relationship of trust?

6.1.1 Method

The study involved ten adolescents, five males and five females, aged between 12 and 17, with a mean age of 15.5 years. The majority of participants (six) had a middle school education level, and the remaining four had a lower school education level. Regarding the main social media used to interact, everyone used both Instagram and WhatsApp, while only one interviewee used Facebook and another one used Snapchat.

In this qualitative research, the socio-cognitive model of trust [1, 2] informed the entire study, from the formulation of the research questions and the definition of the interview outline, to the data analysis and the interpretation of the results. Data were collected using a semi-structured interview, which seemed to be the most appropriate methodology for the research objectives, having the very purpose of understanding the participants' point of view [67], representing a flexible and powerful tool to capture the voices and the ways people make meaning of their experiences [68]. In particular, the interview opened with a question aimed at identifying the social media used by adolescents to interact with others online, followed by an opening question on the trust-specific topic aimed at understanding how young people conceptualize online interpersonal trust and on which elements they base it. Subsequently, the specific components of trust, as theorized in the socio-cognitive model [1, 2], were explored starting from the expectations placed on the other online - which are considered a prerequisite on which trust is based - and following with the competencies recognized in the other, his/her willingness, the perception of unharmfulness in the relationship, the dependence on the other and finally the context in which relationships take place.

As regards the analysis of the collected data, the authors used qualitative content analysis, which aims to promote knowledge and understanding of the phenomenon under study [69]. In particular, on the basis of the purpose of the study, authors chose to use a mixed deductive-inductive approach to content analysis [70]. Therefore, using NVivo10 software [71], they first created a coding system based on the main determinants of trust identified through the socio-cognitive model [1, 2]. Then, they analyzed the text of the interviews within this coding system (deductive approach). Subsequently, guided using an inductive approach, the researchers created a system of categories based on content similarity for each determinant of trust.

6.1.2 Results

The results for each belief in the model - competence, willingness, unharmfulness, dependence, and contex - will be presented below. Regarding the *Competence* belief, the adolescents interviewed highlighted various skills that promote trust in the others with whom they interact online. Some refer to skills related to the management of online communication, others, on the other hand, highlight aspects relating to interaction management. Further skills refer to the ability to use social media constructively. Alongside these, other skills emerge, albeit shared by a smaller number of adolescents, relating to the use of applications connected to social media and self-presentation. As regards this second aspect, young people focus on the ability to be original and to provide a good description of themselves. With reference to the *Willingness* belief, the teens talk about a number of elements that are indicative of how much others are really motivated and willing to accomplish the adolescent's purpose, namely to establish a positive online relationship. Most of them refer to the presence of the other, which can be deduced from a set of elements such as from an interaction that occurs through comments and responses to photos, stories, and surveys, but also through simple likes, or from real and concrete proximity, which manifests itself in times of need, moving from the virtual world to the real world. Some adolescents instead focus on how communication exchanges should take place in the online environment. Someone, for example, underlines how online relationships should be guided by kindness and mutual respect, while others highlight the importance of respecting privacy. Finally, a smaller number of adolescents, refer to the trustworthiness of the person they interact with. As far as it concerns the *Unharmfulness* belief, most adolescents focus on knowledge of the other. In particular, they relate to people who they already know in real life, but also to a knowledge built up over time. Another element that emerges from the interviews is related to a sort of social closeness; in particular it emerges how having elements in common, such as friendships, interests, and ways of thinking, but also a similar age represents a factor that fosters long-distance relationships of trust. Further elements that can be deduced from the profile analysis are relevant; specifically teenagers, interacting online with people they do not know in real life, act with caution at first paying attention to profile analysis, looking, for example, at the quantity and type of shared photos, number of followers, and activity in terms of published posts. Later, the attention shifts to the interaction, in particular to the adopted communication method. Regarding the *Dependence* belief, some of the adolescents interviewed stated that they do not feel dependent. Most, however, depend on others essentially out of a need for visibility. Even the need for help and interaction makes the adolescent feel dependent on the other. In reference to the *Context* belief, among the favoring factors, adolescents mention relational and communicative accessibility and immediacy. However, most adolescents focus on the hindering factors, among which they mainly highlight the difficulty in interpreting the real communicative intentions of the other. Fewer adolescents speak about anti-normative behaviors, which are related to the unauthorized dissemination of information, such as personal photos or confidential conversations.

6.1.3 Conclusion

Overall, the present study provides a detailed photograph of how trust in online relationships between adolescents can be configured, guiding the comprehension of the ingredients that contribute to the attribution of trust in online environments. In this sense, this study represents an attempt to fill the lack of qualitative studies on online interpersonal trust in young people. The usefulness of this study rests on its theoretical basis as well as on its application aspects. At the theoretical level, it sheds light on the social and cognitive factors that contribute to the attribution of trust in virtual environments, enhancing the comprehension of this complex phenomenon. On the one hand, the study contributed to the advancement of the theoretical reference model, showing its potential application in online trust attribution processes. On the other hand, the study contributed to filling the literature gap not only concerning the knowledge of factors on which adolescents base their trust in social media but also concerning what these factors mean for young people. From an applicative point of view, knowing these factors can be useful for supporting young people's relationships on social media. The results of this study also guided the development of a scale to measure adolescents' trust in online interpersonal relationships, with particular attention to its different components and the weight of each in contributing to overall trust.

6.2 Study 2 - Peer Trust and Trust in Professionals in Facebook Support Groups for Mothers: A Qualitative Study

In this qualitative study [72], the trust relationships established among mothers, and between mothers and professionals within Facebook support groups for mothers, were investigated. Specifically, using the socio-cognitive model of trust [1, 2] as the theoretical framework, the various beliefs underpinning trust were explored to identify the elements characterizing peer trust and trust in professionals.

6.2.1 Method

The study was conducted within two Facebook groups dedicated to mothers, created for the purpose of exchanging opinions and advice, sharing experiences, and providing support during pregnancy and motherhood. Specifically, 10 mothers (average age 27) who are members of and regularly participate in at least one online support group were interviewed. A total of 20 interviews were conducted, two for each mother: one focusing on trust in peers and one on trust in professionals.

Data was collected through two in-depth interviews, following interview guides specifically designed for this study. The interview began with a general question about the trust placed in other mothers and the elements on which it is based. Subsequently, each specific belief, as theorized by the socio-cognitive model [1, 2], was explored. After completing this first interview, the same mothers participated in a second interview, which followed the same outline as the previous one but focused on the trust placed in professionals who provide support to mothers within the dedicated Facebook groups.

The interviews were faithfully transcribed, and an inductive content analysis was conducted on the textual material, following the procedure described by Elo and Kyngäs [73]. The analysis was carried out using NVivo10 software [71], which ensured that the complete text associated with each category remained accessible during the analysis. Specifically, after an initial open coding, the identified categories were grouped into higher-order categories, combining similar categories into broader ones. These subcategories were then grouped into general categories and subsequently into main categories, creating a system of subcategories, general categories, and main categories for each investigated trust belief.

6.2.2 Results

Below, the results for each belief of the model will be presented, detailing the findings for each actor in whom trust is placed, namely other mothers and professionals.

Trust in Other Mothers. Regarding the *Competence* belief, one of the most significant aspects is the lived experience as a mother. Another important aspect concerns the communication and language skills possessed by other mothers and the possibility of receiving an alternative perspective on the situation. Additional competencies are related to specific knowledge, such as that possessed by mothers who enjoy researching certain topics, personal characteristics/abilities like creativity, and educational qualifications, as often there are mothers in the groups who work as educators. Regarding the *Willingness* belief, the aspects that emerged include the style of writing and commenting, with particular reference to messages being written politely and in an engaging manner; the frequency of responses, the time dedicated, and the request for private contact details indicating particular interest; an attitude aimed at sharing experiences; the personal characteristics of the mothers, such as sincerity, empathy, patience, tact, consistency, and understanding; and a preference for constructive and peaceful interaction. Regarding the *Unharmfulness* belief, this is primarily linked to the positive and empathetic attitude of other mothers, the sharing of experiences, and having similar ideas and ideals. It also involves the nature of the responses given, which should be polite and avoid unconstructive criticism, the knowledge gained over time, and a profile that is not closed but provides adequate information. With reference to

Dependence belief, in general, mothers do not feel dependent on other mothers. However, some report a dependency related to the need for advice, opinions, or consolation. Finally, regarding *Context* belief, favorable factors include the ability to maintain anonymity during conversations, the possibility of forming small groups where mothers can meet more easily, and the ability to feel close even when physically distant. Hindering factors include the lack of physical contact and the management of remote communication.

Trust in professionals. Regarding the *Competence* belief, mothers seek qualities in professionals such as professionalism, the ability to recognize their own limits and refer to other "non-virtual" professionals, and the capacity to provide appropriate advice and responses. Additionally, the credentials held by the professionals are considered an important aspect of competence. Regarding the *Willingness* belief, mothers seek empathy and a willingness to listen in professionals, as well as humanity, a reassuring attitude, and the sharing of experiences and cases related to their profession and procedures. Additionally, the availability of professionals is perceived through their responses, which should be polite and respectful towards the person they are interacting with, avoiding evasiveness or haste. Regarding the *Unharmfulness* belief, it emerged that mothers feel secure interacting with professionals if they publicly expose themselves, have verifiable references, and demonstrate professionalism. Regarding the *Dependence* belief, it was found that mothers do not feel dependent on the professionals they consult in Facebook support groups. Instead, they generally turn to their trusted professional and seek validation of the advice they have received online. Finally, regarding the *Context* belief, only factors that hinder the relationship between mothers and professionals were identified. Specifically, these factors relate to the lack of direct contact with the professional online and the potential for misunderstandings arising from remote communication.

6.2.3 Conclusion

Overall, this study provided a comprehensive view of the elements underlying trust within Facebook support groups for mothers. Once again, the socio-cognitive theoretical model demonstrated its practical applicability, allowing for a detailed exploration of each belief and identifying the key aspects on which trust in remote relationships is based. Specifically, this analysis revealed that trust among peers, on one hand, and trust between mothers and professionals, on the other, although based on the same beliefs, revolves around different elements. This difference is likely related to the varying roles that mothers and professionals hold within Facebook groups, which inevitably influences the trust relationships that develop. Based on the data collected in this qualitative study, two ad-hoc scales were developed: one to measure trust among mothers and one to measure trust towards professionals. These scales were then administered to 84 mothers. They allowed not only for an assessment of overall trust but also for evaluating the "strength" of the various beliefs as determinants of trust.

6.3 Study 3 - Online psychological support in the COVID-19 era: Social representations, trust and perceived effectiveness from the perspectives of clients and professionals

In this study [74], the socio-cognitive model [1, 2] served as the theoretical foundation for developing two scales: one aimed at therapists and another at patients, with the goal of analyzing the trust invested in remote therapeutic relationships.

6.3.1 Method

The sample consists of 133 Italian licensed professionals qualified to provide psychological support (Age M=46), along with 716 Italian individuals who have had direct experience with psychological support in their lives (patients, Age M=33), selected regardless of the type (in-person, remote, or mixed) and duration of these experiences. A section of the questionnaire used for data collection focused on trust in the online therapeutic relationship, assessed using two 5-point Likert scales (1=strongly disagree; 5=strongly agree), specifically designed based on the trust construct as conceptualized in the socio-cognitive model [1, 2]. Trust in remote therapy was subsequently examined in relation to other factors, including general trust in online interactions and perceived effectiveness of remote psychological support. Regarding the scales on trust in the online therapeutic relationship, Principal Component Analysis (PCA) was first conducted, confirming the multifactorial nature of the scale for both therapists and patients. The calculation of Cronbach's alpha verified internal consistency, highlighting good reliability indices for both scales, both as a single indicator (α of .92 and .94, respectively) and in the explored individual dimensions, namely competence, willingness, and unhurmfulness (α of .92, .83, .84 for therapists and .86, .94, .82 for patients). To reduce the complexity of the items characterizing the trust scale, PCA was then conducted, applying Varimax axis rotation for each item group related to each dimension of trust. Table 1 summarizes the components identified for therapists, while Table 2 summarizes the components identified for patients.

Trust Dimension	Items Considered	Principal Components Identified	Cronbach's Alpha	Cumulative Variance (%)
Competence	31, 32, 33, 34, 35	Expression of Emotions (31)	n.a.	91%
		Relationship Management (32, 33)	.87	
		Metacognitive (34, 35)	.86	
Intentionality		Active Presence	.83	78%
		Physical Presence	.70	
Unhurmfulness		Setting Suitability	.84	76%

Table 1: Principal Component Analysis Results – Therapists. The numbering used for items follows the order of presentation in the survey: the relevant items are from 7 to 68, since item 1 was the informed consent, whereas items 2–6 asked for demographic information.

From the identified dimensions, synthetic indices were constructed (e.g., high/moderate/low competence). After constructing the different trust indices and recoding some variables such as age, to investi-

Trust Dimension	Items Considered	Principal Components Identified	Cronbach's Alpha	Cumulative Variance (%)
Competence	25, 26, 27, 28, 29	Communication Management (25)	n.a.	94%
		Boundary Management (26, 27)	n.a.	
		Distance Adaptation (28, 29)	.87	
Intentionality		Patient Attention	.92	93%
		Presence/Proximity	.93	
Unhurmfulness		Setting Suitability	.79	88%
		Profile Verifiability	n.a.	

Table 2: Principal Component Analysis Results – Patients. The numbering used for items follows the order of presentation in the survey: the relevant items are from 5 to 58, since item 1 was the informed consent, whereas items 2–4 asked for demographic information.

gate the relationships between various variables, correlation analyses were conducted, and statistical significances were verified by calculating χ^2 and Pearson's R coefficient.

6.3.2 Results

The results of this study revealed that, regarding trust in the online therapeutic relationship, 51.1% of therapists and 49.9% of patients reported having high trust. Concerning therapists, trust in the online therapeutic relationship appears to increase with experience in the online modality ($R = .232$, $p < .05$). Additionally, trust in the remote therapeutic relationship seems to enhance satisfaction with the online modality ($R = .548$, $p < .01$). Regarding patients, trust in the online therapist decreases with age ($R = . - 127$, $p < .01$). However, increased trust in online interactions with others correlates with increased trust in the online therapist ($R = .128$, $p < .01$). Furthermore, trust in the professional offering online psychological support is primarily high for those who experienced remote or mixed therapy, and low for those who had in-person psychotherapy ($\chi^2=57.30$,

$p < 0.05$). Finally, for patients, satisfaction with the online modality seems to increase with growing trust in the remote therapeutic relationship (R=.416, p<.01).

Examining the components on which trust is based, it is noticeable that both therapists and patients primarily base their trust on *willingness* (interpreted by therapists as active presence of patients during sessions and physical presence, and by patients as attention shown by the professional and presence/proximity during the session). This is followed by *unhurmfulness* (interpreted by therapists as suitability of the setting, and by patients as suitability of the setting and verifiability of the professional's profile) and, lastly, competence (interpreted by therapists as patients' ability to manage the relationship, followed by metacognitive skills and the ability to express their emotions; interpreted by patients as the professional's ability to manage and process communicative exchanges through the computer tool, followed by the ability to set boundaries to their willingness and adapt to the distance). Regarding *dependence*, most therapists believe that the relationship with patients offering online psychological support depends little or not at all on the virtual environment, and most patients think they depend little or not at all on the professional offering them online psychological support. Finally, both therapists and patients mostly consider the virtual environment as not hindering the therapeutic relationship (according to some therapists, the virtual context would enhance the relationship due to the familiarity of the connecting location and the reduction of travel that promotes relaxation for both patient and therapist; according to some patients, the virtual context would favor the relationship due to the distance that allows continuity in adverse circumstances).

6.3.3   Conclusion

Results of this study highlight the critical role of online interpersonal trust in the adoption of remote therapy. Trust in online relationships significantly facilitates engagement in remote therapy for both therapists and patients. Both therapists and patients who exhibit high levels of trust in online relationships, particulary in remote therapeutic context, are more prevalent. This finding diverges from previous studies, such as Haberstroh and colleagues[75], which suggested that the absence of traditional interpersonal cues in remote therapy might limit trust in therapist. Instead, results of this study align more closely with the findings of Nguyen and Canny (2007) on general trust in online relationships, emphasizing that trust levels do not significantly differ between online and face-to-face encounters when adequate technical support is provided. Increased exposure to online therapy positively correlates with trust in remote therapeutic relationship for both therapists and patients. This underscores the role of experiential learning in fostering trusting therapeutic connections, echoing the insights of Békés and Aafjes-van Doorn [76]. Furthermore, findings of this research indicate that younger patients and individuals with higher trust levels in online interactions generally place more trust in online therapists. Examining the components of trust in remote therapeutic relationship, competencies appear to be less influential compared to the therapist's willingness and unhurmfulness. This trend aligns with broader healthcare research during the pandemic, where patients valued positively a physician's willingness to provide care over their specific expertise [77]. In times of uncertainty like the pandemic, reliance on the therapist's presence and and responsiveness outweighs their technical abilities. Therapist place greater trust in the therapeutic relationship when patients consistently attend sessions, actively participate, and connect from a suitable, private location conducive to honest expression. Patients, on the other hand, trust therapist who demonstrate genuine interest, attempt to foster emotional closeness despite the virtual barrier, and present a credible professional profile. Notably, the virtual environment is generally perceived as facilitating openness among patients, as the perceived distance can encourage them to express their problems and emotions more freely. This finding resonates with Saladino et al.'s [78], emphasizing how a familiar context can bridge trust gaps among individuals less inclined to embrace remote therapy.

In conclusion, this study underscores the application potential of the socio-cognitive theoretical model, which guided the development of research tools enabling measurement of trust levels and analysis of its components. In particular, this study stands out as the first to thoroughly analyse trust in remote therapeutic relationships. By examining trust in detail, this study offers valuable insights into the factors influencing the development of trust between therapists and patients in the context of online therapy.

Based on these findings, the study holds strong potential for practical application in the field of remote therapy. It provides valuable insights for defining and cultivating a trustworthy remote therapeutic relationship, which is essential for the success of a psychological interventions.

6.4 Study 4 - Trust and Attachment in the World of Dating Apps: An Exploratory Study

In this study [79], the socio-cognitive model [1, 2] served as the theoretical foundation for developing a scale to assess trust in relationships formed through dating apps.

6.4.1 Method

The study involved 81 participants (73% female, 27% male, average age = 38), all of whom were adults and users of dating apps. Of these, almost all (93%) reported using the apps moderately and, in 83% of cases, considered their use to be widely accepted in contemporary society. Only 16% of the sample viewed using dating apps as something to be ashamed of. Regarding their expectations from dating apps, while 26% of the sample stated that they used them explicitly for casual relationships, the majority (58%) reported having no clear expectations in this regard. Meanwhile, 15% expected to establish a lasting relationship, and 1% hoped to make friends.

A section of the questionnaire used for data collection focused on trust in relationships established through online dating apps. Specifically, trust was measured using an ad-hoc scale based on the socio-cognitive model [1, 2]. In addition to assessing the overall level of trust, the scale, consisting of 29 items rated on a 5-point Likert scale

(1=strongly disagree; 5=strongly agree), examined the dimensions of competence, willingness, unharmfulness, dependence, and context.

Regarding the data analysis, specifically concerning the trust scale, a Principal Component Analysis was first conducted, which confirmed the multifactorial nature of the scale. Subsequently, internal consistency was assessed using Cronbach's alpha, which demonstrated good overall reliability ($\alpha = .889$), as well as reliability within individual dimensions, particularly for competence ($\alpha = .647$), willingness ($\alpha = .902$), and safety ($\alpha = .735$). Finally, to compare the different components of trust, the means of the items related to competence, willingness, unharmfulness, dependence, and context were calculated.

6.4.2 Results

The results revealed that, concerning overall trust in relationships established through dating apps, just over half of the sample (54%) reported having little or no trust in the person they met online, 41% claimed to trust somewhat, and the remaining 5% reported having a lot or complete trust. When considering the various beliefs that constitute trust, the most significant for the respondents were competence beliefs (M = 3.76), followed by willingness beliefs (M = 3.74) and unharmfulness beliefs (M = 3.44).

Delving into the specific beliefs, regarding *competence*, the sample attributed the greatest importance to the ability to respond in chats with attention to dialogue (M=4.30), followed by the ability to share personal stories without showing off (M=3.63) and the creation of a photo gallery showcasing different aspects of one's life (M=3.37). Among *willingness* beliefs, the most significant elements identified by respondents were kindness and courtesy (M=4.19), willingness to deepen the relationship in person (M=4.01), and dedicating time to the interaction (M=3.99). Moving to *unharmfulness* beliefs, respondents placed the highest importance on respecting each other's boundaries and needs (M=4.27), followed by the ability to verify the profile on various social media (M=3.83) and sharing passions and interests (M=3.47). Regarding *dependence* beliefs, participants predominantly expressed the need for a relationship, followed by the need to combat loneliness, receive attention, and increase self-esteem and confidence. Finally, concerning *context* beliefs, factors that facilitate building a relationship online included the ability to overcome spatial and temporal barriers and the functionality of apps to select suitable profiles and create increasingly appropriate matches. On the other hand, factors that hinder relationship building included the ability of apps to maintain multiple relationships simultaneously, the ease of manipulating information, and the difficulty of expressing emotions.

6.4.3 Conclusion

Overall, the results of this study revealed a predominantly distrustful sample regarding relationships established through dating apps, although a significant portion expressed moderate trust. Competence and willingness emerged as the key components that seem to most favor the attribution of trust. The dating world appears to encompass communication aspects that need to be understood and mastered. Among the most important skills, the ability to respond thoughtfully in chats, the ability to share personal stories, and the skill to create an appropriate photo gallery to present oneself stood out. The willingness of the other person seems to be conveyed not only through courteous and kind manners but also by showing genuine interest in the other person and dedicating time to the interaction. Similar to the previously discussed studies, this research demonstrated how the socio-cognitive model of trust [1, 2], originally a theoretical construct, can be applied to the study of social phenomena. In this specific case, it provided a detailed snapshot of the elements on which trust in online dating is based.

7 Conclusion

This chapter does not aim to be an exhaustive work on a topic that is quite complex and multifaceted, but rather to provide a point of reflection, a starting point for engaging in an open and continuously evolving debate. The intricate issue of online trust, in particular, requires further study and in-depth analysis, potentially contributing to the expansion of literature on online interpersonal trust, whose theoretical and research advancements remain secondary compared to studies in other areas, such as online e-commerce. An emerging field that requires particular attention in terms of trust is artificial intelligence, a new cognitive challenge that should be addressed and integrated into the current scientific debate. In this regard, it would be useful to consider whether, and how, the socio-cognitive model of trust [1, 2] could find practical application, thereby contributing to the expansion of knowledge in this field as well.

References

[1] R. Falcone and C. Castelfranchi, "The socio-cognitive dynamics of trust: Does trust create trust?," in *Trust in Cyber-societies: Integrating the Human and Artificial Perspectives*, pp. 55–72, Springer, 2001.

[2] C. Castelfranchi and R. Falcone, *Trust theory: A socio-cognitive and computational model.* John Wiley & Sons, 2010.

[3] A. Ficorilli, "La relazione di fiducia: un approccio bioetico alle questioni della cura," 2014.

[4] T. Yamagishi and M. Yamagishi, "Trust and commitment in the united states and japan," *Motivation and emotion*, vol. 18, no. 2, pp. 129–166, 1994.

[5] R. J. Lewicki and B. B. Bunker, *Trust in relationships: A model of development and decline.* Jossey-Bass/Wiley, 1995.

[6] D. H. McKnight and N. L. Chervany, "What is trust? a conceptual analysis and an interdisciplinary model," p. 382, 2000.

[7] M. Koller, "Risk as a determinant of trust," *Basic and applied social psychology*, vol. 9, no. 4, pp. 265–276, 1988.

[8] P. M. Doney, J. P. Cannon, and M. R. Mullen, "Understanding the influence of national culture on the development of trust," *Academy of management review*, vol. 23, no. 3, pp. 601–620, 1998.

[9] D. M. Rousseau, S. B. Sitkin, R. S. Burt, and C. Camerer, "Not so different after all: A cross-discipline view of trust," *Academy of management review*, vol. 23, no. 3, pp. 393–404, 1998.

[10] B. L. Connelly, T. R. Crook, J. G. Combs, D. J. Ketchen Jr, and H. Aguinis, "Competence-and integrity-based trust in interorganiza-

tional relationships: which matters more?," *Journal of Management*, vol. 44, no. 3, pp. 919–945, 2018.

[11] M. Y. Ahn and H. H. Davis, "Sense of belonging as an indicator of social capital," *International Journal of Sociology and Social Policy*, vol. 40, no. 7/8, pp. 627–642, 2020.

[12] V. Salmi, M. Smolej, and J. Kivivuori, "Crime victimization, exposure to crime news and social trust among adolescents," *Young*, vol. 15, no. 3, pp. 255–272, 2007.

[13] B. A. Randall, K. J. Rotenberg, C. J. Totenhagen, M. Rock, and C. Harmon, "A new scale for the assessment of adolescents' trust beliefs," *Interpersonal trust during childhood and adolescence*, p. 247, 2010.

[14] H. Bulińska-Stangrecka and A. Bagieńska, "Investigating the links of interpersonal trust in telecommunications companies," *Sustainability*, vol. 10, no. 7, p. 2555, 2018.

[15] F. Krueger and A. Meyer-Lindenberg, "Toward a model of interpersonal trust drawn from neuroscience, psychology, and economics," *Trends in neurosciences*, vol. 42, no. 2, pp. 92–101, 2019.

[16] R. D. Putnam, "Bowling alone: The collapse and revival of american community," *Simon Schuster*, 2000.

[17] J. B. Rotter, "Interpersonal trust, trustworthiness, and gullibility.," *American psychologist*, vol. 35, no. 1, p. 1, 1980.

[18] Y. Bart, V. Shankar, F. Sultan, and G. L. Urban, "Are the drivers and role of online trust the same for all web sites and consumers? a large-scale exploratory empirical study," *Journal of marketing*, vol. 69, no. 4, pp. 133–152, 2005.

[19] A. Beldad, M. De Jong, and M. Steehouder, "How shall i trust the faceless and the intangible? a literature review on the antecedents of online trust," *Computers in human behavior*, vol. 26, no. 5, pp. 857–869, 2010.

[20] V. Shankar, G. L. Urban, and F. Sultan, "Online trust: a stakeholder perspective, concepts, implications, and future directions," *The Journal of strategic information systems*, vol. 11, no. 3-4, pp. 325–344, 2002.

[21] S. L. Jarvenpaa and V. S. Rao, "Trust in online consumer exchanges: Emerging conceptual and theoretical trends," in *E-Commerce and the Digital Economy*, pp. 255–272, Routledge, 2015.

[22] C. L. Corritore, B. Kracher, and S. Wiedenbeck, "On-line trust: concepts, evolving themes, a model," *International journal of human-computer studies*, vol. 58, no. 6, pp. 737–758, 2003.

[23] Y. D. Wang and H. H. Emurian, "An overview of online trust: Concepts, elements, and implications," *Computers in human behavior*, vol. 21, no. 1, pp. 105–125, 2005.

[24] Y.-H. Chen and S. Barnes, "Initial trust and online buyer behaviour," *Industrial management & data systems*, vol. 107, no. 1, pp. 21–36, 2007.

[25] Y.-H. Fang, C.-M. Chiu, and E. T. Wang, "Understanding customers' satisfaction and repurchase intentions: An integration of is success model, trust, and justice," *Internet research*, vol. 21, no. 4, pp. 479–503, 2011.

[26] K. Hassanein and M. Head, "Manipulating perceived social presence through the web interface and its impact on attitude towards online shopping," *International journal of human-computer studies*, vol. 65, no. 8, pp. 689–708, 2007.

[27] D. J. Kim, D. L. Ferrin, and H. R. Rao, "Trust and satisfaction, two stepping stones for successful e-commerce relationships: A longitudinal exploration," *Information systems research*, vol. 20, no. 2, pp. 237–257, 2009.

[28] M. Tagliaferri, "Reviewing the case of online interpersonal trust," *Foundations of Science*, vol. 28, no. 1, pp. 225–254, 2023.

[29] F. N. Koranteng, I. Wiafe, F. A. Katsriku, and R. Apau, "Understanding trust on social networking sites among tertiary students: An empirical study in ghana," *Applied Computing and Informatics*, vol. 19, no. 3/4, pp. 209–225, 2023.

[30] J. Nesi and M. J. Prinstein, "In search of likes: Longitudinal associations between adolescents' digital status seeking and health-risk behaviors," *Journal of Clinical Child & Adolescent Psychology*, 2019.

[31] J. Fox and B. McEwan, "Distinguishing technologies for social interaction: The perceived social affordances of communication channels scale," *Communication Monographs*, vol. 84, no. 3, pp. 298–318, 2017.

[32] C. Flavian, M. Guinalíu, and P. Jordan, "Antecedents and consequences of trust on a virtual team leader," *European journal of management and business economics*, vol. 28, no. 1, pp. 2–24, 2019.

[33] M. Turilli, A. Vaccaro, and M. Taddeo, "The case of online trust," *Knowledge, Technology & Policy*, vol. 23, pp. 333–345, 2010.

[34] T. Yamagishi, M. Kikuchi, and M. Kosugi, "Trust, gullibility, and social intelligence," *Asian Journal of Social Psychology*, vol. 2, no. 1, pp. 145–161, 1999.

[35] F. N. Egger, ""trust me, i'm an online vendor" towards a model of

trust for e-commerce system design," in *CHI'00 extended abstracts on Human factors in computing systems*, pp. 101–102, 2000.

[36] T. Stratford, "Etrust: building trust online," *Journal on Integrated Communications*, vol. 10, no. 1999, pp. 75–81, 1999.

[37] Y. A. Kim and M. A. Ahmad, "Trust, distrust and lack of confidence of users in online social media-sharing communities," *Knowledge-Based Systems*, vol. 37, pp. 438–450, 2013.

[38] S. L. Jarvenpaa, T. R. Shaw, and D. S. Staples, "Toward contextualized theories of trust: The role of trust in global virtual teams," *Information systems research*, vol. 15, no. 3, pp. 250–267, 2004.

[39] M. J. Hoff, ""i don't conversate with those i don't know": the role of trust/distrust in online engagement.," *Digital Culture & Education*, vol. 8, no. 2, 2016.

[40] K. Koidl and K. Kapanova, "Interpersonal trust within social media applications: A conceptual literature review," *The Psychology of Trust*, 2022.

[41] E. T. Lim, C.-W. Tan, D. Cyr, S. L. Pan, and B. Xiao, "Advancing public trust relationships in electronic government: The singapore e-filing journey," *Information Systems Research*, vol. 23, no. 4, pp. 1110–1130, 2012.

[42] D. Gefen, E. Karahanna, and D. W. Straub, "Trust and tam in online shopping: An integrated model," *MIS quarterly*, pp. 51–90, 2003.

[43] S. Kim and H. Park, "Effects of various characteristics of social commerce (s-commerce) on consumers' trust and trust performance," *International journal of information management*, vol. 33, no. 2, pp. 318–332, 2013.

[44] T. U. Daim, A. Ha, S. Reutiman, B. Hughes, U. Pathak, W. Bynum, and A. Bhatla, "Exploring the communication breakdown in global virtual teams," *International journal of project management*, vol. 30, no. 2, pp. 199–212, 2012.

[45] E. C. Kasper-Fuehrer and N. M. Ashkanasy, "Communicating trustworthiness and building trust in interorganizational virtual organizations," *Journal of management*, vol. 27, no. 3, pp. 235–254, 2001.

[46] P. Palvia, "The role of trust in e-commerce relational exchange: A unified model," *Information & management*, vol. 46, no. 4, pp. 213–220, 2009.

[47] S. L. Jarvenpaa and D. E. Leidner, "Communication and trust in global virtual teams," *Journal of computer-mediated communication*, vol. 3,

no. 4, p. JCMC346, 1998.

[48] A. L. Drolet and M. W. Morris, "Rapport in conflict resolution: Accounting for how face-to-face contact fosters mutual cooperation in mixed-motive conflicts," *Journal of Experimental Social Psychology*, vol. 36, no. 1, pp. 26–50, 2000.

[49] J. M. Wilson, S. G. Straus, and B. McEvily, "All in due time: The development of trust in computer-mediated and face-to-face teams," *Organizational behavior and human decision processes*, vol. 99, no. 1, pp. 16–33, 2006.

[50] H. Wichman, "Effects of isolation and communication on cooperation in a two-person game.," *Journal of Personality and Social Psychology*, vol. 16, no. 1, p. 114, 1970.

[51] V. J. Dubrovsky, S. Kiesler, and B. N. Sethna, "The equalization phenomenon: Status effects in computer-mediated and face-to-face decision-making groups," *Human-computer interaction*, vol. 6, no. 2, pp. 119–146, 1991.

[52] S. P. Weisband, "Group discussion and first advocacy effects in computer-mediated and face-to-face decision making groups," *Organizational behavior and human decision processes*, vol. 53, no. 3, pp. 352–380, 1992.

[53] N. Bos, J. Olson, D. Gergle, G. Olson, and Z. Wright, "Effects of four computer-mediated communications channels on trust development," in *Proceedings of the SIGCHI conference on human factors in computing systems*, pp. 135–140, 2002.

[54] D. T. Nguyen and J. Canny, "Multiview: improving trust in group video conferencing through spatial faithfulness," in *Proceedings of the SIGCHI conference on Human factors in computing systems*, pp. 1465–1474, 2007.

[55] K. Koidl, O. Conlan, W. Reijers, M. Farrell, and M. Hoover, "The bigfoot initiative: An investigation of digital footprint awareness in social media," in *Proceedings of the 9th International Conference on Social Media and Society*, pp. 120–127, 2018.

[56] F. Sabatini and F. Sarracino, "Online social networks and trust," *Social Indicators Research*, vol. 142, pp. 229–260, 2019.

[57] C. E. Beaudoin, "Explaining the relationship between internet use and interpersonal trust: Taking into account motivation and information overload," *Journal of Computer-Mediated Communication*, vol. 13, no. 3, pp. 550–568, 2008.

[58] S. Henderson and M. Gilding, "'i've never clicked this much with anyone in my life': trust and hyperpersonal communication in online friendships," *New media & society*, vol. 6, no. 4, pp. 487–506, 2004.

[59] S. Valenzuela, N. Park, and K. F. Kee, "Is there social capital in a social network site?: Facebook use and college students' life satisfaction, trust, and participation," *Journal of computer-mediated communication*, vol. 14, no. 4, pp. 875–901, 2009.

[60] X. Cheng, S. Fu, and G.-J. de Vreede, "Understanding trust influencing factors in social media communication: A qualitative study," *International Journal of Information Management*, vol. 37, no. 2, pp. 25–35, 2017.

[61] B. Friedman, P. H. Khan Jr, and D. C. Howe, "Trust online," *Communications of the ACM*, vol. 43, no. 12, pp. 34–40, 2000.

[62] A. B. Seligman, *The problem of trust.* Princeton University Press, 2000.

[63] H. Nissenbaum, "Securing trust online: Wisdom or oxymoron," *BUL Rev.*, vol. 81, p. 635, 2001.

[64] P. de Vries, "Social presence as a conduit to the social dimensions of online trust," in *Persuasive Technology: First International Conference on Persuasive Technology for Human Well-Being, PERSUASIVE 2006, Eindhoven, The Netherlands, May 18-19, 2006. Proceedings 1*, pp. 55–59, Springer, 2006.

[65] P. Papadopoulou, "Applying virtual reality for trust-building e-commerce environments," *Virtual Reality*, vol. 11, pp. 107–127, 2007.

[66] E. Colì, M. Paciello, E. Lamponi, R. Calella, and R. Falcone, "Adolescents and trust in online social interactions: A qualitative exploratory study," *Children*, vol. 10, no. 8, p. 1408, 2023.

[67] S. Kvale, "The 1,000-page question," *Qualitative inquiry*, vol. 2, no. 3, pp. 275–284, 1996.

[68] S. E. Rabionet, "How i learned to design and conduct semi-structured interviews: An ongoing and continuous journey," *The Qualitative Report*, vol. 14, no. 3, pp. 203–206, 2009.

[69] B. Downe-Wamboldt, "Content analysis: method, applications, and issues," *Health care for women international*, vol. 13, no. 3, pp. 313–321, 1992.

[70] P. Mayring, "Qualitative inhaltsanalyse [28 absätze]," in *Forum Qualitative Sozialforschung/Forum: Qualitative Social Research*, vol. 1, pp. 2–00, 2000.

[71] Q. I. P. Ltd, "Nvivo qualitative data analysis software," *Version 10,*

2012.

[72] E. Colì, A. Petillo, and R. Falcone, "Fiducia tra pari e fiducia verso i professionisti nei gruppi facebook di supporto dedicati alle mamme: uno studio qualitativo." Unpublished manuscript, 2024.

[73] S. Elo and H. Kyngäs, "The qualitative content analysis process," *Journal of advanced nursing*, vol. 62, no. 1, pp. 107–115, 2008.

[74] E. Colì, L. Gavrila, D. Cozzo, and R. Falcone, "Online psychological support in the covid-19 era: Social representations, trust and perceived effectiveness from the perspectives of clients and professionals," *Counselling and Psychotherapy Research*, vol. 00, pp. 1–17, 2024.

[75] S. Haberstroh, T. Duffey, M. Evans, R. Gee, and H. Trepal, "The experience of online counseling," *Journal of Mental Health Counseling*, vol. 29, no. 3, pp. 269–282, 2007.

[76] V. Békés and K. Aafjes-van Doorn, "Psychotherapists' attitudes toward online therapy during the covid-19 pandemic.," *Journal of Psychotherapy Integration*, vol. 30, no. 2, p. 238, 2020.

[77] E. Colì, G. Pavanello, and R. Falcone, "La relazione tra paziente con artrite reumatoide e reumatologo durante la pandemia da covid-19: una ricerca esplorativa su rappresentazione sociale e ruolo della fiducia," *Psicologia della salute: quadrimestrale di psicologia e scienze della salute: 1, 2022*, pp. 137–160, 2022.

[78] V. Saladino, D. Algeri, and V. Auriemma, "The psychological and social impact of covid-19: new perspectives of well-being," *Frontiers in psychology*, vol. 11, p. 577684, 2020.

[79] E. Colì, F. Musotto, and R. Falcone, "Fiducia e attaccamento nel mondo delle dating app. una ricerca esplorativa." Unpublished manuscript, 2024.

Trust in the Rational and Social Bands of Cognitive Architectures

Antonio Lieto

University of Salerno (DISPC), Cognition Interaction and Intelligent Technologies Laboratory (CIIT Lab), and ICAR-CNR

alieto@unisa.it

Abstract

In this short chapter I propose some possible research directions addressing the problem of how cognitive, formal and computational models of trust (i.e. one of the main areas of Rino Falcone's contributions) can play a major role in the development of cognitive modeling and AI research community in the context of what Allen Newell called the rational and social bands.

1 Introduction

In the context of cognitive systems and architectures, a relevant distinctions between the different types of actions and capabilities to be modeled in a general AI systems is based on the schema between different time-scales, types of operations and bands of cognition proposed by Allen Newell (the so-called Newell's time scale of human actions). In the Newell framework [1], each band captures different types of human experience and represents different types of information processing mechanisms required to describe the levels within them. In

The work is partially supported by the project «Cognitive Architectures for Intelligent Interactive Systems» (2024-2026) founded by the University of Salerno.

particular: the neural band is described in terms of cellular biology, the cognitive band in terms of symbolic information processing, the rational band in terms of knowledge, reasoning and goals, and the social band in terms of distributed, multi-agent processing.

In the present work I argue that the research on Trust carried out by Rino Falcone and his group at the Institute of Cognitive Science and Technologies of the CNR (in particualar with Cristiano Castelfranchi [2, 3, 4]) represents an important contribution that can be leveraged both for the development of the so called Trustworthy AI systems as well as for the development of simulative models of rational interactions in the context of cognitive modelling. In order to better contextualize the context where, in my opinion, Rino Falcone's research can play an important role I will use the the Newell's time scale of human actions and will suggest a space for future developments in the context of cognitive artificial agents whose behavior is governed by cognitively grounded software instantiantions: cognitive architectures (CAs).

2 Cognitive Architectures

The expression "Cognitive Architectures" (as reported in [5]) indicates "both abstract models of cognition, in natural and artificial agents, and the software instantiations of such models".

During the last few decades many different cognitive architectures have been realized ([6]) and there have been attempts to unify the main insights coming from 40 years of research in the realization of a Common Model of Cognition. Overall, most of the work carried out on such systems have been focused on the deliberate act level of the Cognitive band (see Figure 1). The interest of the cognitive models of trust provided by Falcone's research lies in the fact that the phenomenon he analyzed, on the other hand, address issues involving the rational and social bands (and can potentially involve processes at higher bands) and, in addition, involves multiple agents. Given this state of affairs, such models could be integrated in different ways with existing cognitive architectures extending, de facto, their knowledge

Table 1
Newell's Time Scales of Human Action

Scale (sec)	Time Units	System	World (theory)
10^7	months		
10^6	weeks		Social Band
10^5	days		
10^4	hours	Task	
10^3	10 min	Task	Rational Band
10^2	minutes	Task	
10^1	10 sec	Unit task	
10^0	1 sec	Operations	Cognitive Band
10^{-1}	100 msec	Deliberate act	
10^{-2}	10 msec	Neural circuit	
10^{-3}	1 msec	Neuron	Biological Band
10^{-4}	$100\mu s$	Organelle	

Figure 1: Newell's Time Scales of Human Action.

processing capabilities when dealing with macro-cognitive (and meta-cognitive) phenomena. In particular, Falcone's models could either work as a blueprint for modelling functional trustworhy AI agents or could be internally integrated with the micro-level mechanisms of current cognitive architectures. Alternatively, an approach can be chosen in which the higher level of abstraction of Falcone's model is externally linked to the knowledge representation and inference mechanisms of the different memory modules of cognitive architectures with the possibility of being then reduced to the representation and processing mechanisms of a given CA when agent's actions require to operate with processes at the cognitive or biological band. Any of these integration choice will have its own advantages and limitations that will impact the possible use of such models as Functional or Structural models of Trust. While this aspect will be further analyzed later, in the following, I provide additional elements of discussion for what concern the current state of affairs for both the Rational and the Social Band since they represent the major areas of interest for the possible integration and development of Rino Falcone's work in cognitive architectures.

3 Rational Band

Newell equates the rational band with the knowledge level [7]. The knowledge level refers to the level at which knowledge becomes abstract and can be treated largely independently from the physical systems that process it. Currently, the idea that specialized architectural modules able to perform activities belonging to the rational band (e.g. planning, language processing and Theory of Mind) should be developed is not majoritarian. The underlying assumption is that all such activities should arise based on the composition of processes executed during different cognitive cycles according to specific computational models. Despite this position, however, few cognitive architectures have formulated computational approaches to this effect, in particular to the Theory of Mind (ToM)[1]. Among these few works, there is that one by [8] where the SIGMA cognitive architecture [9] is used to demonstrate two distinct mechanisms (automatic processing vs. controlled reasoning) for ToM using as an example several single-stage simultaneous-move games, in particular, the well-known Prisoners Dilemma. Authors left open the possibility of using SIGMA's learning capability to allow the agents to learn models of each other in a repeated game setting. ACT-R [10] has also been used to build several models of false belief and second-order false belief task [11]. Additionally, in [12] several scenarios were set up using ACT-R to show how ToM ability can improve the quality of interaction between a robot and a human by predicting what a person will do in different situations; e.g., that a person may forget something and may need to be reminded or that a person cannot see everything the robot sees.

Despite such efforts, however, the modeling attempts of such rational aspects remains still limited (while there are many more models developed for other phenomena concerning, for example, planning, or natural language processing etc.). In the following section, I similarly introduce some of the main issues concerning the Social Band

---

[1]I focus here mainly on ToM since this aspect is a crucial one also in models of trusts concerning risks, beliefs, predictive and interactive reasoning abilities in artificial cognitive agents.

reminding to [13] for further details.

4 Social Band

Events and processes happening at the Social Band can take place over longer time scales (days, weeks, and months). However, they are composed of social events/processes happening at much shorter time scales (seconds, minutes, and hours) and, as evidenced by Anderson [14] may be supported by cognitive and rational processing in individuals at lower time scales. To discuss the potentially controversial topic of motivations, some of our needs, such as physiological needs, come from the bottom up and may explain why we developed high level cognition, i.e., to solve those problems. The social level can also result in goals that people are intrinsically motivated to pursue and our cognition needs to be able to account for that. Our biological and social needs provide goals and may have resulted in innate cognitive capabilities. For example, we seem to have some innate capabilities associated with social cognition, such as perceiving other people, processing direction of gaze, determining intentions of others, limitations on knowing others as individuals (Dunbar's Number), and theory of mind. Related areas such as cooperation, trust, collective action are also tasks or behaviors that arise at the social level.

Concerning the modeling of macro-scale or macro-cognition events there are, as partially anticipated before, two different perspectives currently debated in the literature. On the one hand such elements are seen as too high level to be included within the minimal information processing mechanisms of a general cognitive architecture and, as such, are left to specific computational models to be developed on the top of such architectures. ACT-R, for example, has been used to study social behaviors and distributed collective decision-making processes which must balance diverse individual preferences with an expectation for collective unity. Romero et al. [15, 16] proposed a multi-agent approach where cognitive agents have to reach global consensus while opposing tensions are generated by conflicting incentives, so agents have to decide whether to follow the most influential agent, follow the majority, negotiate with others, come to an agreement when conflicting interests are present, or keep a stubborn position.

Alternatively, the PolyScheme cognitive architecture [17] applies ToM to perspective taking in the human-robot interaction scenario, that is, the robot can model the scene from the human's perspective and use this information to disambiguate the command when moving in a scenario with multiple occluding elements [18]. However, although there are computational models of social interactions in a practical sense, there are no current CAs supporting research in social cognition, at best there are frameworks (as examples, BDI [19] and PECS [20]). I think that it is in this space that Falcone's work can be leveraged to extend the current (limited) macro-cognitive capabilities of agents governed by cognitive architectures. Apart from the, above mentioned, different possibilities concerning the integration choices, however, another important design decision concerns the ultimate goal with which models of Trust can be included in artificial systems. They could be included, indeed, as Functional or Structural Models.

5 Functional vs Structural Models of Trust

In the literature on cognitive and intelligent systems design an important difference is the one between functionalist and structural models. Fnctionalist models are based on a weak equivalence (i.e. the equivalence in terms of functional organization) between cognitive processes and AI procedures, and - as such - cognitive architectures (and the agents that use them) built with such design perspective are used as blueprint for building integrated AI systems having no explanatory role with respect to the way in which a social phenomenon like trust is built and processed in humans (or other animals). In other words: functionalist systems are not concerned about whether or not a given behavioral model is a faithful model of cognition. On the other hand, "structural" models are based on a more constrained equivalence between AI procedures and their corresponding cognitive processes. This aspect leads them to be used, under certain conditions, as simulative tools of our mental and interactiive abilities. Even if, by definition, Falcone's models are certainly more suited to be implemented as higher order structural cognitive models, and therefore are more suitable to be integrated in architectures like ACT-R, they could nonetheless be used also in Functional AI systems where the modular components of the architectures do play a very limited simulative role (this could be the case, for example of their integration in architectures like SOAR [21]). This versatility represents an advantage since the two research agendas pursued by such different stances are not necessarily conflicting since they are beneficial for different important purposes: advancements in science and in engineering. In the next decade it will be important to include such type of macro-cognitive abilities in AI systems and, whatever will be the different goals and choices of human designers, I think that the work developed by Rino Falcone along his research trajectory could be very beneficial.

6 Conclusion

I have proposed a possible space of application of the work of Rino Falcone on cognitive and computational models of trust in the context of cognitive architectures. In particular, by highlighting the relative low development and integration of macro-cognitive models within such systems, I have proposed different type of integration with existing cognitive architectures and different epistemic purposes through which researchers can use and re-use the models developed by Rino Falcone.

References

[1] Newell, A. (1990). Unified theories of cognition, HUP.
[2] Castelfranchi, C., & Falcone, R. (2010). Trust theory: A socio-cognitive and computational model. John Wiley & Sons.
[3] Falcone, R., & Castelfranchi, C. (2001). The socio-cognitive dynamics of trust: Does trust create trust?. In Trust in Cyber-societies: Integrat-

ing the Human and Artificial Perspectives (pp. 55-72). Springer Berlin Heidelberg.

[4] Falcone, R., & Castelfranchi, C. (2001). Social trust: A cognitive approach. Trust and deception in virtual societies, 55-90.

[5] Lieto, A. (2021). Cognitive Design for Artificial Minds, Routledge.

[6] Kotseruba, Iuliia, and John K. Tsotsos. "40 years of cognitive architectures: core cognitive abilities and practical applications." Artificial Intelligence Review 53, no. 1 (2020): 17-94.

[7] Allen Newell. The knowledge level. Artificial intelligence, 18(1):87–127, 1982.

[8] David V Pynadath, Paul S Rosenbloom, Stacy C Marsella, and Lingshan Li. Modeling two-player games in the sigma graphical cognitive architecture. In Int. Conf. on AGI, pages 98–108, 2013.

[9] Paul S Rosenbloom. Rethinking cognitive architecture via graphical models. Cognitive Systems Research, 12(2):198–209, 2011.

[10] John R Anderson, Daniel Bothell, Michael D Byrne, Scott Douglass, Christian Lebiere, and Yulin Qin. An integrated theory of the mind. Psychological review, 111(4):1036, 2004.

[11] Lara M Triona, Amy M Masnick, and Bradley J Morris. What does it take to pass the false belief task? an ACT-R model. In Cognitive Science Society, volume 24, 2002.

[12] J Gregory Trafton, Laura M Hiatt, Anthony M Harrison, Franklin P Tamborello II, Sangeet S Khemlani, and Alan C Schultz. ACT-R/E: An embodied cognitive architecture for human-robot interaction. Human-Robot Interaction, 2(1):30–55, 2013.

[13] Lieto, A., Kennedy, W. G., Lebiere, C., Romero, O. J., Taatgen, N., West, R. L. (2018). Higher-level knowledge, rational and social levels constraints of the common model of the mind. Procedia computer science, 145, 757-764

[14] John R Anderson. Spanning seven orders of magnitude: A challenge for cognitive modeling. Cognitive Science, 26(1):85–112, 2002.

[15] Oscar J Romero and Christian Lebiere. Cognitive modeling of behavioral experiments in network science using ACT-R architecture. In MAS and Agent-Based Simulation, pages 239–251, 2014.

[16] Oscar J Romero and Christian Lebiere. Simulating network behavioral dynamics by using a multi-agent approach driven by ACT-R cognitive architecture. 2015.

[17] Nicholas Louis Cassimatis. Polyscheme: a cognitive architecture for in-

tergrating multiple representation and inference schemes. PhD thesis, Massachusetts Institute of Technology, 2001.

[18] J Gregory Trafton, Alan C Schultz, Magdalena Bugajska, and Farilee Mintz. Perspective-taking with robots: experiments and models. In Robot and Human Interactive Comm., pages 580–584, 2005

[19] Anand S Rao and Michael P Georgeff. Modeling rational agents within a BDI-architecture. KR, 91:473–484, 1991.

[20] Christoph Urban. PECS: A reference model for the simulation of multi-agent systems. In Tools and techniques for social science simulation, pages 83–114. Springer, 2000

[21] Laird, J. E. (2019). The Soar cognitive architecture. MIT press.

Adjustable Autonomy in AI and Assistive Robotics: Experiences and Lessons Learned

Amedeo Cesta

CNR, Istituto di Scienze e Tecnologie della Cognizione, Rome, Italy

amedeo.cesta@cnr.it

Gabriella Cortellessa

CNR, Istituto di Scienze e Tecnologie della Cognizione, Rome, Italy

gabriella.cortellessa@cnr.it

Angelo Oddi

CNR, Istituto di Scienze e Tecnologie della Cognizione, Rome, Italy

angelo.oddi@cnr.it

Abstract

This article explores the research in the development and deployment of adjustable autonomy in AI and Robotics. By examining a series of projects across two main application domains, it highlights the experiences, challenges, and solutions proposed. It emphasises the critical lessons learnt, providing insight into how adjustable autonomy can be effectively leveraged to enhance the capabilities and performance of AI systems in the real world. The focus is also provided on the applications in robotics, where the integration of AI systems is crucial to the

Authors' work is partially supported by FOCAAL - FOg Computing in Ambient Assisted Living. "ACCORDI DI INNOVAZIONE - Bando" Decreto MiSE 05/03/2018 e Decreto Direttoriale 27/09/2018, CLEVERNESS - "Tecnologie a supporto delle fasce più fragili: giovani e anziani". Finanziamento Progettualità Straordinaria - FOE 2020, and PNRR MUR project PE0000013-FAIR.

advancement of autonomy and interaction. The challenges discussed include determining the appropriate level of autonomy as a trade-off among user preferences, safety and robustness, increasing system effectiveness and user trust, adapting control based on contexts, and ensuring that users do not become overly dependent on the technology.

1 Introduction

Autonomy is a key focus in Artificial Intelligence and Cognitive Sciences, being a fundamental trait of artificial *agents* [1]. As stated in [2], autonomous agents are computational systems designed to sense and act independently within their environment to achieve specific goals or tasks. Likewise, [3] defines autonomous agents as artificial entities capable of operating without direct human intervention and maintaining some level of control over their actions and internal state.

We can consider two types of interaction to study *autonomy*. The first is the interaction between a human and an artificial agent. In this type of interaction, autonomy refers mainly to how an agent is able to acquire and to adapt to human preferences and guidance. The second type of autonomy considers a different type of interaction, which is among a group of agents. Hence, autonomy can be seen as a social notion and can be linked to the area of social theories, for example delegation theory [4].

In this paper, we mainly consider the first kind of autonomy. In this context, an agent is autonomous when it considers both the human directives and preferences, and performs actions accordingly. The agent is said to be fully autonomous when it has access to the complete set of choices and preferences of its human *user*. Hence, the user is a separate entity that could judge or change an agent's autonomy. The idea that the autonomy of an agent can be adjusted to match the pace of the human is called *adjustable autonomy* [5]. *Initiative* is a close notion to autonomy, and in some sense is the agent's ability to act first or on one's own in order to move towards the achievement of a goal. Indeed, to the extent that we build robotic agents that exhibit non-trivial initiative, we build autonomous agents. Sharing initiative with agents is also the motivation we are concerned with *mixed-initiative* in human-agent interaction, or about adjustable autonomy.

This article explores research on the development and implementation of adjustable autonomy in AI and robotics by examining a series of projects across two main application domains and highlighting experiences, challenges, and solutions proposed in the assistive and space domains. It also emphasises the importance of AI integration in robotics to advance autonomy and interaction. It highlights key challenges such as finding the right balance of autonomy considering user preferences, safety, and robustness, enhancing system effectiveness and user trust, adapting control to different contexts, and preventing user over-dependence on the technology.

The paper is organised as it follows. Section 2 briefly revises the main motivations and requirements for using adjustable autonomy in the proposed domains. Section 3 and Section 4 describe respectively the main autonomy issues in assistive and space domains, two quite different fields sharing the proposed vision described in Section 2. Finally, Section 5 describes the main lessons learnt and experiences, whereas Section 6 traces some conclusions.

2   Adjustable Autonomy and its Requirements

In this section we briefly revise the main motivations and the requirements in using adjustable autonomy, which has guided our research in assistive and space domains. As introduced above, we study autonomy as the interaction between an human and an artificial agent in pursuing a common goal, such that the initiative is shared between the two actors and the level of autonomy can be adjusted by the human.

First of all, despite adjustable autonomy is necessary for any application of intelligent control technology, humans who rely on autonomous control systems want to be able *to take control of it at various times and at various levels*. This requirement is very important, at least for two motivations: (i) to overcome the technology limitations of the artificial agent (e.g., in case of navigation of a robotic agent in a domestic environment); (ii) to allow the human user to

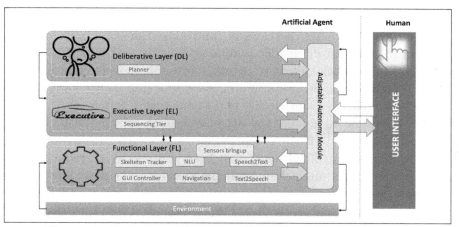

Figure 1: Concept architecture for adjustable autonomy.

maintain the responsibility of the control actions. Second, humans will also want insight into what the system is doing even when all is going well, this requirement is also fundamental in order to share the initiative and implementing a mixed-initiative approach. Third, a robotic agent should have autonomous reasoning and learning capabilities. In particular, the artificial agent should have a set of basic skills (e.g., navigation, communication, grasping), the ability of reasoning on these skills, in order to combine them in more complex plans to reach different goals, and last but not least, the ability of learning new skills. Over more than twenty years of research and the successful completion of several projects, the previous idea of adjustable autonomy has taken the form of the concept architecture of Figure 1. The framework is based on the three-layer robot control architecture [6, 7, 8] commonly accepted in autonomous robotics to support the *Sense-Plan-Act (SPA)* autonomous deliberation and execution paradigm. In this architecture, the *decisional layer (DL)* implements the artificial agent's high-level planning, plan execution, and (re)planning capabilities; the *executive layer (EL)* controls and coordinates the execution of the functions distributed in the software, according to the requirements of the high-level tasks that have to be executed; lastly, the *functional layer (FL)* implements all the basic, built-in agent sensing, perception, and actuation capabilities. The system includes an *adjustable autonomy module* that allows both the user and the artificial agent to regulate autonomy levels. Through the *User Interface module*, users can increase or decrease autonomy at the three levels: the user can modify plans or change priorities (DL); users might intervene to change the actions through which the high level plans are executed (EL); to replace the sensors and actuators (FL). Conversely, the artificial agent can lower autonomy to request user intervention. This might happen if the planner encounters scheduling conflicts, sensors detect unexpected obstacles or dangerous conditions, or if a task repeatedly fails during execution. In these situations, the agent seeks human input to resolve issues and maintain safe operation. In the rest of the paper we will describe several projects both in assistive and space domains, each one focusing on a subset of the previous requirements, but all them based on the framework of Figure 1. The description will detail the main contributions/lessons learned in deploying adjustable autonomy in real world settings.

3   Adjusting Autonomy in Assistive Domain

The article [9], describes an evolution of projects integrating robotics, sensor technology, and artificial intelligence to provide assistance to older adults. Initially, research focused on developing these technologies in controlled laboratory conditions, as seen in early initiatives like the ROBOCARE project, which aimed to demonstrate the feasibility of creating support services through AI and robotics integration. Over time, this field has advanced through more sophisticated projects such as GIRAFFPLUS, EXCITE, SI-ROBOTICS and FOCAAL, which have shifted towards real-world applications. This evolution underscores the increasing complexity and ambition in developing robotic and AI systems to support older adults in the wild and the different role of autonomy to ensure safe, accepted and effective assistive services. The shift from *technical* solutions to *socio-technical* systems [10] highlights the central role of humans, including older adults and caregivers, and

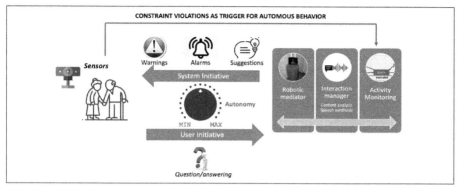

Figure 2: Adjustable Autonomy in ROBOCARE

emphasizes the need to understand how to adjust autonomy to meet their needs and preferences.

ROBOCARE. The project developed a prototype smart home integrating sensors, a mobile robot, and an intelligent stereo camera to support elderly individuals by monitoring their daily activities. The system uses automated reasoning (Deliberative Layer of Figure 1) to ensure that the person's actions align with predefined behavioral patterns set by caregivers, based on medical needs. It can detect anomalies and provide verbal warnings if inconsistencies occur. The primary focus is to promote adherence to "good living" habits through continuous monitoring and intelligent intervention. Interactions are the result of an Artificial Intelligence based reasoning process enabling *mixed-initiative* and are facilitated by a robotic assistant, which serves as the primary interface between technology and older adults. At the core is the Interaction Manager, a rule-based system that assesses situations and triggers specific actions. It evaluates the environment to determine the *timing* and *content* of interactions, using a speech synthesizer to respond to user requests or initiate proactive communication. The system features adjustable autonomy, allowing it to adapt its level of independence based on user needs and context. The activity monitor uses data from sensors to track user behavior against caregiver recommendations. Two key elements guide the system's interaction: a) *Constraint Violations*: these act as triggers for system intervention, determining when the system should engage with the user; b)*Semantic Interpretation*: this shapes the content of interactions, with the system synthesizing verbal *warnings*, *suggestions*, or *alarms* based on the situation. Interactions occurs in two main directions: a) *User Initiative*: users initiate interactions to engage in question-and-answer sessions. The system provides answers based on the activity history or current state, assisting with daily planning and queries. b) *System Initiative*: the system initiates interactions in dangerous or warning scenarios. Warnings occur when constraint violations are detected, prompting the system to engage the user proactively. Overall, adjustable autonomy is used to balance system and user-initiated interactions, adapting to the user's needs and ensuring timely support and intervention. The ROBOCARE system was evaluated [11] through a video-based assessment that distinguishes between scenarios of full autonomy and those where the user takes the initiative. Older users recognize the difference between important and unimportant activities at home. For critical tasks, especially related to safety and healthcare, users value the robot's autonomy in managing the environment and making decisions. However, there is a concern about becoming overly dependent on the system, with a preference for a balance where the robot addresses key issues while also encouraging users to remain proactive.

ExCITE. The ExCITE project primarily focused on testing technology to facilitate social interaction for isolated individuals, such as the elderly in their homes or healthcare institutions, with the goal of increasing social participation and reducing loneliness. The project utilized a remotely controlled robot called Giraff, equipped with a teleconferencing system. Unlike autonomous robots, Giraff is operated by a secondary user—such as a relative, caregiver, or friend—who remotely navigates the robot within the primary user's environment (see Figure 3). The primary user, typically a senior, benefits from this interaction by experiencing increased connectivity with loved ones and caregivers. The secondary user can maneuver the robot to different

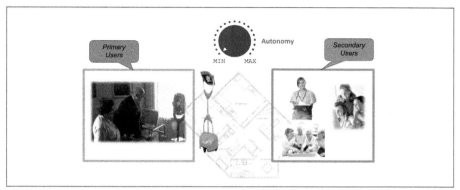

Figure 3: Low Autonomy in ExCITE

locations within the senior's living space using simple software on a computer. In this project, the level of autonomy was deliberately kept low, with full control given to the human operator. This approach ensures safety in delicate settings like the homes of elderly individuals and supports long-term experimentation. The project performed a long-term evaluation to assess whether such technology could be effectively and appreciated beyond the initial "novelty effect" and the "controlled conditions" of a laboratory. This evaluation focused on measuring *acceptance*, *usability*, and *perceived social support*. Users tend to judge technology as worthwhile if it demonstrates intelligence and usefulness in real-world situations, offering valuable services that enhance daily life beyond merely fulfilling basic functions.

GiraffPlus. The GiraffPlus project focuses on using advanced technology to support older adults through continuous monitoring and long-term experimentation. It builds upon the ExCITE project, integrating more mature technology within an Intelligent Environment framework. The system collects data on daily behaviors and physiological measures using distributed sensors, enabling context recognition and long-term trend analysis. This data is processed by AI algorithms, which translate it into personalized services for both primary users (older adults) and secondary users (caregivers and healthcare professionals). The primary users benefit from access to health data,

120

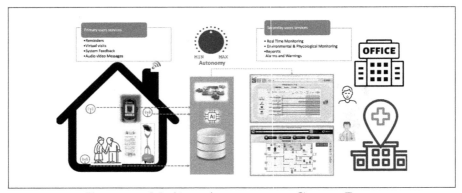

Figure 4: Medium Autonomy in GIRAFFPLUS

reminders, and telepresence communication, enhancing their autonomy and sense of security. They also appreciated the possibility of an increased autonomy for the robotic platform (see Figure 4), such as returning to charging bases and sending personalized messages. Secondary users gain an efficient monitoring tool that provides alerts, reports, and the ability to manage multiple patients remotely. The system employs automated planning to create adaptive reminders, promoting healthier lifestyles based on the users' conditions and objectives. The project was tested in long-term trials across 15 sites, focusing on *acceptance, adoption, domestication, psychological impact*, and *telepresence*. Results showed high acceptance, with users appreciating the system's integration into daily life. Older adults felt more autonomous and protected, while secondary users valued the system's ability to optimize care and provide insights into the assisted person's health. The trials also highlighted a desire for additional features like voice commands and increased autonomy for the robotic platform. Overall, GIRAFFPLUS proved to be a valuable tool for enhancing the care and quality of life for older adults in their homes.

SI-ROBOTICS **and** FOCAAL. Both projects aim to develop advanced solutions for assisting seniors, though they approach the challenge with different focuses and technologies. SI-ROBOTICS seeks to create innovative AI-based robotic systems to provide a variety of assistance services across different environments, such as daily home living and hospitals. Its goal is to support older adults in managing their health through robotic assistants that offer empathetic communication and reliable assistance, tailored to individual needs. FOCAAL, on the other hand, is distinguished by implementing the architecture of Figure 1 using a distributed fog-edge computing approach. This project integrates sensors, actuators, home automation, robotics, and AI to offer innovative monitoring and assistance services that help frail people stay in their own homes, maintain a healthy lifestyle, and receive personalized care. By leveraging local data processing through fog computing, FOCAAL optimizes in-person visit times and resource allocation for healthcare workers, enabling more effective and personalized care. Both projects emphasize the importance of adjustable autonomy. For repetitive tasks and warning signals, autonomy is crucial, but it must be adjustable to allow for human oversight. This flexibility ensures that while the systems can operate independently for routine tasks, human control is crucial for managing more complex or sensitive aspects of care.

Currently the authors are working on Cleverness, exploring how different technologies can create intelligent services that migrate between devices to support the elderly also studying how adjustable autonomy is influenced by the embodiment of the intelligent agent.

4 Adjusting Autonomy in Space Domain

This section concisely describes a selection of projects regarding space domains and to which the authors of this article have actively worked over the past two decades. As matter of facts, these projects demonstrate a consolidated expertise in the development of controlling architectures for autonomy in the space domain. The topics of adjustable autonomy and mixed-initiative are transversal to the considered projects. According to the vision proposed in Section 2, these projects exhibits an increasing level of autonomy on the side of the artificial agent, whereas the role of the human changes from a tight coordination of the artificial agent [12], to the role of prompting it to reach a goal [13] or discovering new domain knowledge [14].

MEXAR2. The project MEXAR2 [12] proposes an advanced AI software tool for continuous support of the mission planning operations of Mars-Express. It generates the activity plan used to download the satellite data from the on-board memory to ground. The problem is solved on a periodic basis at ESOC, the Mission Operations Centre of the European Space Agency (ESA). The tool is still in use and supports on regular basis the work of mission planners within the Mars-Express mission. The space probe continuously produces a large amount of data resulting from the activities of its payloads and from on-board device monitoring and verification tasks. All these data are to be transferred to Earth during bounded downlink sessions. On-board data generally requires be first stored in an on-board memory and then transferred to Earth. In case new data are produced before the previous are dumped to Earth, older data are *overwritten*. MEXAR2 is based on the interactive problem-solving architecture introduced in Figure 1 devoted to the support of data dumping activities of Mars-Express. In particular, it synthesizes sequences of spacecraft operations (dump plans) that are necessary to deliver the content of the on-board memory during the available downlink windows. The idea is to integrate human strategic capabilities and automatic problem-solving algorithms (see Figure 5) to find solutions (dump plans) with the right compromise among different and contrasting goals under the full control of the mission planners. A dump plan is evaluated wrt a set of objectives, which can be classifies as primary and auxiliary: the *volume of lost data* (if any) – primary; *robustness* (i.e., the max percentage usage of the on-board memory) - primary; the *size of the dump plan* (i.e., the number of dump commands contained in a dump plan); the *average delivery delays of the stored data*. The design of the user interaction style allows to obtain different solutions for the same period (or dump interval). In particular, it allows to recognizes one of the main sources of brittleness for a downlink schedule, i.e., peaks of data volumes stored in the on-board

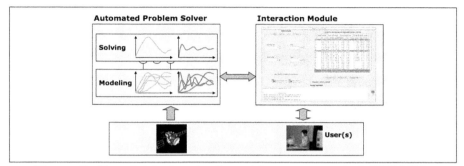

Figure 5: User initiative / adjustable autonomy in MEXAR2

memory. Hence, a user can perform a set of *what-if* analysis steps for exploring a set of different solutions for the same dump interval, by: tuning the solving algorithms; changing the input parameters; compare them according to different metrics; choosing the best candidate for execution.

GOAC. The objective of GOAC (Goal Oriented Autonomous Controller) project [13] was to design, build and test a viable on-board controller to demonstrate key concepts in fully autonomous operations for ESA missions. For example, deep space and remote planetary exploration missions, characterized by severely constrained communication links (e.g., short communication window duration and low data transmission rates) and such that round-trip light time delays and lack of models of remote planetary environments exacerbate the mission control problem. GOAC is an autonomous controller using a *Sense-Plan-Act* paradigm to provide increasing levels of autonomy for robotic task achievement. In particular, the project proposed a three-layer architecture [7] as an integrative effort to bring together three different technologies: a verification and validation system, a planning engine and a controller framework for planning and execution for goal oriented autonomy. Indeed, GOAC is organized as a hierarchy of deliberative and command-dispatcher *reactors*, such that each deliberative reactor uses an automated planner. GOAC follows a divide-and-conquer approach, by splitting the deliberation prob-

124

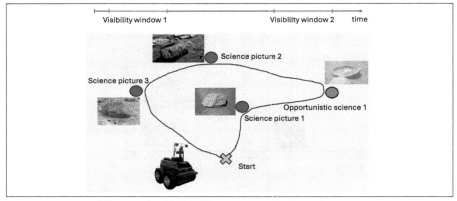

Figure 6: GOAC: opportunistic science.

lem into sub problems, thus making it more scalable and efficient. Each sub problem is assigned to a deliberative reactor. One of the main contribution of the project is a *full* autonomous control system, where the human (i.e., the mission planner) has the role of setting the mission goals and verifying the quality of the achieved results. In addition, the mission scenario is characterized by the presence of a mission goal (i.e., taking a picture), the possibility of adding on the fly additional mission goals (opportunistic science), storing the data on-board and communicating them during the available communication windows. The GOAC control architecture has been tested with a real scenario, using a DALA robot, simulating an extraterrestrial planetary environment (see Figure 6).

IMPACT. The IMPACT project aims at investigating the possibility of employing Artificial Intelligence (AI) techniques to increase operational *autonomy* of artificial agents (e.g., robotic platforms) targeted at the space domain. The idea is based on the creation of a *learning loop* in which the agent increases its capabilities (skills) through the direct interaction with the real environment, and then exploits the autonomously acquired knowledge to execute activities of increasing complexity: this process is cumulative and virtually *open-*

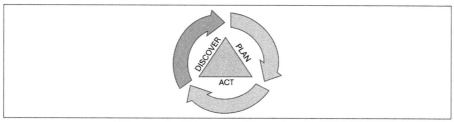

Figure 7: *Discover-Plan-Act* cycle.

ended [15] as the information and abilities acquired up to a certain time are employed to further increase the agent's knowledge of the application domain, as well as the skills to adequately operate in it. This self-induced tendency towards autonomously learning new skills, based on *intrinsic motivations* (IM) [16], will enable the artificial agent to face situations and solve problems not foreseeable when the agent is designed and implemented, especially because of the limited knowledge on the environment the agent will operate in.

IMPACT [14] implements a *Discover-Plan-Act* (DPA) cycle (see Figure 7), which directly extends the *Sense-Plan-Act* (SPA) [6, 7, 8] cycle (see Section 2 for more details) with a more general open-ended learning step (*Discover*) acquiring new knowledge from the external environment. In particular, the IMPACT project extends the three-layer architecture by implementing the following functionalities: (i) extension of the functional layer with a new module for learning additional *skills* based on self-generated goals, a process driven by intrinsic motivations; (ii) extension of the executive layer with an automatic translation functionality of the newly acquired skills, from a low-level sub-symbolic representation to a high-level symbolic representation (e.g., expressed in Planning Domain Definition Language - PDDL); (iii) finally, extension of the deliberative layer with a Long-Term Autonomy (LTA) module allowing to an user the possibility of setting different learning strategies and setting the strategy of mixing intrinsic and extrinsic goals (or mission goals). The ideas proposed has been demonstrated through the development of a proof-of-concept software prototype implementing two simulated scenarios.

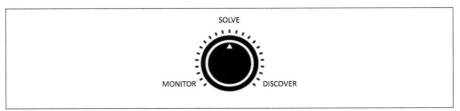

Figure 8: *Levels of autonomy.*

A *Rover* scenario, which demonstrates how the IMPACT system can discover new ways to reach an already known effect by applying an abstraction procedure (see [14]) for the automatic translation of the newly acquired skills from the sub-symbolic level to the PDDL (symbolic) level. This scenario describes a situation where a planetary rover's antenna orientation mechanism is damaged, preventing stable communication. The proposed technology enables the rover to adapt by using its locomotion capabilities to orient the antenna. The rover can enrich its planning domain with the necessary knowledge to adjust its entire body position, using terrain features like slopes or small rocks, to achieve and maintain communication.

A *Robot Arm* scenario, which demonstrates the ability of the IM-PACT system to acquire new ways to interact with the environment and integrate them in its planning domain. In this scenario, a robot equipped with a gripper actuator attached to a manoeuvrable arm tries to grasp a "vase shaped" rock whose size exceeds the max opening span of the gripper. The robot is thus not able to pick-up the rock with its basic grasping skill - however, upon failure, the IM-PACT system will automatically trigger the learning of a new skill and the robot will at the end be able to pick-up the "vase shaped" rock by grasping it from its edge.

5   Lessons Learned and Experiences

Given the project review above and in line with the model proposed in Section2, the following paragraphs summarise the key lessons learned.

Balancing autonomy and human control. Across the set of proposed projects the level of autonomy changes significantly. Figure 8 shows an intuitive representation of the set of considered levels of autonomy according to the complexity of the tasks requested to the artificial agent. In particular, starting from *monitoring* domain activities (e.g, in part of the activities of the ROBOCARE project), to the request for the artificial agent to take the initiative for *solving* some relevant sub-problems (e.g., in the GOAC project), till the requirement of increasing the agent initiative for *discovering* new knowledge about the considered domain, this is the case of the IMPACT project.

It is worth noting that many of the lessons learned regard the possibility of changing the level of autonomy within a single project activity. The evaluation of the ROBOCARE system highlights several key lessons about the use of adjustable autonomy in assistive technologies. One significant insight is that full autonomy is highly valued in critical situations where safety is a top priority or when it helps compensate for age-related deficits. In such scenarios, the robot's ability to operate independently and make decisions enhances the quality of life by addressing urgent needs and supporting users in maintaining their independence and safety. However, despite autonomy is required to deal with critical situations or to compensate *primary* user's deficits, less autonomy is advisable for encouraging *primary* users to stay active and utilise their own abilities. In addition, a key insight from the ExCITE project is the necessity of incorporating adjustable autonomy in telepresence robots. Adjustable autonomy involves balancing the robot's ability to operate independently with the user's need for manual control in certain situations. While increased autonomy can enhance the robot's ability to perform tasks proactively, such as navigating to different locations or initiating interactions, there are scenarios where users prefer to retain control. For example, users may want to manually direct the robot during sensitive interactions or when navigating unfamiliar or complex environments.

The GOAC project offers a different perspective on adjustable autonomy. Indeed, it fosters an operational concept in which operators can focus on the *domain level* (i.e., the Deliberative Layer in

Figure 1) rather than lower-level aspects of the mission. In addition to potentially reducing the costs of the mission as a consequence of goal-oriented operations, a high-level of autonomy also improves the performance of the robotic platform in two different ways. First, *scientific return* is potentially improved, because the high-level guidelines for nominal as well as *opportunistic* scientific exploration (which increases the level of autonomy if activated) allows being more efficient in doing science than relying on pre-calculated plans. Second, *robustness* of the robotic system is also potentially improved. Indeed, when a robotic system reaches off-nominal situations it can be safer and more efficient to autonomously react to the environmental conditions and re-plan rather than wait for ground instructions or rely on fixed pre-programmed alternative plans.

Flexible interaction and human awareness. *Flexible interaction* for developing intuitive interfaces that enable users to easily switch between autonomous and manual control is key issue in making the level of autonomy adjustable. According to the general architecture of Figure 1, through the *User Interface (UI)*, users can increase or decrease autonomy at all the three levels. Indeed, this flexibility helps users feel more comfortable and in control of the technology, enhancing their overall experience. By providing users with the ability to adjust the level of autonomy based on their specific needs and preferences, developers can create more effective and user-friendly systems that integrate seamlessly into daily life. Human-awareness is essential for effective robotic solutions that interact closely with end-users. In the projects SI-ROBOTICS and FOCAAL, which share the objective of enhancing support for elderly individuals and both highlighted that adjustable autonomy is a key element. For tasks that are repetitive or involve alerting users, having a system that can operate autonomously while still allowing for human intervention is crucial. This balance ensures that while technology can handle routine functions efficiently, human control remains available to oversee and adjust care as needed.

Knowledge discovery and update. In this section we consider a complementary perspective about adjustable autonomy. Indeed, the level of autonomy can be also changed with respect the possibility for the artificial agent of discovering and learning autonomously new knowledge about the reference domain. This idea is not new in the literature, *Long Term Autonomy (LTA)* [17] is known as the ability of an artificial agent (e.g., a robotic system) to (i) perform reliable operations for long periods of time under changing environmental conditions, (ii) increase over time its knowledge about the environment. LTA is clearly challenge for a robotic systems. Our experience is mainly based on the results of IMPACT project and the ongoing FAIR project. The integration of automated planning and learning is a promising approach to increase the level of autonomy for an intelligent agent. In particular, we have explored the possibility of increasing the level of autonomy, such that the artificial agent is (i) able to self-generate goals (guided by intrinsic motivation) and to autonomously learn the *skills* necessary to achieve them, (ii) use the new skills in the symbolic planning process. The possibility of learning new skills, increases the knowledge of the system one the one hand, but increases also the risk of failure on the other hand. A possibility for future research is to mitigate the risk of failure by using a digital twin of the robotic agent to test the new learned skills.

6 Conclusions

This article reviews research on the development and deployment of adjustable autonomy in AI and robotics, highlighting key projects in two main application areas: assistive and space domains. It discusses experiences, challenges, and solutions, emphasising critical lessons for effectively leveraging adjustable autonomy to enhance AI systems' capabilities in real-world scenarios, for example robotics applications, where integrating AI is vital for advancing autonomy and interaction. One key contribution of the paper is a revision of the main requirements for a concept architecture for adjustable autonomy. In particular, the well-known three tier architecture, for controlling an artificial agent, is extended in two different ways: (i) it allows the interaction between an user (human expert) and an artificial agent in pursuing a common goal, such that the initiative is shared between the two actors and the level of autonomy can be adjusted by the user at different levels of abstraction; (ii) the artificial agent's initiative can be expanded till the possibility of discovering and learning new skills, which can be integrated in the knowledge of the artificial agent. The authors of the paper are actively working in the previous two extensions as a part of their future research work.

References

[1] Hexmoor, H., Castelfranchi, C., & Falcone, R. (2003). Agent Autonomy. Springer New York, NY. https://doi.org/10.1007/978-1-4419-9198-0

[2] Maes, P. 1994. Modeling Adaptive Autonomous Agents, Artificial Life Journal, C. Langton, ed., Vol. 1, No. 1 & 2, MIT Press.

[3] Wooldridge, Michael and Nicholas R. Jennings, 1995. Agent Theories, Architectures, and Languages: a Survey, In Wooldridge and Jennings Eds., Intelligent Agents, Berlin: Springer-Verlag, 1-22

[4] Castelfranchi, C., Falcone, R., 1998. Towards a Theory of Delegation for Agent-based Systems, Robotics and Autonomous Systems, Special issue on Multi-Agent Rationality, Elsevier Editor. Vol. 24, pp. 141-157.

[5] Musliner, D. and Pell, B., 1999. Call for Papers, 1999 AAAI Spring Symposium on Agents With Adjustable Autonomy, March 22-24, 1999, Stanford University.

[6] R. Peter Bonasso, R. James Firby, Erann Gat, David Kortenkamp, David P. Miller, and Mark G. Slack, 'Experiences with an architecture for intelligent, reactive agents', Journal of Experimental & Theoretical Artificial Intelligence, 9(2-3), 237–256, (1997).

[7] Erann Gat, 'Three-layer architectures', in Artificial Intelligence and Mobile Robots, eds., David Kortenkamp, R. Peter Bonasso, and Robin Murphy, 195–210, MIT Press, Cambridge, MA, USA, (1998).

[8] Félix Ingrand, Simon Lacroix, Solange Lemai-Chenevier, and Frederic Py, 'Decisional autonomy of planetary rovers', Journal of Field Robotics, 24(7), 559–580, (2007).

[9] Cortellessa, Gabriella, De Benedictis, Riccardo, Fracasso, Francesca, Orlandini, Andrea, Umbrico, Alessandro and Cesta, Amedeo. "AI and

robotics to help older adults: Revisiting projects in search of lessons learned" Paladyn, Journal of Behavioral Robotics, vol. 12, no. 1, 2021, pp. 356-378. https://doi.org/10.1515/pjbr-2021-0025

[10] G. Baxter and I. Sommerville, "Socio-technical systems: From design methods to systems engineering," in Interacting with Computers, vol. 23, no. 1, pp. 4-17, Jan. 2011, doi: 10.1016/j.intcom.2010.07.003.

[11] A. Cesta, G. Cortellessa, R. Rasconi, F. Pecora, M. Scopelliti and L. Tiberio. Monitoring elderly people with the robocare domestic environment: Interaction synthesis and user evaluation. Computational Intelligence, 27(1):60–82, 2011.

[12] Amedeo Cesta, Gabriella Cortellessa, Michel Denis, Alessandro Donati, Simone Fratini, Angelo Oddi, Nicola Policella, Erhard Rabenau, and Jonathan Schulster. Mexar2: AI Solves Mission Planner Problems. IEEE Intelligent Systems, 22(4): 12-19, 2007.

[13] A. Ceballos, S. Bensalem, A. Cesta, L. De Silva, S. Fratini, F. Ingrand, Felix, J. Ocòn, A. Orlandini, F. Py, K. Rajan, R. Rasconi, M. Van Winnendael. A Goal-Oriented Autonomous Controller for Space Exploration. 11th Symposium on Advanced Space Technologies in Robotics and Automation. Proceedings (ASTRA 2011), pp. 1–8, Noordwijk, the Netherlands, 12-14 April 2011.

[14] Angelo Oddi, Riccardo Rasconi, Vieri Giuliano Santucci, Gabriele Sartor, Emilio Cartoni, Francesco Mannella, Gianluca Baldassarre (2020). "An Intrinsically Motivated Planning Architecture forCuriosity-driven Robots." Proceedings of the 24th European Conference on Artificial Intelligence https://ebooks.iospress.nl/doi/10.3233/FAIA200373

[15] Stephane Doncieux, David Filliat, Natalia Diaz-Rodriguez, Timothy Hospedales, Richard Duro, Alexandre Coninx, Diederik M. Roijers, Benoit Girard, Nicolas Per- rin, and Olivier Sigaud. Open-ended learning: A conceptual framework based on representational redescription. Frontiers in Neurorobotics, 12:59, 2018.

[16] Todd Hester and Peter Stone. Intrinsically motivated model learning for developing curious robots. Artificial Intelligence, 247:170 – 186, 2017. Special Issue on AI and Robotics.

[17] L. Kunze, N. Hawes, T. Duckett, M. Hanheide, and T. Krajník. Artificial intelligence for long-term robot autonomy: A survey'. IEEE Robotics and Automation Letters, 3(4), 4023–4030, (Oct 2018).

HOW WILL ARTIFICIAL INTELLIGENCE AFFECT OUR LIVES? SOME REFLECTIONS IN THE AREAS OF LABOR AND HEALTH

PIERLUIGI ZOCCOLOTTI

Department of Psychology, Sapienza University of Rome, Rome
Institute of Sciences and Technologies of Cognition, CNR, Rome
Tuscany Rehabilitation Clinic (CRT), Montevarchi (AR), Italy
`pierluigi.zoccolotti@uniroma1.it`

FABIO LUCIDI

Department of Psychology, Sapienza University of Rome, Rome
`fabio.Lucidi@uniroma1.it`

Abstract

Here we analyze labor market changes associated with the introduction of new technologies, particularly Artificial Intelligence (AI). Throughout history, the advent of new technologies has profoundly modified the labor market. While automation may reduce jobs directly related to the economic sector from which they derive, other economic forces may generate new tasks that reinstate the demand for labor. Historical series indicate that although this reinstatement effect proved efficient in the decades following World War II, it has proved less effective in more recent decades. These changes might be due to the prevalent introduction of "so-so" technologies that tend to substitute human labor with machines instead of exploring the full potential of AI in restructuring the production process in ways that generate new tasks with a high level of productivity. A harmonious development of AI applications does not emerge automatically due to market forces but may take different paths depending on the industrial policies that are activated. Thus, this analysis underscores the importance of fostering an interchange between active researchers and their political counterparts, a legacy of Rino Falcone, to whom this chapter is dedicated.

1 Introduction

We know that new knowledge deriving from scientific and technological developments can significantly impact our everyday lives. Rino Falcone, to whom this chapter is dedicated, has always shown great awareness and sensitivity to the implications of scientific and technological advances for policymaking, i.e., the ability to plan and modulate society's development processes, and for the need, as researchers, to seek dialogue with social and political institutions to foster mutual understanding. In fact, at the urging of Rino Falcone in October 2017, a conference was organized at the Senate of the Republic on the potential impact of the development of artificial (AI). The aim was to create a dialogue between experts and representatives of the political forces. This was a cooperative project between the Italian Association of Psychology (AIP) and the Institute of Science and Technology of the CNR; the latter was represented by Rino Falcone and Olga Capirci, respectively, director and researcher at CNR. The experiences over the next few years increasingly confirmed the belief that the development of AI would fundamentally change our lives. Nevertheless, there are still many uncertainties regarding the future development of this process. Thus, increasingly, the main topic of interest at meetings, conferences, and speeches, as well as in scientific journals and the press, is "How will artificial intelligence affect our lives?". For example, "How will it change job opportunities?" "What impact will it have on our healthcare?".

In a subsequent target article, which was published in the Giornale Italiano di Psicologia [1],we described the aforementioned conference and especially some of the potential implications related to the development of artificial intelligence (see also [2], in which we commented on various responses to the target article). At that time, there was considerable uncertainty about the impact of the introduction of intelligent technologies. For example, in the labor market, we noted the presence of alarm bells that emphasized both the severe potential risk of job losses and signs of interest in the potential of new forms of professionalism. This ambivalence is not entirely surprising.

Indeed, we know from economic theory that there is no direct correspondence between the evolution of productivity and labor demand at the sector level and the evolution of labor demand at the aggregate level (e.g. [3]). This underlying uncertainty is, to some extent, still current. However, in recent years there has also been an increase in labor market dynamics associated with the growing introduction of automation, in particular AI (e.g. [4, 5, 6, 7]). Another relevant area is the introduction of intelligent technologies for healthcare (e.g. [8]), which is associated with the increasing use of AI in training healthcare workers(e.g. [9]). Nevertheless, there are still various complexities in this context about understanding the economic impact of AI on healthcare(e.g. [10]).

Although there are still many elements of uncertainty regarding the future, at present, there is a greater understanding of the mechanisms through which the labor market changes as a function of the introduction of new technologies and automation processes, and this knowledge provides a relevant basis for interpreting recent changes related to the increasing introduction of AI. Although we are aware of the rapidity of change in the development of AI, in this chapter, we will briefly consider the changes in the world of labor and healthcare related to the development of intelligent technologies.

2 The impact of automation on aggregate labor demand

Autor and Salomons [7] carried out an in-depth analysis of the impact of automation on the aggregate demand for labor in the period between 1970 and 2007, based on data from the major Western industrialized nations, as well as Australia, Canada, Japan, South Korea, and the United States. These authors note that a balanced assessment of the influence of technological development requires both an evaluation of the (direct) impact of the development of automation in the industries it originates from and an examination of a range of (indirect) changes in aggregate labor demand. For these purposes, they combine a microeconomic analysis that captures well-directed changes

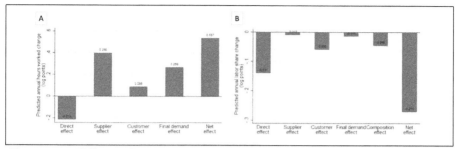

Figure 1: Effect of TFP growth on aggregate employment (A) and "labor-share displacing" (B). Data refer to the period from 1970 to 2007 (modified from [7]).

with a macroeconomic analysis necessary to highlight indirect changes that are needed to have an overall picture of the labor situation. Using this approach, it is possible to examine (1) direct effects at the industrial level, (2) indirect effects of customers and suppliers in related industries, (3) effects on final demand resulting from the effect of productivity growth on aggregate value-added, and (4) effects on composition resulting from productivity-induced changes in industry shares of value added.

In their analysis, Autor and Salomons [7] used total factor productivity growth (TFP) to measure technological progress. It should, however, be noted that this measure is not only sensitive to technological innovations but is also affected by other factors that influence productivity; therefore, further investigation is required to identify the origin of the observed effects. However, the authors note that their main conclusions are substantially confirmed using measures more directly related to technological advancement (although less precise), such as the flow of patents by the production sector.

Autor and Salomons [7] first found that economic sectors with more robust productivity gains (mainly related to increased automation) showed a corresponding reduction in the aggregate employment share. Thus, the direct effect of increased productivity was to reduce employment in the sectors where it originated. However, two indirect effects offset this direct effect. First, the increase in TFP in the

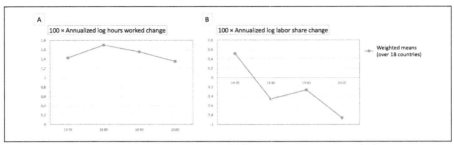

Figure 2: Trends in hours worked (A) and labor share (B) by decade (modified from [7]).

supplying industries catalyzes robust offsetting employment increases in the downstream customer industries; second, TFP growth in each industry sector contributes to aggregate growth in real value added and thus increases final demand, which in turn stimulates further employment growth in all industries (See Figure 1 a).

However, in addition to this effect, an increase in TFP also produces a labor-share displacement, reducing the value-added brought by labor in the production cycle. In this case, the direct effect of introducing automation processes in the sectors where they originated is not counterbalanced by indirect effects related to input-output linkages, changes in composition or increases in final demand (See Figure 1 b).

Autor and Salomons [7] analyzed the development of these two effects over the period under consideration; they noted that the influence of TFP on aggregate labor demand was relatively stable over the period being considered (Figure 2 a). By contrast, the effect of labor-share displacing changed appreciably over time. Thus, in the 1970s, the effect was positive but became progressively more and more negative over the following decades and was highest in the first decade of this century (Figure 2 b). The latter figure underscores the relevance of examining the impact of technological innovation as a function of the historical period.

Acemoglu and Restrepo [5] also carried out an articulate analysis of how automation processes impact the labor structure. According to these authors, the effect of automation in general and, more specifically, of AI and robotics is to reduce the need for workers in various types of tasks by producing what they call a "displacement effect" [5]. However, opposing forces that promote job creation counterbalance this trend. For example, the replacement of labor with machines can reduce production costs ("productivity effect"), and the resulting economic expansion can produce an increase in the demand for labor both in sectors not subject to the automation process and in the same one in tasks that cannot be easily automated.

According to Acemoglu and Restrepo [5], however, the most relevant factor that can counteract the reduction in the need for work associated with the displacement effect is the fact that automation processes act by creating new tasks in which human labor has a relative advantage over the use of machines. Thus, over time progressive labor automation is accompanied by changes in the labor structure, which creates new job opportunities ("reinstatement effect" [5]).

The presence of opposing forces creates the conditions for the automation process to find an equilibrium in the labor market. However, it should also be noted that the economic adaptation of an economy at a given historical moment is not necessarily painless. For example, the automation process may create outplacement difficulties for workers who do not have the necessary expertise to perform the new tasks associated with automation. In other words, critical discrepancies might originate from the mismatch between workers' skills and new technologies [5].

3 Evolution of sources of labor demand

The formalization proposed by Acemoglu and Restrepo [5] lends itself effectively to a temporal analysis of trends in sources of labor demand. For at least some economies, particularly the United States, data are available that document the evolution over time of the impact of introducing new technologies and automation processes. Acemoglu and Restrepo [5] reported data on the evolution of labor demand sources related to the period after World War II. These authors compared the

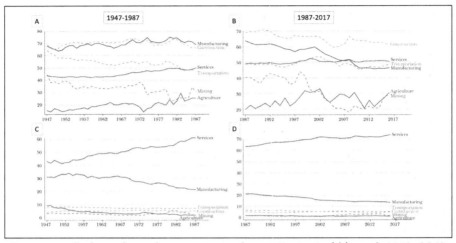

Figure 3: Labor share by sector in the 1947-1987 (A) and 1987-2017 (B) periods. Share of gross domestic product by sector in the 1947-1987 (C) and 1987-2017 (D) periods (modified from [5]).

period of the four decades immediately following the conflict (1947-1987) with that of the subsequent three decades (1987-2017). The first four decades were characterized by a high per capita wage bill increase, which grew by 2.5 per cent per year. This rapid and steady growth was largely due to an increase in productivity (2.4 per cent per year on average). Over this period, the share related to labor costs by sector was rather constant (See Figure 3 a), except for a decline in the mining and transportation sectors. Figure 3 c, shows the weight of each sector on the gross domestic product of the U.S. economy. There is a gradual decline in the manufacturing (and mining) sectors and some growth in the service sector. What is interesting here is to see how the evolution in the content of work tasks has mostly remained stable over time (Figure 4 a). This result is related to the fact that the robust "displacement effect" (estimated at 0.48 per cent per year) was almost entirely offset by a strong "reinstatement effect" (estimated at 0.47 per cent per year). In the decades after World War II, major technological advancements led to a significant increase

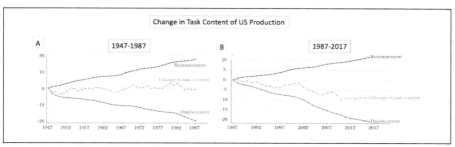

Figure 4: Changes in production task content in the 1947-1987 (A) and 1987-2017 (B) periods (modified from [4]).

in automation. However, these advancements were accompanied by changes in work tasks that almost balanced out the labor reduction directly caused by automation processes.

The next three decades (1987-2017) saw lower economic growth (1.33 per cent per year) mainly related to a reduction in productivity growth (1.54 per cent) compared to the previous period, with substantial stagnation after 2000. The relative labor shares showed a downward trend for the construction and manufacturing sectors as well as the mining sector, which continued its downward trend (Figure 3 b). The relative weight of sectors in gross domestic product indicated a continuation of the trend of decreasing weight in the manufacturing sector and a corresponding growth in the service sector (Figure 3 d).

Compared with the previous period, the balance between reinstatement and displacement effects became negative. Thus, an increase in the displacement effect (amounting to 0.7 per cent per year) more robust than in the previous period was associated with a significantly smaller reinstatement effect (0.35 per cent per year; Figure 4 b). Overall, changes in the content of work tasks (mostly related to the introduction of automation processes) reduced labor demand by 10 per cent (which becomes 30 per cent when considering only the manufacturing sector).

Overall, the labor demand in the United States has gradually declined over the past 30 years. On the one hand, this is due to a flattening of productivity growth and, on the other, because shifts in the content of productive tasks were not offset by the creation of new tasks. This analysis of U.S. data presents a pattern generally consistent with the one (illustrated above) by [7] on data from major industrialized nations.

Technological innovation creates profound changes in the labor structure that can trigger a relevant mismatch between workers' skills and the new skills required by introducing new technologies with increasing automatization. This mismatch can be problematic at given historical moments, but the evolution of the labor market can also gradually absorb these issues.

The historical analysis by Acemoglu and Restrepo [4] indicates how the outcome of this process can be different at different historical moments. The differential outcomes are partly related to the spontaneous evolution of the labor market but also to the policies introduced to modulate the process of introducing technological innovations. This consideration underscores the interest in reasoning about what can be put in place in periods of robust technological innovation to facilitate the redistribution of labor by developing new tasks and duties.

4 Evolution of income distribution since the 1980s

The analyses by Acemoglu and Restrepo [4] and Autor and Salomons [7] highlight a gradual erosion of the contribution of labor to production processes since the 1980s. It is interesting to see how these changes in labor have impacted the distribution of household incomes during this time. Figure 5 shows the evolution of the Gini coefficient [11] between 1985 and 2013 in various industrialized nations, including Italy [12]. This coefficient represents a concentration index to measure inequality in the distribution of income or even wealth. Higher values (up to a maximum of 1) indicate a more unequal distribution. In the 30 years or so considered by the Organization for Economic Cooperation and Development (OECD), inequality has grown consistently in almost all nations, including Italy. A few nations show

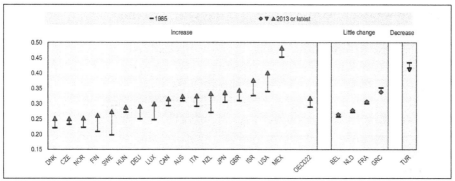

Figure 5: Change in the Gini coefficient [11] between 1985 and 2013.
Source: OECD [12].).

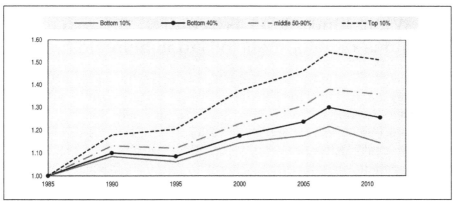

Figure 6: Trends in real household incomes with low, medium, and
high values over the 1985-2013 period (OECD average); 1985 = 1.
Income refers to disposable household income adjusted for household
size. OECD is the unweighted average of 17 countries (Canada, Germany, Denmark, Finland, France, the United Kingdom, Greece, Is-
rael, Italy, Japan, Luxembourg, Mexico, the Netherlands, Norway,
New Zealand, Sweden, and the United States).

a predominantly stable trend (Belgium, the Netherlands, France, and
Greece), and only one shows an appreciable reduction (Turkey).

The increase in inequality appears to be related to differences in relative wage growth. OECD [12] data on real incomes in the 1985-2013 period show limited growth for low and lower-middle incomes (see Figure 6) compared with much more robust growth for higher incomes. Overall, these data indicate how the period with the most substantial changes in the structure of labor was associated with reduced growth in low and middle incomes with a consequent increase in inequality that predominantly affected the most industrialized nations. Thus, these observations underscore the relevance of understanding how the development of new technologies (particularly AI) impacts the structure of the world of work and whether it is possible to actively intervene to try to steer this process in directions with the harmonious development of the more industrialized nations.

5 What kind of AI is required for a harmonious development of labor demand?

In a recent paper, Acemoglu and Restrepo [13] provocatively asked whether the AI developed in recent years is the "right kind". The authors point out that the evolution of technologies does not have a homogeneous impact on the labor structure and that AI represents a technological platform that can be developed with very different goals.

In general, we have seen that automation processes lead to the replacement of humans with (inherently cheaper) machines by reducing the value-added, labor-related component. Automation produces an increase in productivity, and the strength of this effect has a critical impact on labor demand.

A non-obvious implication, according to Acemoglu and Restrepo [13], is that the critical risks to the world of labor do not come from introducing brighter and more innovative technologies but, conversely, from those that only moderately increase productivity. Acemoglu and Restrepo [13] refer to them as "so-so technologies" to emphasize the idea that they are convenient enough to be developed and used but not productive enough to create the opportunity for new tasks with potentially high levels of added value for human labor. One example is the introduction of industrial robots with the sole purpose of replacing humans with machine labor. In general, industrial data indicate that there is a production advantage with the introduction of robots, but the productivity gains are not so advantageous as to induce a consistent reinstatement effect.

If one applies this logic to AI, one can see that there has been a prevailing attempt to use it to progressively automate more and more tasks (such as, for example, image recognition, speech recognition, translation, accounting, and recommendation). However, despite the often-surprising effectiveness of these applications, AI itself has the potential to go beyond automating known tasks and to be used to restructure the production process in ways that generate new tasks with a high level of productivity.

Acemoglu and Restrepo [13] propose the field of education as an example. In general, AI has played a marginal role in teaching up until now. The potential given by AI might be used to foster teaching focused on the learning styles (and possibly specific difficulties) of individual students. This individualized teaching, which is impossible in a traditional setting, could thus represent a structural change in teaching methods and not merely an automation of processes already in place.

Another area of interest concerns applications in the field of health care. Here, AI penetration is growing rapidly but has been predominantly limited to the application of automation processes. However, according to Acemoglu and Restrepo [13], it is possible to think of a much more pervasive use of the opportunities provided by AI. Further considerations on the recent evolution of the use of AI in healthcare are developed below.

This analysis emphasizes the importance of closely monitoring the development of technologies from a policy standpoint. Thus, applications of AI from the perspective of automating tasks already predominantly performed by humans are stimulated directly by the market to the extent that they lead to predictable production benefits in a relatively short time. Conversely, the development of technological applications, and AI in particular, that have a profound impact on the structure of the work process does not have an immediate eco-

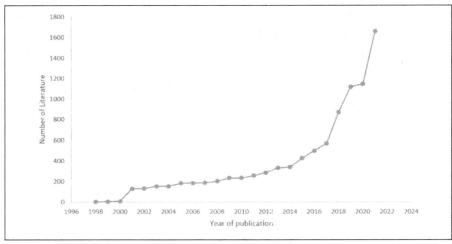

Figure 7: Number of papers on AI+medical/medical education in the last two decades (from [9]).

nomic payoff and may require a struggle to gain traction without appropriate incentives. Acemoglu and Restrepo [13] point out that the United States has traditionally been active in incentivizing research with social spillovers through different public-private partnership forms. However, the momentum toward this type of intervention has subsided significantly in the recent period, partly due to economic circumstances.

6 AI in the area of healthcare

The most impressive way to show growth in this area is probably to show the trend of papers containing, among the keywords, "artificial intelligence and medical or medical education" (see Figure 7). This indicates the growing scientific interest in AI applications in the medical field in recent decades.

As AI technology becomes more advanced, the role of physicians may also be changing. The updating of decision-making processes in healthcare, as well as the design of AI-based management standards and decision-making processes, portends the future direction of AI, which analyzes large amounts of data to inform diagnostic and therapeutic decisions [14]. As of late 2022, the Large Language Models underlying, for example, Chat GPT, allowing conversational interaction, in just a few months have been widely used in medical education, and their potential for application development has already become evident.

At the same time, as the findings of the aforementioned review suggest [9]: (a) AI is not yet part of medical student education; (b) there is a lack of accreditation standards for AI applications in health care, and the effectiveness of AI is not certain; (c) using the Oxford Evidence Scale to evaluate studies, most of them were ranked in the categories with evidence of the lowest scientific quality, indicating the poor quality of current research.

Against the current reality, it is certainly possible to foreshadow a much more pervasive use of the opportunities offered by AI. For example, it is possible to think about the collection and analysis of information that would enable healthcare providers to offer a broader range of individualized services, both diagnostic and treatment. Again, then, it is possible to think of uses that do not simply automate processes already envisioned under human control but represent genuine innovation in real-time health services in which the enhancement of the role exercised by different operators remains central [13]. Already, the application of AI, for example, in imaging, seems to show the potential to improve the accuracy and speed of the diagnostic process while posing issues related to responsibility in the diagnostic process or trust in the clinician-patient relationship.

7 Closing remarks

Overall, the brief analysis presented here underscores the importance, often stressed by Rino Falcone, of fostering an interchange between active researchers and their political counterparts. In the case of AI, Falcone, before others, realized the importance of bringing people to understand the importance of activating incentive policies that, when used appropriately, might have beneficial medium-term effects on the labor market. AI development is not predetermined or magically harmonized by market forces but can take different paths with very different outcomes for the labor market depending on the industrial policies that are activated. In the words of Acemoglu and Restrepo [13], "The considerable promise of AI implies that we need to devote care and serious thought to its implications and to the question of how best to develop this promising technological platform - before it is too late."

We believe that Falcone's value has always been to approach scientific issues by devoting constant care and serious thought to their implications for the future. However, Rino did this without neglecting his basic research interests. He was, in fact, aware that it is the accuracy of the questions that changes the future, not the usefulness of the answers. Foreshadowing the scenarios that might develop by formulating correct and constantly topical questions is the task of researchers; it is then up to politics to build the conditions so that these scenarios can be a source of shared, sustainable, and accessible growth and development for all. The many questions that Rino Falcone has posed to us on every occasion of scientific exchange and collaboration, as well as friendship, will inevitably continue to inspire the actions of many of us as scholars and as individuals.

References

[1] Falcone, Rino, et al. Prospettive di intelligenza artificiale: mente, lavoro e società nel mondo del machine learning. Giornale italiano di psicologia 45.1 (2018): 43-68.

[2] Falcone, Rino, et al. Riflettere sull'IA rivela la varieta e complessita delle competenze coinvolte. Risposta ai commenti. Giornale italiano di psicologia 45.1 (2018): 173-186.

[3] Foster, L.S., Grim, C., Haltiwanger, J., and Wolf, Z. (2017). Macro and micro dynamics of productivity: From devilish details to insights. NBER Working Paper No. 23666, August.

[4] Acemoglu, D., & Restrepo, P. (2019a). Automation and new tasks: How technology displaces and reinstates labor. Journal of Economic Perspectives, 33(2), 3-30.

[5] Acemoglu, D., Restrepo, P. (2019b). Artificial Intelligence, automation, and work. In A. Agrawal, J. Gans, and A. Goldfarb (Editors) The Economics of Artificial Intelligence: An Agenda (p. 197-236). University of Chicago Press.

[6] Acemoglu, D., & Restrepo, P. (2022). Tasks, automation, and the rise in US wage inequality. Econometrica, 90(5), 1973-2016.

[7] Autor, D., & Salomons, A. (2018). Is automation labor-displacing? Productivity growth, employment, and the labor share (No. w24871). National Bureau of Economic Research.

[8] Kocakoç, I. D. (2022). The role of artificial intelligence in health care. In S.B. Kahyaoğlu (Ed.) The Impact of Artificial Intelligence on Governance, Economics and Finance, Volume 2 (pp. 189-206). Singapore: Springer Nature Singapore.

[9] Sun, L., Yin, C., Xu, Q., & Zhao, W. (2023). Artificial intelligence for healthcare and medical education: a systematic review. American Journal of Translational Research, 15(7), 4820. PMID: 37560249; ISSN:1943-8141/AJTR0150419.

[10] Wolff, J., Pauling, J., Keck, A., & Baumbach, J. (2020). The economic impact of artificial intelligence in health care: systematic review. Journal of Medical Internet Research, 22(2).

[11] Gini, C. (1921). Measurement of inequality of incomes. The Economic Journal, 31(121), 124-125.

[12] OECD (2015). In It Together: Why Less Inequality Benefits All. OECD Publishing, Paris.

[13] Acemoglu, D., & Restrepo, P. (2020). The wrong kind of AI? Artificial intelligence and the future of labour demand. Cambridge Journal of Regions, Economy and Society, 13(1), 25-35.

[14] Rimmer, A. (2019). Technology will improve doctors' relationships with patients, says Topol review. BMJ, 364.

Trust, delegation, and cognitive diminishment: the case of generative AI

Fabio Paglieri

Institute of Cognitive Science and Technologies (ISTC), CNR, Rome

fabio.paglieri@istc.cnr.it

Abstract

This paper leverages the socio-cognitive theory of trust developed by Castelfranchi and Falcone to address two urgent and interrelated questions: (1) Should we trust generative AI systems in a variety of tasks? (2) Even if we trust them, is that reason enough to systematically rely on them for performing such tasks? I argue that the answer to the first question is mostly affirmative, whereas the answer to the second conundrum is eminently negative: generative AI systems are, by and large, trustworthy for the execution of various tasks, yet we should resist the temptation of using them too often and too unthinkingly. The theoretical framework pioneered by Castelfranchi and Falcone helps us understand what makes this form of trust without delegation perfectly reasonable: most notably, the need to avoid cognitive diminishment and excessive loss of autonomy.

The author's work is supported by the PRIN 2022 PNRR research project "B-Hu-Well – Boosting human wellbeing with behavioural insights" (PRIN 2022 PNRR, P202227LNS), funded by the European Union, Next Generation EU, Mission 4, Component 2, CUP B53D23030060001.

1 Introduction

One of the key merits of the socio-cognitive theory of trust [1] is the clarification of the complex interplay between trust and delegation. At its core, trusting someone or something entails delegating to that entity either a specific action or an open-ended plan for the realization of some goals of the trustor. The conflicts that such delegation might generate between trustor and trustee have been analyzed in detail [2], as well as the tension and the complementarity between trust and control [3, 4]. A less studied aspect of this dynamic, however, is what negative consequences frequent delegation of crucial cognitive activities to another agent might have for the trustor, *regardless* of the success of their act of trust – or precisely *because* of that success. This dark side of delegation is rarely discussed, since the most prominent effect of successful acts of reliance is an increase in performance (e.g., [5]), whereas the costs in terms of impaired self-sufficiency become apparent only taking a long-term perspective.

In this paper, I offer some thoughts on this topic, arguing that loss of competence (aka *cognitive diminishment*, e.g. [6, 7]) is best understood as an *externality of delegation*, one that cannot be completely avoided, yet it might (and often should) be mitigated with appropriate strategies of *"reasonable trust"*. As a case study, I use the rapidly growing trust that online users exhibit towards generative AI platforms (e.g., ChatGPT, Gemini, DALL-E, Stable Diffusion), both to highlight the dangers of excessive reliance on such systems and to sketch how preserving (or even enhancing) the trustor's competence level and skillset may be factored in the decision to delegate a certain task to another agent or not, preferring instead to do it ourselves – not because we expect the trustee to be faulty, but because we are unwilling to risk a loss of cognitive autonomy and competence due to overdependence from external support. Incidentally, making sense of this suggestion requires leveraging another aspect of the socio-cognitive model of trust developed by Falcone and Castelfranchi: namely, their distinction between trust as an evaluation (what they call "core trust"), as a decision (more properly labelled "reliance"), and as an action (actual "delegation" [1]). Indeed, minimizing or avoiding delegation-induced cognitive diminishment entails deciding not to rely too often on generative AI systems, even though we fully trust their competence in performing the tasks we are tempted to delegate them. This, in turn, provides interesting insight on the complexities of the decision to rely on technological aids in our daily activity.

2 Reliance on generative AI and cognitive diminishment

Advancements in machine learning (ML) techniques, together with access to large amounts of machine-readable datasets, have led to the success of generative AI systems, both for text generation (e.g., ChatGPT, Bard, and Bing Chat, later rebranded as Copilot) and text-to-image art systems (e.g., DALL-E, Midjourney, and Stable Diffusion). In the recent media hype surrounding these applications, it has become commonplace to share anecdotes on their use and its results: since early 2023, social media feeds have been inundated by stories regarding users' personal experiences with AI generative systems like ChatGPT, the one that attracted the most prominence in public discourse. Initially, these narratives tended to be ironically dismissive, pointing out silly mistakes and bizarre hallucinations that these systems were prone to. Soon enough, however, the overall tone of social discourse changed, once we realized how good these technologies are at performing a wide range of tasks: answering queries, summarizing texts, engaging in dialogue, and more generally writing all sorts of documents, just to focus only on text-based Large Language Models (LLMs) like ChatGPT.

As a result, the adoption of such technologies has been massive, in a very short time: ChatGPT reached the mark of 100 million users just 2 months after its release, making it the fastest growing online app to date; for comparison, the previous record was 9 months for Tik Tok, whereas other incredibly popular platforms, such as Instagram, WhatsApp, and Facebook, needed respectively 2.5 years, 3.5

years, and 4.5 years to garner the same level of traffic. Initial use was mostly curiosity-driven and playful, but nowadays these apps are becoming increasingly integrated into educational and professional practices, as well as entering social relationships: social media posts and email messages can be (and often are) written by LLMs, and the tech giants engaged in developing these systems are eager to incorporate them within other services. As a case in point, a recent update of Google's browser, Chrome (August 2024), offers users the option of giving prompts to their LLMs, Gemini, directly in the navigation bar, simply prefacing the search with the tag "@gemini".

Indeed, extensive reliance on LLMs is creating new challenges in several sectors, including education. The World Economic Forum commissioned a report on "Shaping the Future of Learning: The Role of AI in Education 4.0" (April 2024), which early on includes the following statement: "The increasing adoption of AI-driven tools by students for writing assignments and completing assessments has led some educators to question the basic assumptions that classroom work accurately reflects students' cognitive processes" (p. 4)[1]. This is a polite way of saying that traditional assessment tools (e.g., written home assignments) for measuring fundamental cognitive skills (e.g., reasoning and writing) are no longer reliable. Similar concerns have certainly been voiced by several educators, and with good reason. As a case in point, here is a particularly poignant Facebook post by Andrea Rocci, professor of Communication Studies at the University of Lugano, re-

---

[1] Source:https://www.weforum.org/publications/
shaping-the-future-of-learning-the-role-of-ai-in-education-4-0/
An even more ominous passage of the same report is the following: "Generative AI can mimic human logic, writing and even creativity, mirroring some human thought processes and putting into question the relevance of some of the skills, principles, formulas and processes taught in classrooms today, including basics such as writing, grammar and even logic and discourse" (p. 4). Here the 'rationale' endorsed by the World Economic Forum seems to be that, if a cognitive capability can be successfully mimicked or mirrored by an AI, then it is not really important for humans to possess it – after all, you can always rely on the technology for it, so why bother learning it in the first place? The dangers of such carefree attitude towards adoption of cognitive technologies, here expressed in a particularly naïve and crass form, are precisely the topic of this paper.

flecting on the impact of LLMs on university theses: "Students are writing master theses with Chat-GPT. Your students are doing it right now. It's sufficient to pay the access to GPT 4 for one month and you get Consensus to produce a literature review with correct citations and the humanizer to foil detectors. Good luck policing it!" (posted on March 17, 2024, retrieved the same day). I could not agree more, especially with the punchline on the futility of trying to police this kind of behavior: we are already witnessing a classic example of "technological arms race" between AI text generators, AI-generated text detectors, and AI humanizers, and the irony is that it is LLMs all the way down – that is, all these systems are based on the same technology, hence every effort in this direction is bound to increase our dependence from generative AI, rather than curtail it.

Massive reliance on generative AI technologies is problematic for various reasons (among other things, it entails the risk of significant displacement of workers, precisely because it enhances productivity via automation [8, 9]), but also because it pressures towards *technological replacement of cognitive functions and capabilities*. Interestingly, this is a perspective that has engendered significant enthusiasm in the past, at least since the seminal paper on the extended mind by Andy Clark and David Chalmers [10]: in the following decades, the proposed approach, in all its variations,[2] garnered substantial philosophical interest and became one of the pillars of so called "4E cognition", according to which the mind is best understood as being embodied, embedded, enactive, and extended [14]. At the core of the extended mind thesis is the functionalist intuition that what matters to qualify a process as cognitive is the role it plays, not the physical infrastructure that allows it to happen. The original formulation of this tenet is the famous parity principle: "If, as we confront some task, a part of the world functions as a process which, *were it done in the head*, we would have no hesitation in recognizing as part of the cognitive pro-

---

[2]Mapping the checkered history of the extended mind hypothesis is beyond the purposes of this paper, yet it is worth mentioning at least three main varieties of its theoretical tenets: active externalism [10], vehicle externalism [11, 12], and locational externalism [13].

cess, then that part of the world is (so we claim) part of the cognitive process" ([10], p. 8).

Conceptually, I have a lot of sympathy for the extended mind thesis, and it is undisputable that its research program prompted highly relevant debates on the nature of mental process and "the mark of the cognitive", to use an expression popularized both by supporters [15] and critics [16] of this approach. However, it is quite disconcerting to realize that, amidst such a vibrant debate, almost no one paid attention to the obvious (practical) elephant in the room: if mental processes are externalized to artifacts outside of our body, what is to prevent the expropriation of such artifacts, and therefore of whatever cognitive processes they were supposed to perform? If that happens, what are the consequences? Do we have any specific grounds to object against such expropriation, other than invoking standard property rights? Does the role these artifacts play in our mental processes give us any special right over them? Ironically, the issue of ownership came up frequently in the extended mind debate: Rowlands [15], for instance, proposed ownership as a way of insulating the extended mind hypothesis against accusation of cognitive bloat (i.e., being too prone to accept any causally relevant external influence over cognitive processes as integral to them); yet the notion of ownership invoked by Rowlands, and by others after him (e.g., [17, 18]), is not only hard to define, but also phenomenological in nature and related to subject-hood. In contrast, the concept of ownership that ought to preoccupy us is legal. Unfortunately, to the best of my knowledge, the key question "who (legally) owns the extended mind?" has been asked only once in the literature, in a somewhat obscure paper [19], buried in a handbook on intellectual property: not exactly the most likely reading for philosophers of cognitive science, which might explain why the issue was not picked up in the broader debate on the extended mind hypothesis.

In the current technological scenario, I believe the issue must be reconsidered, with more emphasis on political implications and less enthusiasm for conceptual debate: digital technologies in general, and generative AI systems in particular, offer increasing opportunities to offload a rich variety of cognitive processes. This might be good news to vindicate the original externalist intuition, yet it presents all of us (externalists included) with a serious, practical challenge: *the more our mind is extended, the more easily it can be expropriated*, either by private companies or by public powers. A rarely mentioned but highly desirable feature of our good-old-fashioned central nervous system is that, barring illegal clinical procedures, it cannot be taken away from us: the same does not apply to the vast majority of the cognitive artifacts that make our mind extend so seamlessly in the external world. In fact, most of them are not really "ours" to begin with: at best, we own the hardware (your personal computer, your tablet, your mobile phone – unless you are leasing them, as more and more people tend to do nowadays), but what really makes digital artifacts work is the software, and that is typically offered to us as a service for a limited period of time, at a price or for free (i.e., paid with our data and/or contents; for discussion, see [20]). The same applies to the majority of online platforms, including social media: as Mark Bonchek put it on the *Harvard Business Review*, "most social media is rented, not owned. Facebook, Twitter, and LinkedIn are your landlords and you just lease the space"[3]. In this context, cognitive expropriation does not require any manifestly violent act: it would be enough to stop the flow of apps and services (which is something that most providers are legally entitled to do, under current provisions), and we will no longer be able to use those artifacts to perform the cognitive processes that we are now happily delegating them.

Is this reason for concern? Maybe we are better off delegating a lot of cognitive work to technological devices, thus we are genuinely happy to pay for their continued use; and, insofar as we are willing to pay, tech companies will have no reason to expropriate such tools, since that would be self-defeating for them. Now, anyone who believes this path will lead to a happy ending for all involved should first consider two formidable problems. The first is that embracing a societal model in which our growing dependency on technological devices to

---

[3]Source:https://hbr.org/2014/10/making-sense-of-owned-media (last consulted on December 29, 2023).

perform fundamental cognitive activities makes us subordinate to the interests of multinational tech companies is a very risky proposition, since it will further move the balance of power towards private interest and away from public oversight. Nor it would be any better is such private companies were state-controlled (China, as of 2024, provides a reasonable approximation of that scenario), since the power would remain unhealthily concentrated in the hands of a very limited number of individuals, who would just happen to be state officials rather than industry magnates. The second problem concerns what exactly we are so eager to pay for: if it is something that effectively enhances our cognitive powers, then it is easy to see the appeal; but if it is something that *merely replaces pre-existing skills and leads to their progressive loss, in order to induce unnecessary technological dependence*, then it is equally easy to spot that as a rotten deal, if not an outright fraud.

Thus, we are faced with a highly relevant question: are AI-based technologies aimed at extending our mind towards enhancement[4], or are they skewed towards replacement of pre-existing cognitive functions? As Acemoglu and Johnson [23] remind us, the answer to this question is not set in stone: these technologies have the potential to play either role, or both, in our social lives, so it is ultimately up to our collective deliberation to steer them in a direction that will maximize the public good. This deliberation, however, cannot be safely left to the free interplay of market forces, since there is no reason to assume that they will deliver collectively optimal results; in fact, there are grounds to suspect they will not, based both on past experience (the current scenario is severely unbalanced to the advantage of increasingly smaller minorities) and on general principles (*ceteris*

---

[4]Readers should be mindful that there is, in the philosophical and neurological literature, a technical meaning of the expression "cognitive enhancement", which tends to be confined to enhancements resulting in modified brain functioning (for discussion, see [21]), even when such modifications are not pharmacologically induced [22]. However, in the context in this paper, I see no reason to limit the use of "cognitive enhancement" to similar cases, so the expression should be understood in the broader sense of any positive extension of the cognitive capabilities of the agent.

paribus, replacement technologies are easier to market than enhancement technologies, since the latter require active engagement and effort by their users, whereas the former promise the opposite as their main selling point). The shift towards technological enhancement, as opposed to mere replacement, will not happen on its own: it requires active measures at all levels of society.

It is important not to set up a false dilemma, though: enhancement and replacement are not mutually exclusive, as mentioned, and there are cases in which it might be ok to let technological devices replace certain cognitive processes in a stable manner, even if this means losing the capability or the habit of performing them otherwise. The memorization of a large set of phone numbers is a good example[5]: I belong to a generation that grew up without mobile phones, which implied (among other things) that, whenever I wanted to call someone on my cable phone, I had either to find their number on some written repository (my agenda or a phone book, those relics of old...), or simply recall it from memory. The latter was by far the most convenient option, so it was quite common for anyone to memorize the phone number of a variety of people: your close relatives and your love interests, but typically also some of your friends, colleagues, and even the stores or clubs that you frequented the most. At the age of 15, I could easily recall about 50 phone numbers, and that was absolutely normal: today, in the prime of my academic career, that figure is down to 2 – my own number and my wife's, whereas I shamelessly ignore those of my sons. And I have memorized those numbers simply because I am often asked to input them in online forms, where using my smartphone to retrieve them would be impractical. On the plus side, of course, calling others by phone has become much easier, and nowadays my extended mind (me and my smartphone, in this case) can reliably recall several hundreds of phone numbers. Overall, this is

---

[5]There is an important caveat, though: should it be the case that digital technologies make long-term memorization in general increasingly obsolete (which does not seem particularly farfetched), then we might have reasons for concern. No longer recalling many phone numbers is fine, and not having the need to do so is perhaps even pleasant; not being able to memorize anything effectively without a digital support, however, is much less appealing.

an instance where I do not mind permanently offloading the cognitive task of "memorizing phone numbers" to a digital device.

Technological replacement, however, is in general suboptimal, because it generates dependency without empowerment. Comparing it with enhancement clarifies the point: we become dependent on enhancing technologies for achieving results that would be impossible without them, *and always were*. In this case, we are trading off our independence for an extension of our powers: motorized vehicles are an example of physical enhancement, whereas search engines exemplify well cognitive enhancement. Sure, we are all dependent on Google and similar platforms in our ability to search the Internet for information and services, yet such an ability constitutes a significant extension of our mental powers: if the gain is substantial enough, individually and collectively, then the resulting dependency will be justified. With technological replacement, however, we become dependent from an external device to perform a task that was already available to us, which we lose the ability or the inclination to perform as a direct result of the adoption of the new technology: here the loss of independence is not compensated by any extension of our cognitive powers, but rather by comfort and effort reduction. If the replaced skill or process is trivial enough to be forsaken without regret, as in the phone numbers example, then all might be well; but, as soon as the lost competence is a relevant one, we are stuck in an awful situation, where we have made ourselves dependent from proprietary technologies for no good reason.

Similar considerations are relevant overall, yet they become particularly urgent in the context of education, because it is during development that most cognitive skills are acquired through training – or fail to be acquired, if such training is absent due to technological replacement (a danger discussed in the literature under the label "cognitive diminishment"; see [24, 25, 26]). With respect to generative AI, how it will be used in educational settings will have a significant impact on the shape of future societies (for discussion, see [6]). Thus, whether and how to regulate its educational applications should constitute a primary concern for governments and citizens alike. In that regard,

I recently had a chilling experience, although it came with its own silver lining: I was sitting at a cognitive science conference, listening to a talk presenting a field study on the use of ChatGPT for source-based writing in school, when the speaker started lamenting the fact that students were clearly digitally naïve, since they insisted on using ChatGPT as a source, e.g. asking questions about the subject matter of the essay to be written, rather than doing the obviously correct thing, i.e., having ChatGPT write the whole essay on their behalf. The students' tenacity in using generative AI for enhancement rather than for replacement was stigmatized by the researchers as a sign of incompetence in the proper use of ChatGPT, without even contemplating the possibility that it was instead motivated by ethical reasons (i.e., the students did not want to cheat on a school assignment) and by a deep understanding of the rationale of educational activities (where the point is to learn, not to score points by letting someone or something else do the work). As a result, I was personally overjoyed by the students' performance in the task, but rather appalled by the researchers' interpretation of it[6]. Such a blind attitude towards the use of generative AI in education goes beyond mere technological optimism, bordering instead on "technological inebriation": this is something both scholars, practitioners, and policy makers need to steer clear from.

It is also important to clarify how these concerns are affected by a fundamental ongoing debate on extended cognition: namely, the distinction between the constitutive incorporation of a device into the cognitive system (a narrow and stronger sense of "extension"), and the frequent or even continuous use of a device to perform some cognitive function, but without leading to incorporation (a broad and weaker sense of "extension"). This distinction has engendered a significant amount of philosophical work (notable examples include [27, 28, 29]), resulting in a very nuanced taxonomy of cognitive artifacts, based on

---

[6]The whole episode is deliberately narrated without too many details, to protect the identity of the researchers involved: they were relatively young scholars, and these observations were directly discussed with them, thus I believe they will correct course in due time, with no need to expose them to public scorn.

how these artifacts absolve their cognitive functions in relation to the users' capabilities: Fasoli [30], for instance, distinguishes between substitutive, complementary, and constitutive cognitive artifacts. Moreover, this debate has implications on whether some technologies, including generative AI, should be considered as either likely harmful or potentially beneficial for education. Pritchard [31] has framed this as a technology-education tension: on the one hand, introducing such technologies in educational practice might result in deskilling or cognitive diminishment for students, which is a very negative outcome; on the other hand, keeping them out of educational activities will prevent students from learning about and being trained with highly relevant technological tools in the protected and well-structured context of school instruction – that is, another undesirable result. Pritchard appeals to an "extended virtue epistemology" [31] to solve this dilemma: insofar as the new technologies are incorporated in the cognitive processes of their users, there is no diminishment and no deskilling, just a substitution of the physical substrate being used to perform the required function. Cassinadri [6], however, has recently argued against this solution, pointing out that (i) the distinction between cognitive incorporation and mere functional embedding is both debatable in theory [32, 33] and hard to apply in practice [28, 34], plus (ii) actual instances of cognitive extension in educational contexts are likely to be rare, so that most uses of new technologies for learning purposes do not fit that description, yet have great educational potential nonetheless.

Regardless of one's stance towards Pritchard's position, this discussion shows that establishing whether generative AI extends our cognition in the strong or in the weak sense matters only for the second problem raised above: namely, its tendency to replace pre-existing cognitive competences, rather that enhance users with new skills. Insofar as systems like ChatGPT or DALL-E can be described as being fully incorporated in our cognitive processes, to the point that they become constitutive of them, then of course there is no longer need to worry about replacement: much as per Pritchard's suggestion, nothing is being replaced, except the physical substratum carrying out a certain cognitive function – in this case, writing a text or drawing a picture. Of course, it is not at all obvious that such systems can actually meet the rather stringent criteria for cognitive extension in the strong sense, and the burden of proof here falls on those who would like to defend such radical position: however, the option of defusing the replacement/enhancement problem, by reinterpreting generative AI as a constitutive cognitive artifact, is at least theoretically viable.

Yet the first problem discussed in this section, the one of ownership of these technologies, is certainly not defused by appealing to the constitutive nature of AI-based cognitive extensions; in fact, similar appeals only make it *worse*. Let us assume that generative AI is indeed to be considered as a fully incorporated extension of our cognition, no more and no less than any part of our central nervous system: should we therefore rejoice for the fact that now important parts of our basic cognitive capabilities are (literally) owned by private companies? Not at all! If anything, this frames our future as an even bleaker dystopian nightmare. This has an interesting, general implication: the more strongly one believes that AI technologies are (or will soon be) integrated in our cognition, the more worried they should be of the fact that such technologies remain owned by private corporations, and mostly outside of any serious public oversight. Proponents of a strong version of the extended mind hypothesis should be the most vocal advocates against private ownership of cognitive technologies, because, in their view, legal ownership of the tech entails legal rights on somebody's mind, or at least part of it.

Reframing this problem in terms of trust, based on the socio-cognitive model of Castelfranchi and Falcone [1], sheds new light on its implications: the point is not to establish whether generative AI systems counts or not as bona fide cognitive extensions, since the issue of their ownership remains problematic either way; the question is whether we should rely on them, to what extent, for what purposes, and based on what reasons?

3 On the rationality of trust without reliance

Castelfranchi and Falcone [1] individuate two key cognitive ingredients in the core notion of trust (also called "internal trust"): a belief in the *competence* of the trustee (they can do what we are trusting them for) and a belief in their *willingness* to perform the required action (they intend to do what we are trusting them for). These, however, are not sufficient for deciding to rely on the trustee: some *external factors* must also be in place (e.g., we must believe that they will have the opportunity to perform as expected), as well as a belief on *dependence* (we need or prefer to delegate). Given this rich understanding of trust, it is easy to see how the theory can accommodate both instances of *reliance without trust* (i.e., when we are forced by the circumstances to delegate a task to another agent, even if we do not fully trust them), and cases of *trust without reliance* (i.e., when we do trust an agent for a task, yet we avoid delegating it to them, either because we do not think the external conditions are favorable, or because we consider better to do it ourselves).

If we apply this model to generative AI systems, in general they seem adequate to warrant our internal trust for most of the tasks they are designed for: in other words, these systems appear trustworthy in terms of competence and willingness, insofar as they are used appropriately. Competence is hardly in question: indeed, the current success of generative AI is precisely due to the exceptional levels of competence in performing a wide variety of tasks exhibited by these applications, well beyond past performance of similar software. Granted, they are prone to occasional hallucinations and not ideally suited for all types of duties, yet their average performance is more than adequate for most tasks they are used for; besides, they keep improving and are already often capable of outperforming their human counterparts in terms of quality – not to mention the fact that they are much faster. As for willingness, the current functioning of generative AI systems is modeled as service providers: once you give them a prompt, they will deliver their answer. Thus, willingness to perform the required task can be taken for granted, although there is more to be said in terms of second-order goals involved in the use of these technologies: this, however, will be the topic of section 4. For the sake of trusting generative AI systems to do what is asked of them by the user, their willingness is a given.

This is tantamount to saying that, for most ordinary tasks, we are justified in trusting generative AI platforms. If you ask ChatGPT to write a standard email to a potential client for your business, it is reasonable to expect from the system a high-quality output. Similarly, prompting DALL-E to draw a watercolor landscape at sunset will most likely result in a decent image. However, as discussed, having good reasons to trust generative AI platforms in terms of competence and willingness is not enough to justify the choice of delegating some tasks to them: we also need to believe that the external conditions are suitable for them to correctly perform those tasks, as well as being convinced that it is preferable to be dependent from these systems for the achievement of our goals. External conditions do not necessarily constitute a major obstacle against reliance on generative AI platforms, except in specific cases: e.g., if the tasks to be delegated are particularly delicate and entail significant personal responsibilities, or if we have reasons to believe generative AI systems to be particularly likely to make mistakes in those circumstances (due to noise or bias in the training data, for instance).

It is the notion of dependence that needs unpacking, with respect to generative AI. Castelfranchi and Falcone [1] distinguish two versions of this fundamental belief: for trust to prompt actual reliance, we must either believe that delegating that task is the only option available to achieve it (*strong dependence*), or believe that, all things considered, it is better for us to delegate that task to the agent in question, rather than doing it ourselves or delegating it to someone else entirely (*weak dependence* [35]). With generative AI systems, the relevant notion is typically weak dependence: as discussed in section 2, these technologies often frame themselves as a convenient replacement of pre-existing skills – that is, they do on our behalf things that we would be capable of doing ourselves, if so inclined (the very definition of weak dependence). Moreover, from a market perspective, the key benefit of generative AI is to increase productivity by drastically reducing costs, precisely because it allows to perform certain tasks more quickly and with lesser efforts [8]: again, a clear instance of weak dependence.

With weak dependence, the decision whether to rely on a trustworthy technology ultimately hinges on determining whether doing so is better than other available alternatives, *all things considered*. This is where risks of cognitive diminishment should loom large in the mind of potential users of generative AI systems, possibly making it perfectly rational for us *not* to rely on such technologies, even if we trust their capabilities of delivering good results. This does not change the nature of Castelfranchi and Falcone's model of trust, yet it emphasizes how the decision to delegate (or not) a certain task is affected not only by short-term concerns (i.e., is it better for me now to delegate?), but also by long-term considerations of costs and benefits. With respect to generative AI systems, it might be correct to see them as beneficial in the short run, in relation to the specific tasks we use them for: nonetheless, if systematic reliance on such tools ends up making us unable to perform those tasks on our own (the essence of cognitive diminishment), then that choice might be extremely detrimental in the long run. Interestingly, the distinction between weak and strong dependence provides a nice way of framing this problem: cognitive diminishment refers to the danger of *turning weak dependence into strong dependence*, since we start using generative AI platforms to replace us in effectively performing tasks that we could do on our own (weak dependence), until we find ourselves no longer able to successfully undertake them (strong dependence). This entails a significant loss of autonomy for users, however [4].

Once again, stalwart defenders of extended cognition might opine that here any loss of autonomy is illusory, since we are simply updating the physical machinery used to perform our cognitive processes. But this does not change the problem of legal ownership of the technologies we end up being strongly dependent upon (see section 2), nor does it modify our need to rely on them. In fact, the socio-cognitive theory of trust contemplates the possibility of trusting (or not) and relying on (or not) internal processes: for instance, when we think we remember something, but we are not sure enough to act on it, we are expressing a certain degree of trust in our internal memories, and making a decision on whether to rely on them or not. The same applies to generative AI systems understood as "cognitive extensions": even if we accept them as an integral part of our mind, we are still left with the problem of determining their degree of trustworthiness and deciding to what extent we want to rely on them. The latter decision, however, is much more problematic if we become strongly dependent from them for essential activities in our daily life, such as writing or accessing information. These are tasks that we must perform with significant frequency: should delegating them to generative AI be our only option, then our "decision" to rely on them would become merely nominal, as a case of blind delegation. Unfortunately, blind delegation is extremely risky, whether it is directed to an internal process or to an external agency: this danger is compounded by the fact that we would lack significant control over the technology in question, both because we would not be its legal owners (see section 2), and because its workings are radically opaque, even to those who design it (for discussion, see [7]). If my reasoning powers or memory skills appear defective, I can take action to improve them, via training, exercise, and education: but when a generative AI platform fails to perform as expected in a task for which I am strongly dependent on that technology, my options are limited to begging the provider for a better version (typically for a price, one that I can no longer negotiate), or to accept those limitations as insurmountable. Neither scenario is particularly encouraging.

Before summarizing what the socio-cognitive theory of trust reveals on the complex interplay between trust, delegation, and cognitive diminishment associated with the rapid diffusion of generative AI technologies, it is useful to take a brief detour, to reconsider whether willingness to help us in performing some specific tasks can really be ascribed to such technologies, to what extent, and under what conditions.

4 Interlude: local vs. global trust in generative AI

As discussed in section 3, when we rely on a generative AI system for one of the tasks they have been designed for, willingness to comply with our requests to the best of their capacity can be taken for granted: this, together with their objective competence on such tasks, justify trusting them. But trust is always relative to a certain task, and in these instances the task is usually well-specified: write an email, find an information, organize a schedule, summarize a text, and the like. Let us call this "local trust", in the sense that it refers to a well-defined task, in specific circumstances. The willingness to help of generative AI systems in local trust is, I argue, generally justified: we are right in assuming that these systems will do their best to satisfy our immediate requests.

But what about global trust, that is, their willingness to take into consideration our overall wellbeing in the long run? Are they designed to care for that? This is an interesting question, because the goal of taking care of our long-term interests (instead of just satisfying our explicitly stated goals) is akin to the notion of *benevolence*, and benevolence is not, generally speaking, a necessary precondition of trust. Adam Smith was famously adamant about that, when, in *The Wealth of Nations* (1776), noted how "it is not from the benevolence of the butcher, the brewer, or the baker, that we expect our dinner, but from their regard to their own interest. We address ourselves, not to their humanity but to their self-love, and never talk to them of our own necessities but of their advantages". This kind of adversarial reliance is, as remarked by Castelfranchi, still an instance of trust, but one in which benevolence plays no role: yet, "when I order the brewer to send me a box of beer and I send the money, I 'trust' he will give me the beer" ([36], p. 54). Thus, benevolence is not needed for local trust.

However, when we trust someone or something in a deeper and broader sense (global trust), benevolence constitutes a key ingredient: in a healthy family, children trust their parents not only to satisfy their immediate needs and explicit requests, but also to genuinely care for their long-term wellbeing, beyond and possibly even in contrast with short-term gratification of the children's whims; similarly, in a well-functioning state, citizens expect their elected representatives and public institutions to pursue the common good, over and above any petty interest – including those of the parties that directly voted for them. To paraphrase Adam Smith, when it comes to global trust, it is not from the self-interest of the mother, the father, or the public servant, that we expect our well-being, but from their regard to our own objective interest; we address ourselves, not to their self-love but to their humanity, and never talk to them of their advantages but of our own necessities. The question, then, is whether generative AI is worthy of our trust also in this global, loftier sense, or not.

It is worth noticing that the most passionate prophets of generative AI are eager to answer positively to this challenge: in economics, they emphasize benefits for the common good (e.g., increase in productivity [8]) and downplay negative externalities (e.g., workers' displacement and resulting reduction of demands for goods and services in the economy, with risks of recession [9]); in education, enthusiasts of widespread adoption of generative AI technologies insist on opportunities for new and better assessment tools, as well as more personalized and inclusive educational services [37, 38], usually bracketing any concerns in terms of cognitive diminishment [7], data exploitation [20], and privacy risks [39]. Once again, the socio-cognitive theory of trust championed by Castelfranchi and Falcone offers an interesting and highly relevant perspective on this issue, by focusing on the goals we need to attribute to these technologies, in order to assess their willingness to protect our interests in the long run.

These goals are, at present, associated to the private companies responsible for the design, production, and commercialization of generative AI systems, and not to the systems themselves. Whether the agency of complex entities like LLMs is characterized by proper intelligence [40] or not [41] is beside the point here: the pressing concern is not to attribute intentions to these systems, but rather to consider the foreseeable agenda of the gigantic industry behind them. Their business goals are not hard to divine, and doing so does not require attributing them any particularly sinister plan: quite simply, the tech giants involved in the development of generative AI systems are interested in making a significant profit out of them (which should not come as a surprise to anyone, I believe). In contrast, they are not committed to securing the long-term wellbeing of users, beyond what is needed to ensure customer satisfaction and the resulting success of their products. Any proclamation to the contrary, like OpenAI commitment "to ensure that artificial general intelligence (...) benefits all of humanity",[7] is just good marketing, and should be regarded as such.

This has fundamental implications for the kind of trust that we are justified in granting to generative AI systems: local trust in them is reasonable (albeit not necessarily sufficient to warrant reliance, see section 3), insofar as they appear very willing and largely competent to comply with our immediate requests, if properly formulated; however, global trust is a matter of debate, since there is no reason to assume a priori that the interests of the private companies responsible for developing and deploying such technologies will align with the common good; moreover, it remains to be seen whether generative AI's potential for "benefiting all of humanity" will compensate the dangers and costs that it might impose at the individual and collective level. In other words, the jury is still out on the long-term outcomes of the generative AI revolution, thus it would be wise to adopt a precautionary principle in our dealings with such technologies, both personally and publicly, instead of blindly trusting them, out of misplaced enthusiasm for their short-term benefits.

5 Concluding remarks

Hopefully, this paper should have clarified the many ways in which Castelfranchi and Falcone's socio-cognitive theory of trust [1] helps illuminating the complex interplay between trust and delegation, with

---

[7]Source:https://openai.com/charter/

respect to generative AI systems. The following are the main points to keep in mind:

- The key ingredients of internal trust, i.e. competence and willingness, can usually be taken for granted with generative AI systems, when it comes to satisfying our immediate requests: this makes such technologies (locally) trustworthy, under typical circumstances.

- Having good reasons to trust generative AI systems, however, is not sufficient to rationally justify delegating them the tasks they can perform so effectively: for that, we must also believe that making us dependent from these systems is in our best interest, in a long-term perspective.

- Reasons for doubting the wisdom of massive reliance on generative AI include problems of legal ownership of these technologies, fear of cognitive diminishment, and loss of autonomy (among other things, e.g. workers displacement and privacy issues).

- Loss of autonomy, in particular, is best understood as a transition from weak dependence (we can do X on our own, but it is easier to delegate it to generative AI) to strong dependence (we are no longer capable of performing X on our own, so we can only delegate it to generative AI): when the X in question is something as basic as writing skills, memory, or information search, such a loss of autonomy should give us pause.

- Whereas local trust in generative AI systems is generally justified (they will effectively comply with our immediate requests, if well formulated), there are no good reasons to trust them in a more global sense, e.g. assuming the industry behind generative AI to be devoted to the betterment of humankind.

Overall, this analysis suggests *great caution in the systematic adoption of generative AI systems* in our daily practices. This should not, however, lead anyone to enlist me among the apocalyptic intellectuals (to use Eco's categories) who believe generative AI should be banned or severely curtailed. Pointing out risks is not the same as saying that such factors should sway our collective choice on the matter one way or the other: what I am arguing, however, is that we should be aware of those risks and consider them carefully in deciding how to steer the development and diffusion of generative AI technologies. Which is tantamount to insisting that *the future of generative AI is first and foremost a matter of public policy*, one that we should certainly not leave in the hands of private interests (contrary to what has happened so far and is still happening nowadays).

There is, of course, also a more personal version of this conundrum: should each of us privately use these technologies, in view of the risks discussed in this paper? Once again, all I ask is that potential users factor the long-term dangers into their decision-making process: after that, the choice is ultimately left to each of us. As an interesting thought experiment, consider the very act of writing this paper: should I have relied on a LLM to do so, instead of undertaking the task myself? Barring ethical considerations (scholars are not authorized to have LLMs write their articles, and those who are caught in the act face dire consequences) and social decorum (having an AI write an essay in honor of a friend is the ultimate *faux pas*, and I would never do that to Rino!), the option is available, if one cares to consider it: paid versions of the best LLMs are good enough to write decent scholarly work, possibly with some human supervision and usually without too much originality in contents. Or perhaps one might consider asking for some local support from generative AI systems, rather than delegating them the whole thing. What about compiling the references, or maybe choosing a catchy title? Regarding the latter option, I just asked ChatGPT (the basic, free version, not the fancy one that requires payment) to "suggest the title of a scholarly essay on trust, delegation, and generative AI", and this is what I got in response: "Navigating trust and delegation in the age of generative AI: implications for human-AI collaboration". Not bad at all, admittedly: the hint at human-computer interaction is a particularly nice touch, since it was not mentioned in my prompt, yet it makes perfect sense, in context.

Would that have been a better title for this essay, with respect to the one I wrote myself? That is a matter of taste, yet I am honest enough to admit that mine ("Trust, delegation, and cognitive diminishment: the case of generative AI") is less catchy and slightly more boring than the one whipped up by ChatGPT; sure, it is also more informative on the actual contents of the paper, but that is sort of cheating on my part, since I knew where I was going with my arguments, when I selected that title, whereas ChatGPT had no information on that. Be it as it may, the main reason that prompted me to write this whole essay, title included, without any assistance from generative AI technology is not the belief that the outcome would be better that way, but rather the simple fact that I like writing academic prose and I cherish my ability to do so: thus, I am extremely reluctant to let LLMs have the wheel in my place, because... where would be the fun in that?! Thus, at the personal level, my take-home message is simple enough: be careful what you wish to delegate for, lest it is promptly delivered by a generative AI system – letting you unable to do it on your own.

References

[1] C. Castelfranchi and R. Falcone, *Trust theory: A socio-cognitive and computational model.* John Wiley & Sons, 2010.

[2] C. Castelfranchi and R. Falcone, "Delegation conflicts," in *Multi-Agent Rationality: 8th European Workshop on Modelling Autonomous Agents in a Multi-Agent World, MAAMAW'97 Ronneby, Sweden, May 13–16, 1997 Proceedings 8*, pp. 234–254, Springer, 1997.

[3] C. Castelfranchi and R. Falcone, "Trust and control: A dialectic link," *Applied Artificial Intelligence*, vol. 14, no. 8, pp. 799–823, 2000.

[4] R. Falcone and C. Castelfranchi, "Issues of trust and control on agent autonomy," *Connection Science*, vol. 14, no. 4, pp. 249–263, 2002.

[5] R. Falcone and A. Sapienza, "The role of trust in dependence networks: A case study," *Information*, vol. 14, no. 12, p. 652, 2023.

[6] G. Cassinadri, "Chatgpt and the technology-education tension: Applying contextual virtue epistemology to a cognitive artifact," *Philosophy & Technology*, vol. 37, no. 1, p. 14, 2024.

[7] F. Paglieri, "Expropriated minds: On some practical problems of generative ai, beyond our cognitive illusions," *Philosophy & Technology*, vol. 37, no. 2, pp. 1–30, 2024.

[8] S. Noy and W. Zhang, "Experimental evidence on the productivity effects of generative artificial intelligence," *Science*, vol. 381, no. 6654, pp. 187–192, 2023.

[9] J.-A. Occhipinti, A. Prodan, W. Hynes, R. Green, S. Burrow, H. A. Eyre, A. Skinner, G. Ujdur, J. Buchanan, I. B. Hickie, *et al.*, "The recessionary pressures of generative ai: A threat to wellbeing," *arXiv preprint arXiv:2403.17405*, 2024.

[10] A. Clark and D. Chalmers, "The extended mind," *Analysis*, vol. 58, no. 1, pp. 7–19, 1998.

[11] S. Hurley, *Consciousness in Action*. Harvard University Press, 1998.

[12] M. Rowlands, *Body language: Representation in action*. MIT Press, 2006.

[13] R. A. Wilson, *Boundaries of the mind: The individual in the fragile sciences-Cognition*. Cambridge University Press, 2004.

[14] A. Newen, L. De Bruin, and S. Gallagher, *The Oxford handbook of 4E cognition*. Oxford University Press, 2018.

[15] M. Rowlands, "Extended cognition and the mark of the cognitive," *Philosophical Psychology*, vol. 22, no. 1, pp. 1–19, 2009.

[16] F. Adams and K. Aizawa, "The bounds of cognition," *Philosophical Psychology*, vol. 14, no. 1, pp. 43–64, 2001.

[17] S. Gallagher, "The socially extended mind," *Cognitive systems research*, vol. 25, pp. 4–12, 2013.

[18] P. R. Smart, G. Andrada, and R. W. Clowes, "Phenomenal transparency and the extended mind," *Synthese*, vol. 200, no. 4, p. 335, 2022.

[19] J. Dunagan, "Who owns the extended mind?: The neuropolitics of intellectual property law," *The Sage handbook of intellectual property. Los Angeles, CA: Sage*, pp. 689–707, 2015.

[20] N. Couldry and U. A. Mejias, "The costs of connection: How data are colonizing human life and appropriating it for capitalism," 2020.

[21] J. Savulescu and N. Bostrom, eds., *Human enhancement*. Oxford University Press, 2009.

[22] M. Dresler, A. Sandberg, K. Ohla, C. Bublitz, C. Trenado, A. Mroczko-Wąsowicz, S. Kühn, and D. Repantis, "Non-pharmacological cognitive

enhancement," *Neuropharmacology*, vol. 64, pp. 529–543, 2013.

[23] D. Acemoglu and S. Johnson, *Power and progress: Our thousand-year struggle over technology and prosperity*. Hachette UK, 2023.

[24] E. Kasneci, K. Seßler, S. Küchemann, M. Bannert, D. Dementieva, F. Fischer, U. Gasser, G. Groh, S. Günnemann, E. Hüllermeier, *et al.*, "Chatgpt for good? on opportunities and challenges of large language models for education," *Learning and individual differences*, vol. 103, p. 102274, 2023.

[25] D. Mhlanga, "Open ai in education, the responsible and ethical use of chatgpt towards lifelong learning," in *FinTech and artificial intelligence for sustainable development: The role of smart technologies in achieving development goals*, pp. 387–409, Springer, 2023.

[26] A. Shiri, "Chatgpt and academic integrity," *Information Matters*, vol. 3, no. 2, 2023.

[27] R. Clowes, "Thinking in the cloud: The cognitive incorporation of cloud-based technology," *Philosophy & Technology*, vol. 28, pp. 261–296, 2015.

[28] M. Farina and A. Lavazza, "Incorporation, transparency and cognitive extension: Why the distinction between embedded and extended might be more important to ethics than to metaphysics," *Philosophy & Technology*, vol. 35, no. 1, p. 10, 2022.

[29] R. Heersmink, "A taxonomy of cognitive artifacts: Function, information, and categories," *Review of philosophy and psychology*, vol. 4, pp. 465–481, 2013.

[30] M. Fasoli, "Substitutive, complementary and constitutive cognitive artifacts: Developing an interaction-centered approach," *Review of Philosophy and Psychology*, vol. 9, pp. 671–687, 2018.

[31] D. Pritchard, "Intellectual virtue, extended cognition, and the epistemology of education," in *Intellectual virtues and education*, pp. 113–127, Routledge, 2015.

[32] M. Facchin, "Why can't we say what cognition is (at least for the time being)," *Philosophy and the Mind Sciences*, vol. 4, 2023.

[33] S. Varga, "Demarcating the realm of cognition," *Journal for General Philosophy of Science*, vol. 49, no. 3, pp. 435–450, 2018.

[34] R. Heersmink, "Extended mind and cognitive enhancement: Moral aspects of cognitive artifacts," *Phenomenology and the Cognitive Sciences*, vol. 16, pp. 17–32, 2017.

[35] N. R. Jennings, "Commitments and conventions: The foundation of co-

ordination in multi-agent systems," *The knowledge engineering review*, vol. 8, no. 3, pp. 223–250, 1993.

[36] C. Castelfranchi, "Trust and reciprocity: misunderstandings," *International Review of Economics*, vol. 55, pp. 45–63, 2008.

[37] J. Mao, B. Chen, and J. C. Liu, "Generative artificial intelligence in education and its implications for assessment," *TechTrends*, vol. 68, no. 1, pp. 58–66, 2024.

[38] Z. Swiecki, H. Khosravi, G. Chen, R. Martinez-Maldonado, J. M. Lodge, S. Milligan, N. Selwyn, and D. Gašević, "Assessment in the age of artificial intelligence," *Computers and Education: Artificial Intelligence*, vol. 3, p. 100075, 2022.

[39] L. L. Dhirani, N. Mukhtiar, B. S. Chowdhry, and T. Newe, "Ethical dilemmas and privacy issues in emerging technologies: A review," *Sensors*, vol. 23, no. 3, p. 1151, 2023.

[40] F. Bianchini, "Evaluating intelligence and knowledge in large language models," *Topoi*, pp. 1–11, 2024.

[41] L. Floridi, "Ai as agency without intelligence: on chatgpt, large language models, and other generative models," *Philosophy & technology*, vol. 36, no. 1, p. 15, 2023.

Harmonizing Artificial Intelligence, Minds, and Society: A Cognitive Science Perspective

Vieri Giuliano Santucci

Institute of Cognitive Sciences and Technologies, National Research Council, Rome, Italy

vieri.santucci@istc.cnr.it

Giovanni Pezzulo

Institute of Cognitive Sciences and Technologies, National Research Council, Rome, Italy

giovanni.pezzulo@istc.cnr.it

Abstract

Artificial Intelligence (AI) is a rapidly evolving technology, increasingly pervasive both within society and in interfacing with the human mind. Studying the relationships among these three elements – namely, novel AI technologies, minds, and society – and guiding the development of new technologies by ensuring that they are well harmonized with the human social and mental contexts represent urgent challenges. Here, we argue that the field of cognitive science is in a privileged position to tackle these challenges. We focus, in particular, on the relationships between AI and the human mind and emphasize the importance of developing novel AI technologies from a human-centered perspective, which combines both the autonomy of AI systems and their cooperative interactions with humans. We argue that cognitive science studies can offer some fundamental "lessons" about (for example) social interaction and communication, building trust, and balancing autonomy and cooperation, which could help realize future AI systems that are more interconnected with us and our societies—foreseeing novel forms of augmented intelligence for humans.

1 Introduction

Artificial Intelligence (AI) is evolving at an unprecedented pace and has become increasingly intertwined with both human cognition and society. Its applications range from everyday technologies like recommendation systems and virtual assistants to more complex systems operating in critical areas such as healthcare and governance. However, this growing integration raises urgent questions about how AI technologies relate to the human mind and societal structures. It is becoming increasingly apparent that developing AI systems in isolation from the people who use them is too restrictive. An alternative approach consists in developing future AIs from a human-centered perspective, by focusing on the fact that they will have to operate within rich cognitive, social and cultural ecosystems. In this perspective, future AIs could be seen as *agents* in hybrid societies that also include humans – not just tools.

The distinction between agents and tools is key here. So far, our civilization has developed many tools, like cars or computers. Now, for the first time, we have the technology to develop AI systems that are so advanced that they could become agents, which participate in human societies – potentially playing the roles of companions, advisors, collaborators, etc. Defining future AIs as agents (rather than as tools) implies attributing them some *autonomy* – in the sense of freedom to select which goals and plans to pursue – some *adaptivity* – in the sense of a capability to continuously improve and adapt to novel situations through learning – and some capacity for *interaction* with other agents, humans or AIs – possibly, in ways that humans understand and trust (as opposed to developing goals and behaviors that run against human values and are potentially harmful).

Crucially, putting issues like autonomy, adaptivity and interaction at the center stage of future AI development implies that cognitive, social and cultural context in which future AIs need to operate cannot be ignored. In other words, in order to ensure that novel AI technologies are well harmonized with the human social and mental contexts in which they operate, it becomes important to study not just AI in isolation but the three-way relations between novel AI technologies, minds and society. The field of cognitive science might be in a privileged position to tackle this challenge.

Cognitive science, which spans fields such as psychology, neuroscience, cognitive robotics, linguistics, AI, anthropology, and philosophy, offers valuable insights into the ways humans think, learn, and interact. This interdisciplinary approach provides a unique framework for understanding how AI can be harmonized with human cognition and society. In this paper, we explore the role of cognitive science in guiding the development of human-centered AI. We argue that by aligning AI technologies with cognitive models and societal values, we can foster a more cooperative and trustworthy relationship between humans and machines.

2 Towards future, human-centered AIs: three lessons from cognitive science

Cognitive science has long sought to understand how humans solve problems, make decisions, and interact socially. These insights can inform the design of AI systems that are better aligned with human thought processes and societies. For example, AI systems designed to work in healthcare or education must understand not only technical tasks but also the nuances of human interaction and the social and cultural context in which their tasks take place. Understanding how humans collaborate, build trust, and act autonomously under uncertainty can help in the development of AI systems that are more adaptive, empathetic, and capable of complex decision-making in dynamic environments – and that meet the expectations and function more harmoniously with human users.

Here we discuss three "lessons" from cognitive science that could be useful to build future, human-centered AIs. These examples encompass cognitive studies addressing how we interact and cooperate; how we build trust; and how we balance autonomy and cooperation.

2.1 Social interaction and cooperation

Cooperation is an essential component of human society [1], and AI systems must be able to engage in cooperative behaviors. Cognitive studies revealed that humans are endowed with a rich set of cognitive mechanisms – or an *interaction engine* [2] supporting cooperative interactions and joint action and scaffolding linguistic communication. Various studies addressed these cognitive mechanisms and their neural underpinnings, comprising for example the mechanisms permitting humans to align goals and form shared representations [3], understand others' goals and intentions from their actions [4], engage in (verbal and nonverbal) communication and create common ground [5].

Some of these findings have already motivated the development of AI and cognitive robotic models that are capable to solve joint action tasks cooperatively with humans [6], to engage in action understanding and mindreading [7, 8] and to make their behavior more predictable and legible by humans, to help them understand their intentions more easily [9, 10].

However, as future AIs move from relatively simple setups investigated so far to more naturalistic interactions with humans in open-ended settings, their social and communication skills will need to be improved and aligned with human expectations and values. One fundamental challenge is passing from AI and robotic models of isolated social skills, as discussed above, to richer architectures that show similar features as the human *interaction engine* [2, 11].

For this, it might be useful to distill fundamental principles from the large body of studies about human social cognition. Here we briefly discuss three such principles. First, interaction appears to be guided by a fundamental capability to understand how people work at a rather abstract, "agentive" level – perhaps using a (predictive) model of the mental constructs that other individuals (or groups of individuals) use to make decisions, encompassing mental states like action goals, plans and intentions (including shared plans and goals). It is at this rather abstract level – of an *intuitive psychology* [12] or *theory of mind* [13] – that humans seem to coordinate, communicate and understand each other; future AIs would benefit from addressing the same level in order to interact seamlessly with humans [14]. Second, social cognition is increasingly seen as guided by predictive processes – encompassing lower level aspects of the interaction, such as the outcomes of (one's own, others' and joint) actions, higher level aspects such as proximal goals and distal intentions [15], and communicative aspects such as the next sentence in a conversation [16]. Third, there is an increasing tendency to move away from "passive" and "unidirectional" descriptions of social cognition – where the focus is exclusively on one agent *mindreading* another agent – to a truly interactive perspective, in which co-agents cooperate to build-up their interaction, encompassing situations in which not only they predict each other, but also make their behavior predictable and legible, help each other inferring and aligning intentions, form and monitor cooperative goals and common ground, and (in some cases) pursue them even when this implies individual costs [17, 18].

These (and other) principles emerging from cognitive and neural studies of social cognition could guide the development of the next generation of human-centered AIs that could engage in open-ended interactions with humans – therefore, in some sense, become agents in a hybrid society. For example, future autonomous driving AIs will have to behave in ways that meet human expectations in the traffic (including expectations about rule violations), not just drive as efficiently and safely as possible. Co-robots that are supposed to engage in cooperative joint actions with humans, even without "reasoning like humans", could nevertheless be trusted companions to the extent that they "expose" (permit humans inferring) the same types of mental states – goals, intentions, etc. – that support human social inference and alignment. The next generation of AI assistants or chat-bots might learn to cooperate with humans and share goals with them, not just answer queries. They might also come to a better shared understanding of reality with humans if they share the same multimodal context and embodiment with them [19].

2.2 Building trust

For AI systems to be fully integrated into human society, they must not only perform efficiently but also be trustworthy and cooperative. Trust is fundamental in shaping effective human-AI interaction because it determines how much autonomy and responsibility humans are willing to delegate to machines [20]: without trust, human operators are less likely to rely on AI, which limits the potential for such systems to truly assist or augment human capabilities. Cognitive science can inform the design of AI systems by providing insights into how trust operates in human relationships and how AI can emulate these dynamics.

Trust is conceptualised as a complex, dynamic process involving not only relational attributes [21], but specific cognitive phenomena [22, 23] involving beliefs and expectations about another agent's behavior, intentions and competences, and it strongly relies on the ability to predict others' behaviour [24, 25, 26]. Moreover, the concept of trust is a multifaceted construct that might dynamically change over time given the influence of factors such as previous experiences, perceived competence, or situational context [27].

In AI systems, trustworthiness is influenced by several core factors, including transparency, reliability, and ethical alignment. Studies on trust in automated systems suggest that transparency is essential, as users must understand the underlying mechanisms by which decisions are made to build confidence in the system's outcomes [28, 29]. This is especially important in high-stakes environments, such as healthcare and autonomous driving, where trust directly impacts safety and decision-making outcomes. Research indicates that the perception of transparency is linked to how well users comprehend the system's operation and its decision-making processes [30]: explainability and the ability to provide clear, interpretable outputs significantly contribute to trust in AI systems, particularly in real-world applications where they have to tackle complex and sensible tasks [31, 32].

Reliability is another crucial element to transfer the concept of trust to AI systems. Studies show that consistency in performance across various situations helps users develop expectations about how the system will behave, which enhances trust. Systems that deliver predictable, reliable outcomes, eventually because users might exert more control over them [33], are perceived as more trustworthy [20]. However, reliability alone is not sufficient for building trust in human-AI cooperation. Trust also hinges on the concept of delegation, where humans decide to transfer certain tasks or responsibilities to AI systems. According to Castelfranchi and Falcone [34], delegation is at the heart of cooperative interactions between agents. It involves a dynamic relationship in which the user places trust in the AI to perform a task autonomously while maintaining the ability to reassess and adjust the level of control [35]. When a system proves itself capable and reliable, users are more likely to delegate complex tasks, confident that the AI will handle them effectively.

Yet, effective delegation in cooperative contexts also requires balancing control and autonomy. Users need to feel that they retain some level of oversight and can intervene when necessary, rather than completely relinquishing control. Castelfranchi and Falcone introduce the idea that this balance is crucial for trust to grow; too much control or overhelp from the system can lead to distrust, as users may feel the AI is encroaching on tasks they would prefer to handle themselves [33]. Thus, trustworthy AI systems not only deliver reliable outcomes but also support appropriate levels of delegation that respect the user's autonomy, reinforcing the cooperative dynamic between human and machine [36].

On the basis of the previously discussed elements – transparency, reliability, and dynamics of cooperation – it becomes clear that developing models of trust rooted in cognitive science insights is essential for building socially capable AI systems. These models offer a way to formalise the mechanisms through which trust is established, maintained, and lost in human interactions, applying them to AI agents [37]. Translating these principles into AI requires sophisticated computational models of trust that can adapt to the user's evolving expectations and perceptions. This adaptability is critical for AI systems operating in unpredictable environments where trust might need to be recalibrated constantly.

This becomes crucial when we consider artificial agents endowed with increasing levels of operational autonomy. As we will explore in the following section, it is conceivable to design social robots capable of autonomously setting their own goals. In this context, the implementation of mechanisms and functionalities that can ensure transparent and adaptable cooperation is not only desirable but essential. The ability to modulate cooperation dynamically allows AI agents to engage more effectively with humans while retaining the flexibility to adjust their behavior in accordance with evolving tasks and user expectations. This adaptability is vital to foster trust in systems that are progressively gaining more independence in decision-making.

2.3 Balancing autonomy and cooperaton

A defining feature of biological agents, and particularly humans, is autonomy. Autonomy has been conceptualized in numerous ways across different disciplines, and philosophy in particular (see [38] for a comprehensive review). While a definitive or universal definition of autonomy is beyond the scope of this discussion, what is essential for our purposes is the ability of agents to independently choose their goals and the means to achieve them. This aspect of autonomy becomes particularly significant when it is contextualised within social settings, where respect for norms and cooperation must be harmonized with the agent's autonomous decision-making. Cognitive science, in particular, has extensively explored how humans balance autonomy and cooperation in social interactions, decision-making, and norm adherence [39, 40, 1]. These studies offer valuable insights into how we might conceptualise the decision-making and action spaces of autonomous artificial agents, especially social robots, as they navigate the complexities of independent goal-setting within cooperative human environments [41].

In recent years, research in AI and cognitive robotics has pushed the boundaries of what we understand by autonomy in machines [42, 43], moving beyond mere automation to embrace a form of autonomy that includes decision-making, motivational capabilities and open-ended learning processes [44, 45, 46]. This progression is particularly visible in the domain of intrinsically motivated learning [47], where artificial agents are endowed with general goals, such as curiosity, allowing them to explore their environment and autonomously set tasks aimed at increasing their knowledge and skills [48, 49]. These developments open the door to agents that can operate in dynamic and unpredictable environments with minimal supervision, discovering solutions to novel problems along the way.

Typically, in AI and robotics, it is more common to discuss (and implement) automation rather than true autonomy. Automation essentially refers to systems that execute predefined instructions or follow a fixed set of rules [50]. In some cases, the term autonomy is applied to systems using machine learning techniques to find solutions to human-assigned tasks: this is exemplified in classical AI planning techniques [51], where a system, given a set of interaction rules with the environment and a precise description of it, can combine a sequence of actions to achieve a desired outcome. Alternatively, at a higher level of autonomy, we can consider reinforcement learning algorithms [52], in which an agent learns through trial-and-error interaction with its environment, to build a policy, or sequence of actions, that maximizes a reward function defined by a human designer; or (active) inference agents that build generative models of the environment and use them to predict environmental dynamics and to achieve (predefined) preferred outcomes [53]. In all these cases, while the developed systems go beyond mere automation, they are still constrained by externally defined goals and objectives, which limit their scope of action.

This reveals the existence of different degrees of autonomy [54], ranging from zero autonomy (i.e, automation), where machines merely follow predetermined instructions, to intermediate levels, in which goals are predefined but the agent learns autonomously how to achieve them, up to higher levels – what we might call *motivational autonomy* – where systems are able to adapt and learn independently not only discovering unpredicted solutions to assigned tasks but also identifying and pursue self-imposed goals. This form of autonomy is crucial when deploying agents in highly unstructured or unknown environments, where predefining all possible (sub-)tasks is impossible, and the agent needs the capacity to set its own objectives to navigate the complexity. The concept of intrinsic motivation, originally inspired by studies in animal psychology [55], and later expanded through research in neurophysiology [56], cognitive and social sciences [57], has been introduced into AI precisely to address the challenge of developing artificial systems capable of achieving higher levels of autonomy. This approach allows agents to act based on internally generated motivations, rather than relying solely on external goals or rewards. By fostering behaviors driven by curiosity [58], novelty [59], goal-directed exploration [60, 61], or the desire to improve competence [62], intrinsically motivated systems are able to autonomously explore and learn from their environment, adapting to new situations without the need for explicit task assignments.

While motivational autonomy enables artificial agents to navigate uncertain environments effectively, it simultaneously poses significant challenges in social and cooperative settings. Systems endowed with the capacity to set their own goals may diverge from human expectations or established social norms, creating friction in scenarios that require collaboration with humans or other systems. This tension becomes especially problematic in high-stakes environments such as healthcare, autonomous driving, or social robotics, where even slight misalignments between autonomous agents and human values can lead to dangerous or unethical outcomes [63, 64]. On the other hand, imposing pre-defined strict rules and limitations might impair the autonomy that these agents require to properly accomplish the assigned tasks in complex, unstructured and possibly unknown scenarios. Achieving a balance between autonomy and cooperative interaction is therefore crucial [65], making it essential that the design of artificial agents incorporates mechanisms to align their actions with societal norms to ensure safe and effective cooperation while maintaining their versatility and autonomy [66].

Balancing autonomy has been a central concern not only in AI and robotics but also in human social relationships, such as those between teachers and students or parents and children. In these scenarios, autonomy is gradually granted as the learner demonstrates competence and trustworthiness. Insights from cognitive science and social theory suggest that similar dynamics can inform the development of artificial agents [67, 68, 69]. Human autonomy is shaped by external norms and internal motivations, and artificial agents must navigate a comparable balance between independence and cooperative behavior [70].

One of the major challenges with highly autonomous systems is ensuring alignment with human goals without imposing too much control, which would undermine the autonomy itself. In cognitive science, autonomy is understood as a relational concept, emphasizing the interplay of independence and interdependence [71, 72]. This notion applies equally to artificial agents: robots in social environments must learn not only how to act but also what is normatively appropriate [41]. This requires continuous interaction with human agents, making artificial autonomy interdependent and relational [63, 70]. The goal is to design systems that gradually acquire autonomy by learning from their interactions with humans and their surroundings. Not only the environment itself can provide hints and strategies to regulate artificial agents autonomy [73]. In this interdependence model, humans and machines rely on each other: the machine learns from human feedback, while humans increasingly delegate more complex tasks. This autonomy is never absolute but context-dependent, with the robot's autonomy continuously adjusted based on the task, context, and its evolving capabilities [74, 75].

A promising approach to this balance is "adjustable autonomy," where the level of human oversight is modulated based on the system's competence in various situations [35, 76, 36, 77]. This concept allows for flexible delegation: granting the system more freedom in low-risk scenarios while retaining human control in critical contexts. Such a model acknowledges that artificial autonomy is not a static property but a set of adaptable capabilities that change depending on the situation and system performance [78]. At a deeper level, this model of interdependence challenges the common misconception that robotic autonomy is synonymous with independence from human control [79, 70]. Just as human autonomy thrives within social contexts, artificial autonomy must be seen as part of a collaborative relationship between human and machine [80]. Systems driven by intrinsic motivations, like curiosity-driven exploration, might provide a frame to exemplify this interdependence. These systems rely on human programmers to set broad goals, but within those guidelines, they autonomously determine specific tasks to pursue, enhancing their adaptability and versatility in unpredictable environments [66].

However, these capabilities also highlight the importance of ethical frameworks [81]. Relying solely on top-down ethical rules embedded into a robot's code is insufficient [82]. A hybrid approach is needed, where artificial agents learn ethical norms through their interactions with humans and the environment. This "hybrid morality" fuses pre-programmed rules with dynamically learned ethical behaviors, enabling robots to navigate complex social landscapes more effectively [83].

3   Conclusion

The integration of AI, the human mind, and society is one of the most important challenges of our time. Cognitive science, with its focus on understanding human cognition and behavior, is uniquely positioned to guide the development of AI systems that are both autonomous and cooperative. By drawing on insights from cognitive science, we can design AI systems that augment human intelligence, foster cooperation, and align with societal values. As AI continues to evolve, it is essential that we ensure it remains human-centered, supporting not only technological advancement but also the well-being of individuals and society as a whole.

In this short piece, we focused on three "lessons" from cognitive science, which might help designing human-centered AIs capable of more effective and trusted interaction with humans. Cognitive science research could offer many more "lessons" like the three that we addressed in this piece that could be relevant for future, human-centered AIs. For example, one challenge is ensuring that AI systems are adaptable and flexible enough to operate in dynamic, real-world environments. Cognitive science offers insights into how humans learn and adapt to new situations, which can inform the design of AI systems that are more resilient and capable of handling uncertainty. Furthermore, advancements in cognitive science might offer hints about designing AI systems capable to integrate multiple and potentially conflicting sources of information, manage multiple and potentially conflicting drives and goals, learn continuously, and are capable of creative thinking.

For the sake of brevity, we have largely set aside important societal and ethical considerations that are essential in shaping the future of AI. One of the most pressing challenges is ensuring that AI systems align with ethical principles and societal values. Cognitive science can contribute significantly by providing models that explain how humans make moral decisions, offering a foundation for designing AI systems to follow similar principles. By embedding human-centered ethical reasoning into AI systems, we can better ensure that their behavior aligns with human values.

Another fundamental challenge is to understand the roles AI systems might play in our societies if they evolve into true "agents". We are still in the early stages of advanced AI development, and it remains difficult to fully grasp the potential roles and impacts – both positive and negative – that these systems may have. In the introduction, we briefly mentioned the possibility of future AIs serving as companions, advisors, or collaborators, but their roles could extend far beyond these examples. As AI systems become more sophisticated, they may not only assist but also augment human intelligence. Cognitive science offers a framework for developing AI that enhances, rather than replaces, human cognitive abilities. For instance, AI can support human decision-making by providing additional data and insights, allowing for more informed choices. In medicine, AI can assist doctors by analyzing large datasets, identifying patterns, and suggesting potential diagnoses.

Beyond individual enhancement, AI has the potential to improve collective intelligence as well. AI systems that mediate group decision-making can help minimise biases like overconfidence or groupthink, leading to more effective outcomes. In group settings, AI can offer diverse perspectives, aggregate opinions, and flag potential biases, facilitating more robust decision-making processes. In this way, AI serves as a tool for enhancing both individual and collective cognition. While these examples highlight the potential for AI to foster new levels of problem-solving and creativity, they are only a small glimpse of what may be possible. The full range of roles and impacts that future AI could have – including potentially negative consequences, which we do not address in this piece – is yet to be understood. Cognitive studies could shed light on how future AIs might integrate into society by drawing on insights from the history and development of human civilizations.

In conclusion, the future of AI depends on harmonizing technology with the human mind and societal frameworks. Cognitive science offers essential tools and insights for this integration, paving the way for AI systems that enhance human experiences rather than disrupt them.

References

[1] M. Tomasello, *Why we cooperate*. MIT press, 2009.

[2] S. C. Levinson, "On the human "interaction engine"," in *Roots of human sociality*, pp. 39–69, Routledge, 2020.

[3] N. Sebanz, H. Bekkering, and G. Knoblich, "Joint action: bodies and minds moving together," *Trends in cognitive sciences*, vol. 10, no. 2, pp. 70–76, 2006.

[4] C. D. Frith and U. Frith, "Mechanisms of social cognition," *Annual review of psychology*, vol. 63, no. 1, pp. 287–313, 2012.

[5] H. H. Clark, *Using language*. Cambridge University Press, 1996.

[6] D. Maisto, F. Donnarumma, and G. Pezzulo, "Interactive inference: a multi-agent model of cooperative joint actions," *IEEE Transactions on Systems, Man, and Cybernetics: Systems*, 2023.

[7] R. Proietti, G. Pezzulo, and A. Tessari, "An active inference model of hierarchical action understanding, learning and imitation," *Physics of Life Reviews*, vol. 46, pp. 92–118, 2023.

[8] D. M. Wolpert, K. Doya, and M. Kawato, "A unifying computational framework for motor control and social interaction," *Philosophical Transactions of the Royal Society of London. Series B: Biological Sciences*, vol. 358, no. 1431, pp. 593–602, 2003.

[9] A. D. Dragan, K. C. Lee, and S. S. Srinivasa, "Legibility and predictability of robot motion," in *2013 8th ACM/IEEE International Conference on Human-Robot Interaction (HRI)*, pp. 301–308, IEEE, 2013.

[10] G. Pezzulo, F. Donnarumma, and H. Dindo, "Human sensorimotor communication: A theory of signaling in online social interactions," *PloS one*, vol. 8, no. 11, p. e79876, 2013.

[11] G. Pezzulo, "The "interaction engine": a common pragmatic competence across linguistic and nonlinguistic interactions," *IEEE Transactions on Autonomous Mental Development*, vol. 4, no. 2, pp. 105–123, 2011.

[12] S. Carey and E. Spelke, "Domain-specific knowledge and conceptual change," *Mapping the mind: Domain specificity in cognition and culture*, vol. 169, p. 200, 1994.

[13] T. Charman and S. Baron-Cohen, "Understanding photos, models, and beliefs: A test of the modularity thesis of theory of mind," *Cognitive Development*, vol. 10, no. 2, pp. 287–298, 1995.

[14] C. Baker, R. Saxe, and J. Tenenbaum, "Bayesian theory of mind: Modeling joint belief-desire attribution," in *Proceedings of the annual meeting of the cognitive science society*, vol. 33, 2011.

[15] N. Sebanz and G. Knoblich, "Prediction in joint action: What, when, and where," *Topics in cognitive science*, vol. 1, no. 2, pp. 353–367, 2009.

[16] A. Goldstein, Z. Zada, E. Buchnik, M. Schain, A. Price, B. Aubrey, S. A. Nastase, A. Feder, D. Emanuel, A. Cohen, *et al.*, "Shared computational principles for language processing in humans and deep language models," *Nature neuroscience*, vol. 25, no. 3, pp. 369–380, 2022.

[17] G. Pezzulo, F. Donnarumma, H. Dindo, A. D'Ausilio, I. Konvalinka, and C. Castelfranchi, "The body talks: Sensorimotor communication and its brain and kinematic signatures," *Physics of life reviews*, vol. 28, pp. 1–21, 2019.

[18] E. Redcay and L. Schilbach, "Using second-person neuroscience to elucidate the mechanisms of social interaction," *Nature Reviews Neuroscience*, vol. 20, no. 8, pp. 495–505, 2019.

[19] G. Pezzulo, T. Parr, P. Cisek, A. Clark, and K. Friston, "Generating meaning: active inference and the scope and limits of passive ai," *Trends*

in Cognitive Sciences, vol. 28, no. 2, pp. 97–112, 2024.

[20] J. D. Lee and K. A. See, "Trust in automation: Designing for appropriate reliance," *Human factors*, vol. 46, no. 1, pp. 50–80, 2004.

[21] F. D. Schoorman, R. C. Mayer, and J. H. Davis, "An integrative model of organizational trust: Past, present, and future," 2007.

[22] G. Mollering, *Trust: Reason, routine, reflexivity*. Emerald Group Publishing, 2006.

[23] J. A. Simpson, "Psychological foundations of trust," *Current directions in psychological science*, vol. 16, no. 5, pp. 264–268, 2007.

[24] C. Castelfranchi and R. Falcone, *Trust theory: A socio-cognitive and computational model*. John Wiley & Sons, 2010.

[25] D. Straker, *Changing Minds: In Detail. How to Change what People Think, Feel, Believe and Do*. Changing Works, 2014.

[26] R. Falcone, G. Pezzulo, and C. Castelfranchi, "A fuzzy approach to a belief-based trust computation," in *Trust, Reputation, and Security: Theories and Practice: AAMAS 2002 International Workshop, Bologna, Italy, July 15, 2002. Selected and Invited Papers 5*, pp. 73–86, Springer, 2003.

[27] J. Rhim, S. S. Kwak, A. Lim, and J. Millar, "The dynamic nature of trust: Trust in human-robot interaction revisited," *arXiv preprint arXiv:2303.04841*, 2023.

[28] J. Urbano, A. P. Rocha, and E. Oliveira, "A socio-cognitive perspective of trust," *Agreement Technologies*, pp. 419–429, 2013.

[29] B. C. Kok and H. Soh, "Trust in robots: Challenges and opportunities," *Current Robotics Reports*, vol. 1, no. 4, pp. 297–309, 2020.

[30] E. Glikson and A. W. Woolley, "Human trust in artificial intelligence: Review of empirical research," *Academy of Management Annals*, vol. 14, no. 2, pp. 627–660, 2020.

[31] F. Doshi-Velez and B. Kim, "Towards a rigorous science of interpretable machine learning," *arXiv preprint arXiv:1702.08608*, 2017.

[32] W. J. Von Eschenbach, "Transparency and the black box problem: Why we do not trust ai," *Philosophy & Technology*, vol. 34, no. 4, pp. 1607–1622, 2021.

[33] C. Castelfranchi and R. Falcone, "Trust and control: A dialectic link," *Applied Artificial Intelligence*, vol. 14, no. 8, pp. 799–823, 2000.

[34] C. Castelfranchi and R. Falcone, "Towards a theory of delegation for agent-based systems," *Robotics and Autonomous systems*, vol. 24, no. 3-

4, pp. 141–157, 1998.

[35] R. Falcone and C. Castelfranchi, "Levels of delegation and levels of adoption as the basis for adjustable autonomy," in *Congress of the Italian Association for Artificial Intelligence*, pp. 273–284, Springer, 1999.

[36] F. Cantucci, R. Falcone, and C. Castelfranchi, "Human-robot interaction through adjustable social autonomy," *Intelligenza Artificiale*, vol. 16, no. 1, pp. 69–79, 2022.

[37] A. Sapienza, F. Cantucci, and R. Falcone, "Modeling interaction in human–machine systems: A trust and trustworthiness approach," *Automation*, vol. 3, no. 2, pp. 242–257, 2022.

[38] J. Christman, "Autonomy in Moral and Political Philosophy," in *The Stanford Encyclopedia of Philosophy* (E. N. Zalta, ed.), Metaphysics Research Lab, Stanford University, Fall 2020 ed., 2020.

[39] H. Baumann, "Reconsidering relational autonomy. personal autonomy for socially embedded and temporally extended selves," *Analyse & Kritik*, vol. 30, no. 2, pp. 445–468, 2008.

[40] M. Oshana, *Personal autonomy in society*. Routledge, 2016.

[41] R. Falcone and C. Castelfranchi, "Social trust: A cognitive approach," *Trust and deception in virtual societies*, pp. 55–90, 2001.

[42] M. Asada, "Rethinking autonomy of humans and robots," *Journal of Artificial Intelligence and Consciousness*, vol. 7, no. 02, pp. 141–153, 2020.

[43] P. Formosa, "Robot autonomy vs. human autonomy: social robots, artificial intelligence (ai), and the nature of autonomy," *Minds and Machines*, vol. 31, no. 4, pp. 595–616, 2021.

[44] O. E. L. Team, A. Stooke, A. Mahajan, C. Barros, C. Deck, J. Bauer, J. Sygnowski, M. Trebacz, M. Jaderberg, M. Mathieu, *et al.*, "Openended learning leads to generally capable agents," *arXiv preprint arXiv:2107.12808*, 2021.

[45] O. Sigaud, G. Baldassarre, C. Colas, S. Doncieux, R. Duro, N. Perrin-Gilbert, and V.-G. Santucci, "A definition of open-ended learning problems for goal-conditioned agents," *arXiv preprint arXiv:2311.00344*, 2023.

[46] E. Hughes, M. Dennis, J. Parker-Holder, F. Behbahani, A. Mavalankar, Y. Shi, T. Schaul, and T. Rocktaschel, "Open-endedness is essential for artificial superhuman intelligence," *arXiv preprint arXiv:2406.04268*, 2024.

[47] V. G. Santucci, P.-Y. Oudeyer, A. Barto, and G. Baldassarre, "Intrin-

sically motivated open-ended learning in autonomous robots," 2020.

[48] P.-Y. Oudeyer, F. Kaplan, and V. V. Hafner, "Intrinsic motivation systems for autonomous mental development," *IEEE transactions on evolutionary computation*, vol. 11, no. 2, pp. 265–286, 2007.

[49] G. Baldassarre, M. Mirolli, *et al.*, *Intrinsically motivated learning in natural and artificial systems*. Springer, 2013.

[50] F. Galdon, A. Hall, and S. J. Wang, "Designing trust in highly automated virtual assistants: A taxonomy of levels of autonomy," *Artificial intelligence in industry 4.0: a collection of innovative research case-studies that are reworking the way we look at industry 4.0 thanks to artificial intelligence*, pp. 199–211, 2021.

[51] S. J. Russell and P. Norvig, *Artificial Intelligence: A Modern Approach*. Harlow, UK: Pearson Education, third ed., 2016.

[52] R. S. Sutton and A. G. Barto, *Reinforcement Learning: An Introduction*. The MIT Press, second ed., 2018.

[53] T. Parr, G. Pezzulo, and K. J. Friston, *Active inference: the free energy principle in mind, brain, and behavior*. MIT Press, 2022.

[54] M. Simmler and R. Frischknecht, "A taxonomy of human–machine collaboration: capturing automation and technical autonomy," *Ai & Society*, vol. 36, no. 1, pp. 239–250, 2021.

[55] R. W. White, "Motivation reconsidered: the concept of competence.," *Psychological review*, vol. 66, no. 5, p. 297, 1959.

[56] P. Redgrave and K. Gurney, "The short-latency dopamine signal: a role in discovering novel actions?," *Nature reviews neuroscience*, vol. 7, no. 12, pp. 967–975, 2006.

[57] R. M. Ryan and E. L. Deci, "Intrinsic and extrinsic motivations: Classic definitions and new directions," *Contemporary educational psychology*, vol. 25, no. 1, pp. 54–67, 2000.

[58] J. Schmidhuber, "Formal theory of creativity, fun, and intrinsic motivation (1990–2010)," *IEEE transactions on autonomous mental development*, vol. 2, no. 3, pp. 230–247, 2010.

[59] A. Barto, M. Mirolli, and G. Baldassarre, "Novelty or surprise?," *Frontiers in psychology*, vol. 4, p. 907, 2013.

[60] P. Schwartenbeck, J. Passecker, T. U. Hauser, T. H. FitzGerald, M. Kronbichler, and K. J. Friston, "Computational mechanisms of curiosity and goal-directed exploration," *elife*, vol. 8, p. e41703, 2019.

[61] K. J. Friston, M. Lin, C. D. Frith, G. Pezzulo, J. A. Hobson, and S. Ondobaka, "Active inference, curiosity and insight," *Neural computation*,

vol. 29, no. 10, pp. 2633–2683, 2017.

[62] V. G. Santucci, G. Baldassarre, and M. Mirolli, "Grail: A goal-discovering robotic architecture for intrinsically-motivated learning," *IEEE Transactions on Cognitive and Developmental Systems*, vol. 8, no. 3, pp. 214–231, 2016.

[63] C. Tessier, "Robots autonomy: Some technical issues," *Autonomy and artificial intelligence: a threat or savior?*, pp. 179–194, 2017.

[64] J. Ji, T. Qiu, B. Chen, B. Zhang, H. Lou, K. Wang, Y. Duan, Z. He, J. Zhou, Z. Zhang, *et al.*, "Ai alignment: A comprehensive survey," *arXiv preprint arXiv:2310.19852*, 2023.

[65] C. Castelfranchi and R. Falcone, "From automaticity to autonomy: the frontier of artificial agents," *Agent autonomy*, pp. 103–136, 2003.

[66] G. Baldassarre, R. J. Duro, E. Cartoni, M. Khamassi, A. Romero, and V. G. Santucci, "Purpose for open-ended learning robots: A computational taxonomy, definition, and operationalisation," *arXiv preprint arXiv:2403.02514*, 2024.

[67] J. S. Eccles, C. M. Buchanan, C. Flanagan, A. Fuligni, C. Midgley, and D. Yee, "Control versus autonomy during early adolescence," *Journal of Social Issues*, vol. 47, no. 4, pp. 53–68, 1991.

[68] E. L. Deci and R. M. Ryan, "The" what" and" why" of goal pursuits: Human needs and the self-determination of behavior," *Psychological inquiry*, vol. 11, no. 4, pp. 227–268, 2000.

[69] W. Wallach, S. Franklin, and C. Allen, "A conceptual and computational model of moral decision making in human and artificial agents," *Topics in cognitive science*, vol. 2, no. 3, pp. 454–485, 2010.

[70] F. Pianca and V. G. Santucci, "Interdependence as the key for an ethical artificial autonomy," *AI & SOCIETY*, vol. 38, no. 5, pp. 2045–2059, 2023.

[71] C. Mackenzie and N. Stoljar, *Relational autonomy: Feminist perspectives on autonomy, agency, and the social self.* Oxford University Press, USA, 2000.

[72] C. Furrer and E. Skinner, "Sense of relatedness as a factor in children's academic engagement and performance.," *Journal of educational psychology*, vol. 95, no. 1, p. 148, 2003.

[73] A. Baranes and P.-Y. Oudeyer, "Maturationally-constrained competence-based intrinsically motivated learning," in *2010 IEEE 9th International Conference on Development and Learning*, pp. 197–203, IEEE, 2010.

[74] R. Schulz, P. Kratzer, and M. Toussaint, "Preferred interaction styles for human-robot collaboration vary over tasks with different action types," *Frontiers in neurorobotics*, vol. 12, p. 36, 2018.

[75] K. Yuan, Y. Huang, L. Guo, H. Chen, and J. Chen, "Human feedback enhanced autonomous intelligent systems: a perspective from intelligent driving," *Autonomous Intelligent Systems*, vol. 4, no. 1, pp. 1–10, 2024.

[76] S. A. Mostafa, M. S. Ahmad, and A. Mustapha, "Adjustable autonomy: a systematic literature review," *Artificial Intelligence Review*, vol. 51, pp. 149–186, 2019.

[77] F. Cantucci, R. Falcone, and M. Marini, "Redefining user expectations: The impact of adjustable social autonomy in human–robot interaction," *Electronics*, vol. 13, no. 1, p. 127, 2023.

[78] G. Dorais, R. P. Bonasso, D. Kortenkamp, B. Pell, and D. Schreckenghost, "Adjustable autonomy for human-centered autonomous systems," in *Working notes of the sixteenth international joint conference on artificial intelligence workshop on adjustable autonomy systems*, pp. 16–35, Citeseer, 1999.

[79] M. Johnson, J. M. Bradshaw, P. J. Feltovich, C. M. Jonker, M. B. Van Riemsdijk, and M. Sierhuis, "Coactive design: Designing support for interdependence in joint activity," *Journal of Human-Robot Interaction*, vol. 3, no. 1, pp. 43–69, 2014.

[80] M. Johnson, J. M. Bradshaw, P. Feltovich, C. Jonker, B. Van Riemsdijk, and M. Sierhuis, "Autonomy and interdependence in human-agent-robot teams," *IEEE Intelligent Systems*, vol. 27, no. 2, pp. 43–51, 2012.

[81] V. Dignum, M. Baldoni, C. Baroglio, M. Caon, R. Chatila, L. Dennis, G. Génova, G. Haim, M. S. Kließ, M. Lopez-Sanchez, *et al.*, "Ethics by design: Necessity or curse?," in *Proceedings of the 2018 AAAI/ACM Conference on AI, Ethics, and Society*, pp. 60–66, 2018.

[82] T. Hagendorff, "The ethics of ai ethics: An evaluation of guidelines," *Minds and machines*, vol. 30, no. 1, pp. 99–120, 2020.

[83] C. Allen, I. Smit, and W. Wallach, "Artificial morality: Top-down, bottom-up, and hybrid approaches," *Ethics and information technology*, vol. 7, pp. 149–155, 2005.

Reputation Interoperability in Multi-Agent Systems

Jaime Simão Sichman
Laboratório de Técnicas Inteligentes (LTI)
Escola Politécnica (EP)
Universidade de São Paulo (USP)
Brazil
`jaime.sichman@usp.br`

Abstract

In this contribution, we briefly describe some research work carried on during the last 20 years whose goal was to provide artificial agents, possibly developed separately and using different internal reputation models, to exchange reputational information between each other, thus making it possible to use their complementarity in an open multi-agent system scenario. In particular, we present (i) a *functional ontology of reputation*, named **FORe**, to represent reputation knowledge in a formal structured form; (ii) an agent model endowed with a mapping mechanism, which could enable heterogeneous agents to interact with each other using different reputation models; (iii) a *service oriented architecture for reputation interoperability*, named **SOARI**, to compute this mapping externally, to alleviate the agents' processing load.

1 Introduction

In 2021, the International Conference on Autonomous Agents and Multiagent Systems (ICMAS) has reached its 20th edition. AAMAS

Most of the work reported in this paper was totally/partially sponsored by CNPq, CAPES, and FAPESP, Brazil.

2021 (`http://aamas2021.soton.ac.uk`) was organized virtually, and has attracted then over 1,000 registered attendees. The first AA-MAS conference was organized in Bologna (Italy) in 2002, by joining three successful conferences in the area: AA (the International Conference on Autonomous Agents), ICMAS (the International Conference on Multiagent Systems), and ATAL (the International Workshop on Agent Theories, Architectures, and Languages).

In this edition, the Program Chairs - Ulle Endriss (University of Amsterdam, The Netherlands) and Ann Nowé (Vrije Universiteit Brussel, Belgium) - organized a panel session to reflect on the history of our conference and to speculate what the future might bring.

In this panel [1], reputation and trust were mentioned as a seminal concept in the field, both theoretically and in practice. Maria Gini (University of Minnesota, USA) discussed the current issue of people trusting AI systems, and mentioned that "*The agents community has recognized the importance of trust for many years and has proposed ways to create systems that can be trusted.*" She added that "*Machine learning results cannot be easily understood and trusted, especially when one tries to generalize.*" Michael Luck (King's College London, UK) mentioned that "*Back in 2005, the AgentLink roadmap identified six broad technological areas of research and development: .... and trust & reputation.*" Ana Paiva (Instituto Superior Técnico, Portugal) added that "*Agents must be able to interact with humans in transparent and trustworthy ways, promoting and contributing to positive societal changes.*" I was honored to be a member of this panel as well, and pointed out that that the AAMAS community has created several associated workshops that last for decades, for instance, one concerned with "*how to establish coordination, organizations, institutions, and norms in such systems. ... One must recognize that such problems have been discussed in the community since a long time [2].*" In this latter reference, Cristiano Castelfranchi and Rino Falcone introduce the concept of adjustable autonomy, which is based on the notions of reputation and trust.

Another long-standing workshop is the International Workshop on Trust in Agent Societies (TRUST). In 2021, its 22nd edition happened co-located with AAMAS 2021. In particular, Rino Falcone was one of its co-chairs [3]; coincidentally, he was also the co-chair of the workshop co-located with the first AAMAS conference, which happened in 2002 [4].

Therefore, since this volume is dedicated to Rino's honor, I decided to briefly describe some research results conducted by my students, colleagues and me during these years regarding reputation interoperability in multi-agent systems.

Our motivation to address this subject was our belief that it was an essential issue (and so it remains!) since (i) several different reputation models, like HISTOS and SPORAS [5], Socio-Cognitive Model of Trust[6, 7], REGRET [8], Cognitive Reputation Model [9], MMH (Mui, Mohtashemi, and Halberstadt) [10], FIRE [11], REPAGE [12] and L.I.A.R. (Liar Identification for Agent Reputation) [13, 14], among others, were proposed; (ii) each of these models focussed on different and complementary reputation aspects, using distinct representation techniques (numerical values, symbolic values, etc.); (iii) these facts prevented heterogeneous agents, in an open multi-agent system, to exchange reputation values, like humans do, despite the fact of using different reputation models.

In order to achieve this research goal, we adopted the following steps:

1. first of all, we tried to understand which were the main concepts associated to the notion of reputation and to represent these concepts in a structured formal way;

2. we tried then to incorporate within an agent model a mapping mechanism, which could enable heterogeneous agents to interact with each other using different reputation models;

3. finally, we provided an external service to compute this mapping, to alleviate the agents' processing load.

Each of these steps is briefly described in the sequence. The interested reader should refer to the particular references to get a more detailed technical description.

2 Designing FORe

Our first works, developed in 2005, were motivated by the fact that the notion of reputation was used in several computer based models in a rather intuitive way. In general, these models used neither a precise definition of reputation nor the theoretical or empiric bases from disciplines that have worked with reputation concepts much longer than Artificial Intelligence (AI), such as economy, sociology and psychology.

We have then proposed a *functional ontology of reputation*, named **FORe** [15, 16], whose goals were (i) to put together the broad knowledge about reputation produced in some areas of interest such as psychology and artificial intelligence, mainly multi-agent systems; (ii) to represent that knowledge in a structured form.

To achieve this latter, we used the primitive categories of knowledge used in the Functional Ontology of Law proposed by Valente [17]: we claimed that the concepts of the legal world could be used to model the social world, through the extension of the concept of legal rule to social norm and the internalization of social control mechanisms in the agent's mind, so far externalized in legal institutions.

In these works, reputation was considered a social product as well as a social process. It is a product, or property, in the sense that it consists of opinion agreement in some level; on the other hand, it may be seen as a process in the sense that there is a flow of information and influence in the social network. While reputation as a product may be seen as a cognitive representation (or a belief), reputation as a process consists of a set of beliefs' transmission in the social network.

The term Ontology originally designates a philosophical branch dealing with the a priori nature of reality [18]. Although there are many definitions of ontology, the more popular is that proposed by Gruber [19]: *"an ontology consists of an explicit specification of a conceptualization"*.

The Functional Ontology of Reputation contains four main categories: Reputative Knowledge, Responsibility Knowledge, Normative Knowledge and World Knowledge. Besides, we have defined a COMMON KNOWLEDGE category to represent concepts related to common sense that are related to REPUTATIVE KNOWL-EDGE. This ontology contains 85 classes and 40 properties. As we can see in Figure 1, these classes are divided into two main knowl-edge categories: 67 in REPUTATION KNOWLEDGE category and 18 in COMMON KNOWLEDGE category. Each of these five main categories

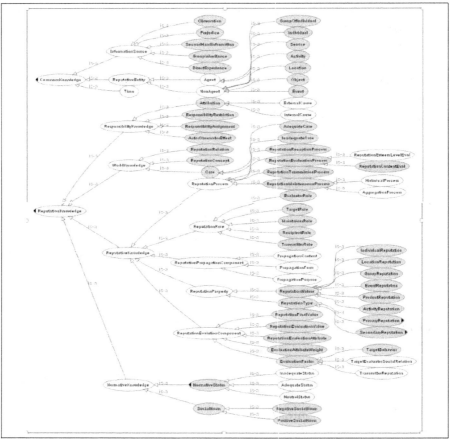

Figure 1: First five layers of **FORe**. Source: [16]

is briefly detailed in the sequence. A more interested reader should refer to [15, 16].

Common Knowledge COMMON KNOWLEDGE category represents the concepts related to the common sense that are related in some way to REPUTATIVE KNOWLEDGE. Among others, it includes the *time concept*. *Information source* represents those facts that act as a source of information for a reputation definition, such as *direct experience, group characteristics, propagated information, observation of behavior* and *prejudice*. *Reputational Entities* represent all things that are able to play at least one reputative role and then take part in a reputation process, such as *individuals* (agents or human being), *group of individuals, objects* and *services*.

Reputative Knowledge REPUTATIVE KNOWLEDGE is the most characteristic category in the Functional Ontology of Reputation, involving 45 classes. The main concepts used to to identify the several aspects of a reputation, both as a product and as a process, were the following:

1. reputation *nature*: distinguishes a reputation according to the entities it is associated to (person, process, etc.);

2. reputation *roles*: inspired by Conte and Paolucci [9], involves four distinct sets of agents: the evaluators, the targets, the beneficiaries, and the propagators;

3. reputation *information sources*: distinguishes reputation obtained by direct interaction (called image in [9]) or by opinions received by others;

4. reputation *evaluation and measurement*: details how reputation is expressed (numerally, by discrete levels, involving attributes and/or context);

5. reputation *maintenance*: described the process that modifies the content and structure of a reputation over time;

6. reputation *scope*: distinguishes reputations according to the manner they are employed (locally or a globally);

7. reputation *propagation* : once more inspired by Conte and Paolucci [9], defines the main aspects involved in an agent's decision on spreading or not a reputation: why to transmit, to whom to transmit, about whom to transmit, what to transmit, how to transmit.

Responsibility Knowledge RESPONSIBILITY KNOWLEDGE category was borrowed from the Functional Ontology of Law[17] and extended to deal with behavior causes. Its main function is to associate a cause to a specific behavior, in order to define whether the reputational entity must be considered responsible for this behavior or, instead, there are circumstances that attenuate and restrict its responsibility. RESPONSIBILITY KNOWLEDGE represents two main notions, namely, the *attribution* notion and the *actor-observer effect* notion, and involves six classes.

Normative Knowledge NORMATIVE KNOWLEDGE category was also borrowed from the Functional Ontology of Law and extended to deal with social norms instead of legal rules. Its main function is to prescribe the agent behavior, through the description of the social norms. The idea is to compare the agent actual behavior with the ideal one, prescribed by the norms, and then conclude whether the behavior is adequate or inadequate. This category contains 7 classes, where *Social Norm* and *Normative Status* are the main ones.

World Knowledge The main function of WORLD KNOWLEDGE is to provide a model of social world which is used by the other categories in order to encapsulate common sense notions. That category involves the following classes: (i) *Reputation Concept*, which represents the class of things in the world; (ii) *Reputation Relation*, which depicts the relations that exist between reputation concepts; (iii) *Case*, which represents a reputative entity behavior in a given circumstance; cases can be compared to social norms in order to define the normative status associated to them.

An overview of **FORe** is shown in Figure 2, where the relation among different knowledge categories are represented. The agent soci-

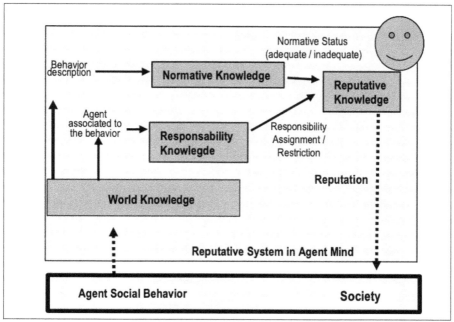

Figure 2: **FORe** Overview. Source: [16]

ety as well as agents' behavior is represented by the inferior rectangle, while the superior rectangle represents the reputative system within an agent's mind. A cycle starts with the interpretation of the agent social behavior by the WORLD KNOWLEDGE category. This category describes the agent behavior in terms of social norms as well as identifies the agent associated to the behavior. After this first step, NORMATIVE KNOWLEDGE category receives this behavior description and matches it with an positive social norm or a negative one, generating the correspondent normative status, Adequate or Inadequate, respectively. RESPONSIBILITY KNOWLEDGE category then defines whether the agent must be held responsible for its behavior (responsibility assignment) or not (responsibility restriction). Finally, using the normative status and the responsibility information REPU-

TATIVE KNOWLEDGE category is able to define the agent reputation, as a reward (good reputation) or a penalty (bad reputation).

Once the ontology has been developed, the next step would be to use it as a basis for interoperability, i.e, allowing agents using different reputation models to exchange reputational information.

3 Applying FORe

In our second step, developed in 2007, we tried to combine several existing technologies to allow agents to interoperate, in particular exchaging notions from their reputation models [20, 21]. We implemented agents on the **ART** testbed [22, 23] that provided an infrastructure for the experimentation of reputation models. We used the **FORe** to allow agents to describe their own reputation model for external purposes and a query language over OWL [24], called nRQL, for agent interactions.

ReGret System	FORe concept
Direct Trust	PrimaryReputation
Witness Reputation	PropagatedReputation
Neighborhood Reputation	StereotypedReputation
System Reputation	CollectiveReputation
Credibility	EvaluatorFactor

Figure 3: Mapping between REGRET model and **FORe**. Source: [20]

Mapping agent reputation models to FORe Visser et al. [25] suggest three different ways to support semantic integration of different sources of information: (i) a centralized approach, where each source of information is related to one common domain ontology; (ii) a decentralized approach, where every source of information is related

203

to its own ontology; (iii) a hybrid approach, where every source of information has its own ontology and the vocabulary of these ontologies are related to a common ontology; this latter organizes the common global vocabulary in order to support the source ontologies comparison. We adopted the hybrid approach by using **FORe** as a common ontology for three reputation models: Cognitive Reputation Model [9], MMH (Mui, Mohtashemi, and Halberstadt)[1] [10] and REGRET [8]. Therefore, considering the ontologies which describe the agent reputation models we can define a mapping between these ontologies and **FORe** whenever the ontologies use a common vocabulary.

As an example, Figure 3 presents the mapping between REGRET model and **FORe**.

Agent architecture   Figure 4 shows an overview of a general agent architecture that allows reputation interoperability. The figure does not represent a full architecture but rather some elements that have to be added to an existing architecture in order to interact about reputation.

This general architecture functions as follow:

1. The interaction module receives a message. If the message is about reputation, it is transmitted to the reputation mapping module.

2. The reputation mapping module analyzes the content of the message and transforms it to comply with the reputation model of the agent. To do so, the reputation mapping module uses **FORe** concepts and the structure of the agent's own reputation model. It must also know how to map information from the ontology into concepts of the agent's reputation model.

3. The translated message is forwarded to the reputation reasoning module. This one can now understand it and handle it as it should. Reputation reasoning is out of the scope of our work and it is not studied further here.

---

[1]This model was referred as *Typology of Reputation* in [16, 20].

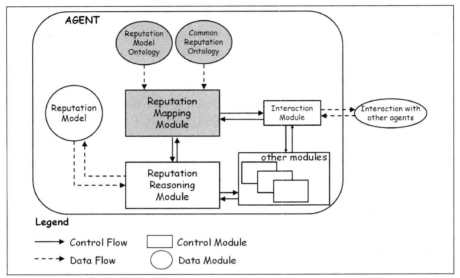

Figure 4: General agent architecture for reputation interaction. Source: [26]

4. If the reputation reasoning module needs to interact with other agents, it formulates a message according to its own reputation model and sends it to the reputation mapping module.

5. In this case, the reputation mapping module translates the content of the message into **FORe** concepts and forwards the resulting query to the interaction module that sends it to other agents.

ART and FOReART testbeds A first version of the **ART** testbed [22, 23] has been developed and used for a competition between trust models during the AAMAS'06 conference. This first version provides a simulation engine on which several agents, using different trust models, are running. The simulation consists in a game where the agents have to decide to trust or not other agents. This game is an art appraisal game in which agents are required to evaluate paintings from different eras to a pool of clients. Therefore, in order to over-

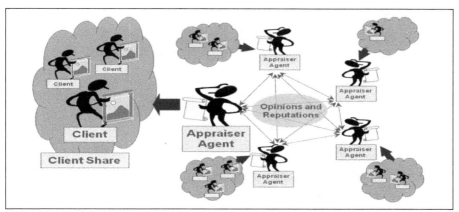

Figure 5: Scenario of **ART** and **FOReART** testbeds. Source: [23, 27]

come their intrinsic knowledge limitation, agents may buy reputation information about third-parties, as well as opinions about paintings, from other agents to produce more accurate appraisals. The scenario of the game can be seen in Figure 5.

In **ART** testbed, interoperability is obtained by mapping the agent's reputation model evaluations into a single value in the domain [0 : 1]. Even though not explicitly defined, it is assumed that 0 refers to the lowest reputation value and 1 to the highest reputation value. This value representation model may incur in loss of expressiveness, since it is required to map usually complex internal reputation models to a single numerical value, hence adopting an efficient, although simplistic, approach.

Since this simplified approach could not serve to our interoperability purposes, we developed an extension of **ART** testbed, called **FOReART** testbed, which enables agents to communicate through symbolic messages. An example of such interaction may be seen in Figure 6. As **FOReART** symbolic messages may refer to internal reputation models concepts (e.g. *DirectReputation* in Figure 6), we consider that it provides a more expressive communication about reputation.

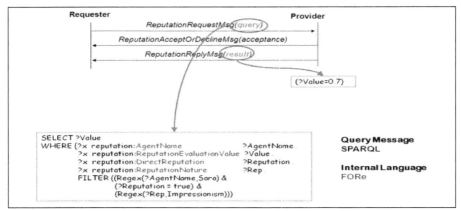

Figure 6: Query path using the ontology concepts and relations. Source: [27]

4 Designing SOARI

In the general agent architecture presented in Figure 4, a new component denominated Reputation Mapping Module (RMM) was integrated to the agent architecture in order to provide the ontology mapping and translation functions. This general agent architecture for reputation interaction facilitates the interoperability of heterogeneous reputation models but it has some main drawbacks from an agent perspective. First, it combines in a single module the ontology mapping and translation functions, hence providing no clear distinction between these two tasks. Second, it requires that each agent represents in ontological terms its internal reputation model and the common reputation domain ontology shared with the others. Finally, each agent must perform the mapping function internally. These are quite important assumptions, and probably would have a big effect in the agents' performance.

We have then proposede a possible solution for these drawbacks, by using a well-know technique in distribued systems: a service oriented approach (SOA).

The main underlying idea in the proposed service oriented archi-

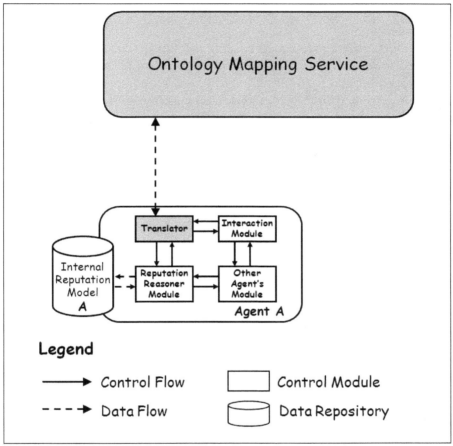

Figure 7: **SOARI** overview Source: [26]

tecture for reputation interaction is that the mapping between different reputation models represented by ontologies may be realized off-line and be available on-line as a service.

The advantage of using a service oriented architecture, from a design/programming perspective, is that the agents become simpler since they do not need to perform the mapping function internally; on the other hand, the advantage of using the hybrid approach as proposed by Visser et al. [25] comes from the fact that the agents do not know other agents internal reputation model, and thus cheating is avoided.

We have therefore, between 2008 and 2014, designed and experimented a *Service Oriented Architecture for Reputation Interoperability*, named **SOARI** [26, 28, 27] that extends the general agent architecture presented in [21, 20] in two ways, as it may be seen in Figure 7:

1. it subdivides the Reputation Mapping Module (RMM) in two distinct and specialized modules: the Ontology Mapping Service (OMS) and the TRANSLATOR module;

2. it performs the ontology mapping function as a service outside the agent architecture.

By defining such extension, we intended to alleviate the agent dynamic workload, since it will not need to perform the mapping function internally and on-line. Moreover, the results of such mapping will be stored in the service and it may be reused by new agents that enter the system and that have an internal reputation model that was already mapped and stored previously. In [26], we have demonstrated the use of the OMS, by mapping two reputation models, L.I.A.R. [13, 14] and REPAGE [12], to **FORe**, which was used as a common reputation domain ontology. The following two steps are required prior to apply the OMS: (i) designing the reputation model ontologies of L.I.A.R. and REPAGE since these reputation models were not described in ontological terms and the OMS needs ontological models as input; (ii) aligning the L.I.A.R. and REPAGE ontologies to **FORe**, since the OMS processes only ontologies that are already described in terms of a common ontology. The interested reader should refer to [26] for a more detailed technical description.

5 Applying SOARI

In [26], we have illustrated the use of the **SOARI** architecture in a simplified electronic commerce scenario, composed of service providers and consumers, both implemented as agents. In this scenario, when a consumer wants to contract a service, it first searches for the providers that have that kind of service available. Secondly, it exchanges reputation information with other consumers to evaluate the reputation of each provider and finally, it decides which provider to contract the service from, based on the reputation values received. The use case was composed of two consumers, which used respectively the REPAGE and L.I.A.R. reputation models.

In a second experiment, described in [28, 27], we were interested in using **SOARI** to answer two questions: (1) is there any improvement in the reputation evaluation accuracy when enabling a more expressive communication? (2) is there any improvement in the reputation evaluation accuracy when considering the heterogeneity of reputation models?

In order to obtain answers to these questions, we have designed and implemented heterogenous agents in the **ART** and **FOReART** testbeds. Out strategy was to develop two types of appraisal agents: *Honest* and *Dishonest*. *Honest* agents answer the requests from other appraisal agents only when they have expertise about the requested painting era and their answer contains information coherent with their internal belief. On the other hand, *Dishonest* agents answer all the requests from other appraisal agents, even when they do not have enough expertise about that painting era, and they never answer the requests with information coherent with their internal belief.

The experiments consist of executing the art appraisal game previously described using the **ART** and **FOReART** testbeds with 20 honest agents and 01 dishonest agent. The main objective of these experiments was to identify the mean value of the reputation assigned by the honest agents to the dishonest agent. Moreover, we have implemented agents with different reputation models, respectively REPAGE, L.I.A.R. and MMH. These experimental scenarios are represented in Figure 8.

When analyzing the effects of the communication expressiveness (question 1), we verified if the mean value of the dishonest agent's reputation model attributes obtained using numerical reputation val-

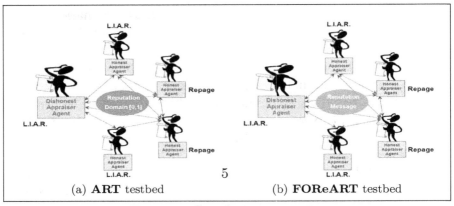

(a) **ART** testbed   (b) **FOReART** testbed

Figure 8: Simulation scenarios

ues (**ART** experiments) were greater than the similar ones obtained in the symbolic testbed (**FOReART** experiments). The underlying idea was that if these results were statistically different, then it would mean that the dishonest agent is better identified if reputation is expressed and exchanged in a more expressive way.

Concerning the analysis of the effect of the reputation model heterogeneity (question 2), we tested if the mean value of the dishonest agent's reputation model attributes obtained from experiments with a homogeneous reputation model were higher than the similar ones obtained from mixed experiments. The underlying idea was that if these results were statistically different, it would then mean that heterogeneous environments, composed of agents with different reputation models, would better identify the dishonest agent, since different aspects of the behavior of the latter could be better captured in the presence of different reputation models.

In our first statistical analysis [28], we have used Wilcoxon ranksum to test our hyphotesis. In our second statistical analysis [27], we used instead a non-parametric multivariate statistical method in order to analyze experimental data considering the possible correlation among reputation models attributes.

Our obtained results were that (i) there is effectively an improve-

211

ment in the accuracy of the agents' reputation evaluation when enabling more expressive communication concerning reputation, except in the homogeneous simulation scenario that considers only agents with L.I.A.R. reputation model (question 1); (ii) we can not conclude such improvement concerning reputation model heterogeinity (question 2). In this case, it would be necessary to perform a more detailed qualitative analysis for better understanding the reputation models interdependence. However, results has shown that reputation heterogeneity have a significant statistical impact on agents using L.I.A.R. and MMH reputation models, which is not he case of those using REPAGE reputation model.

6   Conclusions

In this contribution, we have briefly described some research work carried on during the last 20 years whose goal was to provide artificial agents, possibly developed separately and using diferent internal reputation models, to exchange reputational information between each other, thus making it possible to use their complementarity in an open multi-agent system scenario.

In order to achieve this goal, we have (i) developed a *functional ontology of reputation*, named **FORe**, to represent reputation knowledge in a formal structured form; (ii) designed an agent model endowed with a mapping mechanism, which could enable heterogeneous agents to interact with each other using different reputation models; (iii) developed a *service oriented architecture for reputation interoperability*, named **SOARI**, to compute this mapping externally, in order to alleviate the agents processing load.

As for further work, we believe that this approach could be enhanced by (i) enriching **FORe** with notions of new reputation models; (ii) using the same approach, i.e., designing an external ontological-based service, to interoperate other aspects that could be important in an open multi-gent system scenario, for instance trust, organizations and norms.

As an example, a similar approach was reported in [29], which used other software engineering tecnique - model driven engineering - to interoperate agent organizations models.

Acknowledgements

First of all, I would like to thank Alessandro Sapienza and Filippo Cantucci for their patience in attending this contribution. This paper summarizes the results of a certain number of research projects that were developed during the last 20 years. These works were developed and/or reported by my former students Sara Jane Casare and Luis Gustavo Nardin, and by my colleagues Anarosa Alves Franco Brandão, Laurent Vercouter, and Elisabeti Kira. I would also like to mention Priscilla Avegliano and André Jalbut, former students that have also worked in reputation issues, but whose work could not fit in the length of this contribution.

Last but not least, I would like to express to Rino Falcone my deepest friendship. I first met Rino during an European workshop on multi-agent systems, MAAMAW 1992, which occurred in San Martino al Cimino, Italy. At the time, I was a Phd student, and as a result of the workshop I decided to develop part of my thesis at the Istituto di Psicologia del CNR (IP/CNR), which has been substituted in 2000 by the current Istituto di Scienze e Tecnologie della Cognizione (ISTC/CNR). During that period, my main scientific colaborators were Rosaria Conte and Cristiano Castelfranchi, but I had the chance to interact with Rino, since we were located in nearby offices. During these more than 30 years, I suppose that we were indirectly influenced by each other's work, but I just noticed - after the invitation to write this chapter - that we have never really worked together. Interestingly, this didn't prevent us from becoming true friends, rather than academic colleagues. I have great memories of fancy dinners in Rome or elsewhere in the world, during agent and artificial intelligence conferences. Or when he tried to convince me that "il giallorosso è piu bello che il biancoceleste :-)." Our profession allows us, more than just producing articles and training new researchers, to collect some good friends around the world. Undoubtedly, Rino belongs to the select group of friends that I keep close to my heart.

References

[1] U. Endriss, A. Nowé, M. L. Gini, V. R. Lesser, M. Luck, A. Paiva, and J. S. Sichman, "Autonomous agents and multiagent systems: perspectives on 20 years of AAMAS," *AI Matters*, vol. 7, no. 3, pp. 29–37, 2021.

[2] R. Falcone and C. Castelfranchi, "The human in the loop of a delegated agent: the theory of adjustable social autonomy," *IEEE Trans. Syst. Man Cybern. Part A*, vol. 31, no. 5, pp. 406–418, 2001.

[3] D. Wang, R. Falcone, and J. Zhang, eds., *Proceedings of the 22nd International Workshop on Trust in Agent Societies (TRUST 2021) Co-located with the 20th International Conferences on Autonomous Agents and Multiagent Systems (AAMAS 2021), London, UK, May 3-7, 2021*, vol. 3022 of *CEUR Workshop Proceedings*, CEUR-WS.org, 2021.

[4] R. Falcone, K. S. Barber, L. Korba, and M. P. Singh, eds., *Trust, Reputation, and Security: Theories and Practice, AAMAS 2002 International Workshop, Bologna, Italy, July 15, 2002, Selected and Invited Papers*, vol. 2631 of *Lecture Notes in Computer Science*, Springer, 2003.

[5] G. Zacharia and P. Maes, "Trust management through reputation mechanisms," *Journal of Applied Artificial Intelligence*, vol. 14, no. 9, pp. 881–907, 2000.

[6] R. Falcone and C. Castelfranchi, "The socio-cognitive dynamics of trust: Does trust create trust?," in *Proceedings of the Fourth International Conference on Autonomous Agents, AGENTS 2000, Barcelona, Catalonia, Spain, June 3-7, 2000* (C. Sierra, M. L. Gini, and J. S. Rosenschein, eds.), pp. 55–72, ACM, 2000.

[7] R. Falcone and C. Castelfranchi, "Social trust: A cognitive approach," in *Trust and Deception in Virtual Societies* (C. Castelfranchi and Y.-H. Tan, eds.), pp. 55–90, Dordrecht: Springer Netherlands, 2001.

[8] J. Sabater and C. Sierra, "REGRET: reputation in gregarious societies," in *Proceedings of the Fifth International Conference on Autonomous Agents, AGENTS 2001, Montreal, Canada, May 28 - June 1, 2001* (E. André, S. Sen, C. Frasson, and J. P. Müller, eds.), pp. 194–195, ACM, 2001.

[9] R. Conte and M. Paolucci, *Reputation in artificial societies: Social beliefs for social order*, vol. 6. Springer Science & Business Media,

2002.

[10] L. Mui, M. Mohtashemi, and A. Halberstadt, "Notions of reputation in multi-agents systems: a review," in *The First International Joint Conference on Autonomous Agents & Multiagent Systems, AAMAS 2002, July 15-19, 2002, Bologna, Italy, Proceedings*, pp. 280–287, ACM, 2002.

[11] T. D. Huynh, N. R. Jennings, and N. Shadbolt, "FIRE: An integrated trust and reputation model for open multi-agent systems," in *Proceedings of the 16th European Conference on Artificial Intelligence*, (Valencia, ES), pp. 18–22, 2004.

[12] J. Sabater-Mir, M. Paolucci, and R. Conte, "Repage: REPutation and ImAGE Among Limited Autonomous Partners," *Journal of Artificial Societies and Social Simulation*, vol. 9, no. 2, 2006.

[13] G. Müller and L. Vercouter, "L.i.a.r. achieving social control in open and decentralised multi-agent systems," tech. rep., Ecole Nationale Supérieure des Mines de Saint-Etienne, 2008.

[14] L. Vercouter and G. Muller, "L.I.A.R.: achieving social control in open and decentralized multiagent systems," *Appl. Artif. Intell.*, vol. 24, no. 8, pp. 723–768, 2010.

[15] S. J. Casare and J. S. Sichman, "Towards a functional ontology of reputation," in *4th International Joint Conference on Autonomous Agents and Multiagent Systems (AAMAS 2005), July 25-29, 2005, Utrecht, The Netherlands* (F. Dignum, V. Dignum, S. Koenig, S. Kraus, M. P. Singh, and M. J. Wooldridge, eds.), pp. 505–511, ACM, 2005.

[16] S. J. Casare and J. S. Sichman, "Using a functional ontology of reputation to interoperate different agent reputation models," *J. Braz. Comput. Soc.*, vol. 11, no. 2, pp. 79–94, 2005.

[17] A. Valente, *Legal knowledge engineering: A modelling approach*. IOS Press, 1995.

[18] N. Guarino, "Formal ontology, conceptual analysis and knowledge representation," *Int. J. Hum. Comput. Stud.*, vol. 43, no. 5-6, pp. 625–640, 1995.

[19] T. R. Gruber, "Toward principles for the design of ontologies used for knowledge sharing?," *Int. J. Hum. Comput. Stud.*, vol. 43, no. 5-6, pp. 907–928, 1995.

[20] A. Brandão, L. Vercouter, S. J. Casare, and J. S. Sichman, "Exchanging reputation values among heterogeneous agent reputation models: an experience on ART testbed," in *6th International Joint Conference on Autonomous Agents and Multiagent Systems (AAMAS 2007), Hon-*

olulu, Hawaii, USA, May 14-18, 2007 (E. H. Durfee, M. Yokoo, M. N. Huhns, and O. Shehory, eds.), p. 232, IFAAMAS, 2007.

[21] L. Vercouter, S. J. Casare, J. S. Sichman, and A. Brandão, "An experience on reputation models interoperability based on a functional ontology," in *IJCAI 2007, Proceedings of the 20th International Joint Conference on Artificial Intelligence, Hyderabad, India, January 6-12, 2007* (M. M. Veloso, ed.), pp. 617–622, 2007.

[22] K. Fullam, T. B. Klos, G. Muller, J. Sabater, A. Schlosser, Z. Topol, K. S. Barber, J. S. Rosenschein, L. Vercouter, and M. Voss, "A specification of the agent reputation and trust (ART) testbed: experimentation and competition for trust in agent societies," in *4th International Joint Conference on Autonomous Agents and Multiagent Systems (AAMAS 2005), July 25-29, 2005, Utrecht, The Netherlands* (F. Dignum, V. Dignum, S. Koenig, S. Kraus, M. P. Singh, and M. J. Wooldridge, eds.), pp. 512–518, ACM, 2005.

[23] K. K. Fullam, T. B. Klos, G. Muller, J. Sabater-Mir, A. Schlosser, Z. Topol, K. S. Barber, J. S. Rosenschein, L. Vercouter, and M. Voss, "The agent reputation and trust (art) testbed game description (version 2.0)," 2006.

[24] S. Bechhofer, F. Van Harmelen, J. Hendler, I. Horrocks, D. L. McGuinness, P. F. Patel-Schneider, L. A. Stein, *et al.*, "Owl web ontology language reference," *W3C recommendation*, vol. 10, no. 2, pp. 1–53, 2004.

[25] U. Visser, H. Stuckenschmidt, H. Wache, and T. Vögele, "Enabling technologies for inter-operability," *Workshop on the 14th International Symposium of Computer Science for Environmental Protection*, 2000.

[26] L. G. Nardin, A. Brandão, J. S. Sichman, and L. Vercouter, "SOARI: A service oriented architecture to support agent reputation models interoperability," in *Trust in Agent Societies, 11th International Workshop, TRUST 2008, Estoril, Portugal, May 12-13, 2008. Revised Selected and Invited Papers* (R. Falcone, K. S. Barber, J. Sabater-Mir, and M. P. Singh, eds.), vol. 5396 of *Lecture Notes in Computer Science*, pp. 292–307, Springer, 2008.

[27] L. G. Nardin, A. A. F. Brandão, E. Kira, and J. S. Sichman, "Effects of reputation communication expressiveness in virtual societies - A multivariate statistical analysis," *Comput. Math. Organ. Theory*, vol. 20, no. 2, pp. 113–132, 2014.

[28] L. G. Nardin, A. A. F. Brandão, and J. S. Sichman, "Experiments on semantic interoperability of agent reputation models using the SOARI architecture," *Eng. Appl. Artif. Intell.*, vol. 24, no. 8, pp. 1461–1471,

2011.

[29] L. R. Coutinho, A. A. F. Brandão, O. Boissier, and J. S. Sichman, "Towards agent organizations interoperability: A model driven engineering approach," *Applied Sciences*, vol. 9, no. 12, 2019.

Epistemic Logics Programs: A Survey of Motivations, Recent Contributions and New Developments

Stefania Costantini

DISIM - Università dell'Aquila, via Vetoio, L'Aquila, Italy
Gruppo Nazionale per il Calcolo Scientifico - INdAM, Roma, Italy
stefania.costantini@univaq.it

Andrea Formisano

DMIF - Università di Udine, via delle Scienze 206, Udine, Italy
Gruppo Nazionale per il Calcolo Scientifico - INdAM, Roma, Italy
andrea.formisano@uniud.it

Abstract

Significant time has passed since the initial definition of Epistemic Logic Programs (ELPs) by Gelfond and Przymusinska in 1991. ELPs were developed to extend Answer Set Programming (ASP), as defined by Gelfond and Lifschitz in 1988, with epistemic operators capable of expressing what is known or not known, thereby facilitating affordable epistemic reasoning. However, integrating these new epistemic operators at a semantic level proved to be more complex than anticipated, primarily because they allow introspective examination of a program's semantics, which is defined in terms of its answer sets. Specifically, the operator $\mathbf{K}$ was introduced, which denotes knowledge, where $\mathbf{K}A$ indicates that a ground atom A is true in every answer set of the program P in which $\mathbf{K}A$ appears. The semantics

Research partially supported by the PNRR Project "Enhanced Network of intelligent Agents for Building Livable Environments - ENABLE" CUP E13C24000430006, and by the Interdepartmental Project on AI (Strategic Plan UniUD–22-25), and by INdAM-GNCS project CUP E53C22001930001.

of ELPs are provided through methods that characterise world views, which can be several, where each one is a set of answer sets, as opposed to a single set of answer sets as in traditional ASP. Each world view consistently satisfies the epistemic expressions present in a given program according to a specified semantic approach. Beyond the seminal proposals, various semantic frameworks for ELPs have been introduced. Some of these frameworks determine world views using a modified notion of "reduct", akin to ASP, and are thus referred to as "reduct-based". Other approaches combine non-classical and epistemic logics. In this chapter, we review existing semantic reduct-based frameworks, propose new ones, and mainly discuss how to make it possible to experiment with these semantics. We first propose a method for rapid prototyping of a solver (execution engines) that provides the world views of a given program for the class of semantics we consider. We then propose a method for synthetic characterisation of world views, and for querying world views top-down, in Prolog style.

1 Introduction

The initial formulation of Epistemic Logic Programs (ELPs, referred to simply as "programs" unless specified otherwise) was introduced several years ago in [1] and in [2]. ELPs aimed to extend Answer Set Programs (ASP programs), which are defined under the Answer Set Semantics [3], by incorporating "epistemic operators" to facilitate various forms of epistemic reasoning in a practical yet principled manner. Consequently, there was an early anticipation of extending ASP solvers to accommodate ELPs. However, devising the semantics for ELPs, which should serve as the foundation for solver implementations, proved more intricate than initially anticipated. The obstacles arise from the fact that the new epistemic operators have the capability to introspectively examine a program's own semantics, defined in terms of its "answer sets". In fact, $\mathbf{K}A$, where $\mathbf{K}$ represents "knowledge", is deemed true if the (ground) atom A holds true in every answer set of the program Π where $\mathbf{K}A$ occurs. Derived epistemic operators can be defined: $\mathbf{M}A$, or equivalently $not\,\mathbf{K}not\,A$ (where not is

ASP standard *default negation*), means that A is true in some of the answer sets of Π. The *epistemic negation operator* **not** [4] expresses that atom A *is not provably true*, meaning that A is false in at least one answer set of Π, i.e., **not** A is equivalent to *not* $\mathbf{K}A$. Vice versa, through **not**, one can define both $\mathbf{K}$ and $\mathbf{M}$: in fact, $\mathbf{K}A$ and $\mathbf{M}A$ can be rephrased as **not not** A and **not** *not* A, respectively.

Semantics of ELPs is provided in terms of *world views*, which are sets of answer sets (instead of a unique set of answer sets like in Answer Set Programming). Each world view, in accordance with a specified semantic approach, consistently fulfils the epistemic expressions found within a given program.

Several semantic approaches for ELPs have been introduced beyond the seminal ones, among which those proposed by M. Gelfond [5], M. Truszczyński [6], L. Fariñas del Cerro et al. [7], Y. Shen and T. Eiter [4], P. T. Kahl and A. P. Leclerc [8], E. I. Su [9], and by P. Cabalar et al. [10]. Many of these approaches extend to ELPs what done for ASP, and thus are *reduct-based*, that is, to find world views of a given program they prescribe to:

- start with a candidate world view;

- build the *reduct* of the program with respect to this candidate world view, according to some specific definition of such reduct;

- compute the set of stable models of the reduct;

- check whether the candidate world view is indeed a world view, i.e., coincides with this set of stable models.

While solvers have been devised for some of these approaches, the landscape of semantic proposals continues to evolve, with the likelihood of new ones being introduced in the future. This ongoing evolution is driven by the absence of consensus regarding the "right" semantics.

The paper [11] proposed a fast-prototyping methodology to obtain what we can call a "quick solver" (in the sense that it is quickly and easily obtained) for any reduct-based semantics, with no attention to performance, but with the advantage of being able to experiment with the approach on small/medium programs and not only on very small programs as done so far on paper.

In this chapter, we provide two contributions. In Section 5 we refine the fast-prototyping method and present its implementation in the ASP Chef system [12]. We then illustrate the implementation of some of the seminal semantic approaches and a recently proposed one and show how it is easily customisable to other approaches. So, the demanding process of implementing a performant dedicated solver for a new semantic approach can be postponed after a testing phase is performed by using the quick solver.

In Section 6, we then propose a method for synthetic characterisation of world views, and for querying world views top-down, in Prolog style. The method is based on Resource-Based Answer Set Programming (RAS), that we had previously proposed and prototypically implemented [13, 14, 15].

The ELP approach is aimed at building intelligent cognitive agents in Answer Set Programming or, more generally, in logic programming. So, this work is related to and partly inspired by the wide corpus of research developed by Rino Falcone in this field (see, among others, [16, 17, 18]).

2 Answer Set Programming and Semantics

Answer Set Programming (ASP) [3, 19, 20] is a well-established programming paradigm with many applications, equipped with efficient and formally justified efficient *solvers* for program execution [21, 22, 23, 24]. In ASP, a program can be understood as a collection of statements that define a specific problem. Each answer set represents a solution that aligns with this problem definition. If an ASP program has no answer sets, it is said to be "inconsistent", and "consistent" otherwise. Syntactically, an ASP program Π is a collection of *rules* of the form

$$A_1 \vee \ldots \vee A_g \leftarrow L_1, \ldots, L_n \tag{1}$$

where each A_i, $0 \leq i \leq g$, is an atom, $\vee$ denotes disjunction and the L_is, $0 \leq i \leq n$, are literals (i.e., atoms or negated atoms of the form $not\ A$). The left-hand side and the right-hand side of the rule are called *head* and *body*, respectively. A rule with empty body is called a *fact*. Disjunction can occur in rule heads only (and hence in facts). A *constraint* is a rule with an empty head, '$\leftarrow L_1, ..., L_n$', stating that $L_1, \ldots, L_n$ are not allowed to be simultaneously true in an answer set. The impossibility of fulfilling such requirements is one reason for a program to be inconsistent. Several extensions to the ASP language, including features like "classical negation," aggregates, and weak constraints, have been proposed. For an in-depth discussion, readers can refer to the literature [22, 25]. However, in this paper, we restrict ourselves to basic ASP programs in the above syntactic form, assumed to be "ground", i.e., without variables (thereby implicitly assuming that all atoms are propositional). Note that program *grounding* is the first step any ASP solver performs.

The answer set (or "stable model") semantics (AS semantics, for short) [3] asserts that, if they exist, the answer sets of a program Π are included among the supported minimal classical models of the program, when it is interpreted as a first-order theory in an obvious way. The initial definition [3], which was introduced for programs with rule heads restricted to single atoms, was based on the 'GL-Operator'. Given set of atoms I and program Π, $GL_\Pi(I)$ is defined as the least Herbrand model of the program Π^I, the (so-called) Gelfond-Lifschitz reduct of Π w.r.t. I, obtained from Π by:

1. removing all rules which contain a negative literal $not\ A$, for $A \in I$; and by

2. removing all negative literals from the remaining rules.

The fact that Π^I is a positive program ensures that a least Herbrand model exists and that it can be computed, for instance, by means of the *immediate consequence operator* T_{Π^I} [26]. Then, I is an answer set whenever $GL_\Pi(I) = I$.

3 Epistemic Logic Programming

3.1 Introduction and Intuition

In Epistemic Logic Programming (ELP), one can postulate about what is known, meaning true in every answer set of a program in the program itself. This is done through literals of the form $\mathbf{K}L$, $\mathbf{K}$ intuitively meaning "knowledge", where $\mathbf{K}L$ is called a *subjective literal* in contrast to usual *objective literals*. Note that mentioning literal L in the rules of a program Π means that L is intended to be true in some of the answer sets of Π. In contrast, $\mathbf{K}L$ has a much stronger connotation, meaning that L is unequivocally true in any situation, i.e., true in all answer sets. In turn, $\mathbf{K}\,not\,L$ denotes falsity in every answer set. This makes ELP suitable for all those critical applications in fields such as law, cybersecurity, distributed computing, recommender systems, etc., where one must be able to refer to, so to say, *reliable truth*.

Still, as in ASP, uncertainty or conflict about the truth of some atom gives rise to several answer sets; in addition, in ELP, uncertainty or conflict about knowledge gives rise to radically alternative scenarios called *world views*, that are sets of answer sets, each one stemming from the given program and the assumptions on knowledge expressed therein. For example, the following program Π_1:

$$
\begin{aligned}
a &\leftarrow not\,b \\
b &\leftarrow not\,a \\
e &\leftarrow not\,\mathbf{K}f \\
f &\leftarrow not\,\mathbf{K}e
\end{aligned}
$$

has two world views, under every semantics (each one denoted by using brackets [ ]): one is $[\{a, e\}, \{b, e\}]$, where $\mathbf{K}e$ is true (as e is, in fact, true in both answer sets) and $\mathbf{K}f$ is false, and the other one is $[\{a, f\}, \{b, f\}]$ where $\mathbf{K}f$ is true (as f is true in both answer sets) and $\mathbf{K}e$ is false.

The presence of two answer sets in each world view of Π_1 is due to the cycle on objective atoms a and b. In contrast, the presence of two world views is due to the cycle involving subjective literals (in general, the existence and number of world views are related to such kinds of cycles; see [27] for a detailed discussion).

As a concrete example, consider the following program, expressed in basic non-disjunctive ASP enriched via the **K** operator. The program states (in a hypersimplified way) under which conditions a patient should consult a doctor. In particular, a patient will consult a specific doctor for some problem p on which the doctor is specialised only if the doctor is known to be reliable. Notice here the difference that this brings to the formulation: it is not sufficient that the doctor might be possibly reliable (in some answer set) but must be certainly reliable (in every answer set). Since the first rule must be grounded, in order to show concise world views we consider below only one doctor.

The initial version of the program can be the following:

$$consult(patient, X, p) \leftarrow doctor(X), specialised(X, p),$$
$$good\_reputation(X),$$
$$\mathbf{K}\, reliable(X),$$
$$\mathbf{K}\, good\_reputation(X).$$
$$doctor(d1).$$
$$specialised(d1, p).$$

This embryonic program has only one world view, i.e.,

$$[\{doctor(d1), specialised(d1, p)\}],$$

with just one answer set inside. There is no indication in the program, and thus, from the world view, that the doctor is reliable and has a good reputation and therefore can be consulted. One can then extend the program to outline radical uncertainty by stating that a doctor may be known to be either reliable or unreliable. This is expressed by a negative cycle on $not\,\mathbf{K}$, that generates two world views.

$$reliable(X) \leftarrow doctor(X), not\,\mathbf{K}\, unreliable(X).$$
$$unreliable(X) \leftarrow doctor(X), not\,\mathbf{K}\, reliable(X).$$

Assume that there is also uncertainty about doctor's good reputation, expressed in non-disjunctive ASP as follows:

$$good\_reputation(d1) \quad \leftarrow \quad not\ nogood\_reputation(d1).$$
$$nogood\_reputation(d1) \quad \leftarrow \quad not\ good\_reputation(d1).$$

The program outlined so far has, in fact, two world views, each including two answer sets, reflecting the uncertainty about doctor's good reputation. The doctor is concluded reliable in one world view, W_1, and unreliable in the other one, W_2. Still, even in W_1 it is not concluded that (s)he can be consulted, because of the uncertainty about reputation.

$$
\begin{aligned}
W_1 \quad = \quad & [\{\,doctor(d1), specialised(d1,p), reliable(d1), \\
& good\_reputation(d1)\}, \\
& \{\,doctor(d1), specialised(d1,p), reliable(d1), \\
& nogood\_reputation(d1)\}] \\
W_2 \quad = \quad & [\{\,doctor(d1), specialised(d1,p), unreliable(d1), \\
& good\_reputation(d1)\}, \\
& \{\,doctor(d1), specialised(d1,p), unreliable(d1), \\
& nogood\_reputation(d1)\}]
\end{aligned}
$$

Assume now that a new rule will be established stating that a doctor has a good reputation if (s)he is known to have issued brilliant diagnoses in the past and that evidence about this will also be acquired concerning doctor $d1$.

$$past\_brilliant\_diagnoses(d1).$$
$$good\_reputation(X) \quad \leftarrow \quad doctor(X),$$
$$\mathbf{K}\,past\_brilliant\_diagnoses(X).$$

Note that $\mathbf{K}\,past\_brilliant\_diagnoses(d1)$ immediately follows as $past\_brilliant\_diagnoses(d1)$ is a fact, i.e., it is unconditionally true. The conclusion $good\_reputation(X)$ that can be now drawn rules out the second answer set from both world views, which become:

$$W_1' = [\{doctor(d1), specialised(d1, p), reliable(d1),$$
$$good\_reputation(d1), past\_brilliant\_diagnoses(d1),$$
$$consult(patient, d1, p)\}]$$
$$W_2' = [\{doctor(d1), specialised(d1, p), unreliable(d1),$$
$$good\_reputation(d1), past\_brilliant\_diagnoses(d1)\}]$$

Therefore, according to W_1', the patient can finally consult the doctor about her problem. In addition, since a brilliant doctor can hardly be considered unreliable, one might add the following constraint, stating, in fact, that it is impossible that a doctor issuing brilliant diagnoses in the past is known to be unreliable:

$$\leftarrow \; doctor(X), \mathbf{K}\, past\_brilliant\_diagnoses(X), \mathbf{K}\, unreliable(X).$$

This constraint rules out the second world view, so W_1' will become the unique world view of the program, and no uncertainty is left.

Notice that for this very simple program that we have incrementally constructed, we end up with a single world view, but this is not always the case. In general, there can be several world views, and so uncertainty exists about which scenario (outlined in a world view) is more reliable. Extensions to the basic ELP approach provide epistemic operators ranging over the entire set of world views (cf. [27]).

3.2 Syntax and Semantics

ELPs allow one to express positive and negative *subjective literals* (in addition to *objective literals*, i.e., those occurring in ASP programs, plus the truth constants $\top$ and $\bot$). A positive subjective literal is of the form $\mathbf{K}L$, where $\mathbf{K}$ is the *epistemic operator* of knowledge and L is an objective literal (so, epistemic operators cannot be nested), meaning that L is "known" as it is true in every answer set of the given program Π.

The syntax of rules in ELP is analogous to ASP, save that literals in the body can be either objective or subjective. An ELP program is called *objective* if no subjective literals occur therein; that is, it is an ASP program. A constraint involving (also) subjective literals is called a *subjective constraint*, whereas one involving objective literals only is an *objective constraint*. Let $Body_{subj}(r)$ be the (possibly empty) set of subjective literals occurring in the body of rule r. *Subjective rules* are those whose body is only composed of subjective literals.

Let a semantics S be a function mapping each program into sets of belief views, that is sets of sets of objective literals, where S has the property that, if Π is an objective program, then the unique member of $S(\Pi)$ is the set of stable models of Π. Each member of $S(\Pi)$ is called an S-*world view* of Π. One usually writes "world view" instead of "S-world view" whenever mentioning the specific semantics is irrelevant.

As usual, for any world view W and any subjective literal $\mathbf{K}L$, we write $W \models \mathbf{K}L$ if and only if for all $I \in W$ the literal L is satisfied by I (i.e., if $L \in I$ for L atom, or $A \notin I$ if L is *not A*). W satisfies a rule r if each $I \in W$ satisfies r.

Cabalar et al. [28, 10] propose an interesting attempt to establish useful properties that ELP's semantics should obey, analogous to the ones which have been defined over time for ASP. They propose, in fact, a notion of *epistemic splitting*, where the top and bottom of a program are defined w.r.t. the occurrence of epistemic operators. They also consider *subjective constraint monotonicity*, stating that, for any ELP program Π and any subjective constraint r, W is a world view of $\Pi \cup \{r\}$ iff both W is a world view of Π and W satisfies r. *Foundedness* is achieved if what is known in a world view complains with what can be derived in a founded way from the program. Practically, this means that atoms occurring in sets within a world view cannot have been derived through cyclic positive dependencies, where, to define such dependencies, $\mathbf{K}A$ is the same as A; thus, the world view $[\{a\}]$ of the simple program $a \leftarrow \mathbf{K}a$ is *unfounded* and should be discarded.

4 Proposals for ELP Semantics

In this section we report some of the most relevant semantic definitions for ELPs, concentrating on the reduct-based ones. To compare the behaviour of the various semantics on practical tiny examples, the reader may refer to Tables 1, 2, and 3 in the Appendix. At present, there is no consensus about which is the 'right' semantic approach, and of which results a semantics should return on controversial examples (for instance, on those in Table 3). A debate is quite lively about what the 'intuitive' outcome of these examples should be. This constitutes a strong motivation for our approach, that is, providing a way of quickly and easily constructing a solver so as to be able to make experiments with new semantic approaches or with variations of existing ones.

We start with the seminal definition of the first ELP semantics, introduced in [2], that we call, for short, G94. In what follows, let r be a rule in an ELP program Π.

Definition 4.1 (G94-world views). *The G94-reduct of* Π *w.r.t. a non-empty set of interpretations* W *is obtained by:*

(i) *replacing by* $\top$ *every subjective literal* $L \in Body_{subj}(r)$ *such that* L *is of the form* $\boldsymbol{K}G$ *and* $W \models \boldsymbol{K}G$, *and*

(ii) *replacing all other occurrences of subjective literals of the form* $\boldsymbol{K}G$ *by* $\bot$.

A non-empty set of interpretations W *is a G94-world view of* Π *iff* W *coincides with the set of all stable models of the G94-reduct of* Π *w.r.t.* W.

This definition was then extended in [5] introducing the semantics G11:

Definition 4.2 (G11-world views). *The G11-reduct of* Π *w.r.t. a non-empty set of interpretations* W *is obtained by:*

(i) *replacing by* $\bot$ *every subjective literal* $L \in Body_{subj}(r)$ *such that* $W \not\models L$,

(ii) *removing all other occurrences of subjective literals of the form* $not\ \boldsymbol{K}L$,

(iii) *replacing all other occurrences of subjective literals of the form* $\boldsymbol{K}L$ *by* L.

The set W is a G11-world view of Π iff W coincides with the set of all stable models of the G11-reduct of Π w.r.t. W.

In the paper [28] it is noticed that the semantics K15 introduced in [29] and recalled by the following definition, slightly generalises the semantics G11:

Definition 4.3 (K15-world views). *The K15-reduct of Π w.r.t. a non-empty set of interpretations W is obtained by:*

(i) *replacing by $\perp$ every subjective literal $L \in Body_{subj}(r)$ such that $W \not\models L$, and*

(ii) *replacing all other occurrences of subjective literals of the form $\boldsymbol{K}L$ by L.*

The set W is a K15-world view of Π iff W coincides the set of all stable models of the K15-reduct of Π w.r.t. W.

Semantics G11 and K15, which are refinements of the original G94 semantics, have been proposed over time to cope with new examples that were discovered on which existing semantic approaches produce unwanted or not intuitive world views.

K15 can be seen as a basis for the semantics proposed in [4], called S16, for short. In particular, S16 treats K15 world views as candidate solutions, to be pruned in a second step. In this step, some world views are removed by applying the principle of keeping those that maximise what is not known (or, equivalently, minimise what is known). For brevity, the definition of S16 world views is not reported here.

All the above semantics, in order to check whether a belief view $\mathcal{A}$ is indeed a world view, adopt some kind of reduct, reminiscent of that related to the AS semantics (cf. Section 2), and $\mathcal{A}$ is a world view if it is *stable* w.r.t. this reduct.

The F15 semantics [7, 30] is based on very different principles. It is based on a combination of Equilibrium Logic [31] with the modal logic S5. Differently from F15, FAAEL [10] is based on the modal logic KD45. Cabalar et al. [10] proved that FAAEL world views coincide with *founded* G94 world views; this is an interesting connection, as a solver for G94, enhanced with a post-processing phase, also works for FAAEL.

We refer the reader to the references above for formal definitions of F15 and FAAEL, which we do not report here. Also, we apologise to readers and authors because we do not consider other recent semantics, such as the one proposed in [9, 32].

Another reduct-based semantics has been proposed in [33] and is still under refinement, and discussed in a related submitted paper; the last version is reported below. This approach considers subjective literals $\mathbf{K}G$ and $\mathbf{K}notG$ as new atoms, called *knowledge atoms*. Negation *not* in front of knowledge atoms is assumed to be the standard default negation. So, instead of ELPs proper, we here consider ASP programs possibly involving knowledge atoms. We make a difference with previous approaches where we allow an epistemic interpretation $\mathcal{W}$ to be empty.

Definition 4.4 (CF24F-adaptation). *The CF24F-adaptation $\Pi^{\mathcal{W}}$ of a program Π with respect to an epistemic interpretation $\mathcal{W}$ is a new program, obtained by modifying Π as follows:*

(i) *whenever $\mathcal{W} \models \mathbf{K}G$, in all non-unit rules with head G substitute head G with the knowledge atom $\mathbf{K}G$, and add new rule $G \leftarrow \mathbf{K}G$ and*

(ii) *whenever $\mathcal{W} \models \mathbf{K}not\, G$, add new rule $\mathbf{K}notG \leftarrow not\, G$*

Let $F_{\Pi^{\mathcal{W}}}$ be the set of the newly added rule heads of the form $\mathbf{K}G$.

The CF24F Adaptation is nothing other than our definition of a reduct, though extended and more involved than previously known ones, in order to achieve *foundedness*; such property was previously enjoyed only by FAAEL, and by none of the reduct-based approaches. We have the following notion of world view.

Definition 4.5 (CF24F world view). *An epistemic interpretation $\mathcal{W}$ is a CF24F world view of a program Π if $\mathcal{W}$ equals the set of sets obtained from the stable models of $\Pi^{\mathcal{W}}$ by cancelling knowledge atoms.*

231

Rephrasing the minimality criterion proposed for S16 [4] in terms of our approach, we have:

Definition 4.6 (S16 Criterion - CF24F+S16C). *Each world view $\mathcal{W}$ as defined in Def. 4.5 is considered to be a* candidate world view. *A candidate world view $\mathcal{W}$ is indeed a world view under CF24F+S16C if no other candidate world view $\mathcal{W}'$ exists, where $F_{\Pi \mathcal{W}'} \subset F_{\Pi \mathcal{W}}$.*

5 Fast Prototyping of a Solver for Reduct-based Semantics

An obstacle to developing efficient solvers for ELPs resides in the computational complexity [34]. The central decision problem, that is, checking whether an ELP has a world view, is Σ_P^3-complete [4, 35], unless some particular assumptions on program structure are done.

Table 4 in the Appendix shows some solvers developed for various semantics.

The method introduced in [11] and refined and implemented here allows a solver for any reduct-based semantics to be quickly obtained with little implementation effort. Thus, researchers proposing new semantic approaches (as often happens) or extending/modifying an existing one can exploit the method to conduct experiments. It must be noticed, however, that the method provides the correctness of the solver with respect to a given semantics but does not cope with efficiency issues. Thus, the prototypical solver obtained via our method can only be used to experiment with programs of small-medium size.

5.1 The Method

Notice that, for checking whether an epistemic interpretation $\mathcal{W}$ is a world view for program Π according to a semantics $\mathcal{S}$ based upon a definition $\mathcal{R}$ of a reduct, one has to apply $\mathcal{R}$ to Π, then find the set of answer sets of the reduced program, then possibly perform some post-processing thus obtaining a final set of answer sets, then perform a final comparison to see whether the result coincides with $\mathcal{W}$.

In the case of CF24F, for instance, the post-processing cancels knowledge atoms. If applying a minimality criterion as defined in S16 and adopted in CF24F+S16C, the resulting 'candidate' world views must be filtered, and this, as discussed in [4], adds further complexity.

To find all the world views of Π, one has to devise and check all the epistemic interpretations. This, given the set At_Π of the atoms occurring in Π, one has to find all the subsets of At_Π, and all the sets of such subsets. Clearly, this process is highly complex. Basically, it 'absorbs' most of the complexity of the entire method. Then, all the epistemic interpretations must be checked (possibly with some simplifications, for instance, to exclude those epistemic interpretations containing two sets, one included in the other, etc.). Below we formalise the various steps of the proposed method.

Fast ELP-solver pipeline Given $\mathcal{S}$, $\mathcal{R}$, and $\mathcal{P}$, a 'quick solver' which returns all world views of a given program Π, is obtained by running on Π the following pipeline of modules:

1. A module $M_\mathcal{W}$ that computes all epistemic interpretations $\mathcal{W}_1$, ..., $\mathcal{W}_k$ for Π (i.e., all sets of subsets of At_Π);

2. A module M_{red} that applies, for each $\mathcal{W}_i$, the reduct $\mathcal{R}$ to Π, and generates the reduct program Π_i (which is an ASP program);

3. A module M_{ASP} that computes the set SMs_i of answer sets of Π_i, for $i = 1, \ldots, k$;

4. In case a post-processing $\mathcal{P}$ is required, a module $M_\mathcal{P}$ applying $\mathcal{P}$ to select the desired candidate SMs_i's;

5. A module M_{chk} that checks each SMs_i produced by the previous step and selects those which are world views w.r.t. $\mathcal{S}$, as they coincide with the corresponding $\mathcal{W}_i$.

Correctness of the solver depends upon the correct implementation of the various modules, that, however, should not be difficult to ensure. In fact, each module copes with a single aspect and will thus be sufficiently transparent and reasonable in size.

233

5.2 The ASP Chef System

The ASP Chef system [12] aims to provide a framework, in which users can employ ASP directly by specifying a set of logic rules but also indirectly via mapping data from one format to the format accepted by ASP engines and mapping the output produced by ASP engines to some other format suitable to be presented to the end-user or further processed in a pipeline. In ASP Chef there is a notion of ASP *recipe* as a chain of *ingredients* that are the instantiation of different operations, where an operation can be one of the "traditional" ASP computational tasks, or some data manipulation procedure or data visualisation procedure. In order to enable the composition of linear chains of ingredients, sequences of interpretations (i.e., sequences of sets of atoms) have been adopted as a uniform format for the input and the output of all operations. Each operation can have parameters to customise its behaviour and side output, to enable inspection and visualisation of intermediate states of the evaluation of recipes. The adopted uniform format allows several operations implemented in ASP Chef to be combined in any order and new operations to be easily accommodated.

A web app (`https://asp-chef.alviano.net/`) is available to implement even long pipelines involving ASP as a core engine to perform several computational tasks, putting into practice the notion of ASP recipe. The default ASP engine exploited in ASP Chef is `clingo-wasm`, a web-accessible version of `clingo` [24].

Several operations are already available in ASP Chef. They include searching for the models of programs, filtering and sorting atoms within each model according to their lexicographical ordering, merging and splitting interpretations.

5.3 Implementation in ASP Chef

We implemented the pipeline of modules as an ASP Chef recipe, which we illustrate below. The complete recipe is accessible, usable, and modifiable through the ASP Chef web app, at this clickable link: ELP in ASP Chef .

Let us remark that this recipe is not optimal because we chose to implement the pipeline for expository purposes rather than aiming for an optimised implementation. Also, we chose to implement each step of the pipeline separately, not necessarily using a minimal number of ingredients. Hence, a more compact recipe could be obtained, for instance, by merging different ingredients or by skipping some steps that perform some inessential processing (such as filtering of atoms that are useless for the subsequent ingredients to reduce the size of clingo-wasm input).

Let us now describe the main parts of our recipe. Regarding the input format, since ASP Chef ingredients expect to be fed with a sequence of sets of atoms, we encoded the ELP program Π as a set of atoms. A rule of the form (1) is encoded as the fact

```
rule(head(A_1,...,A_g), body(L_1,...,L_n)).
```

A subjective literal $\mathbf{K}G$ is encoded by the term $\mathtt{k}(G)$, while negation is represented by the functor neg. So, we reserved the symbols neg and k and they cannot occur in the input program. For example, the program Π_1 of Section 3 is encoded as the set of four facts:

```
rule(head(a), body(neg(b))).
rule(head(b), body(neg(a))).
rule(head(e), body(neg(k(f)))).
rule(head(f), body(neg(k(e)))).
```

This single set of facts is the input sequence for the recipe. The following ASP program processes the input set to detect atoms and literals, both objective and subjective, occurring in the ELP rules. This ASP program is evaluated by a SEARCH MODELS ingredient and uses some functions (i.e., @functor, @arity, and @argument) introduced by a specific INTROSPECTION TERMS ingredient (not detailed here) exploited to introduce some Lua functionalities useful to decompose input facts and to access their sub-terms.

```
rule_head(rule(H,B), @argument(H,I)) :- rule(H,B),
    I=1..@arity(H).
rule_body(rule(H,B), @argument(B,I)) :- rule(H,B),
    I=1..@arity(B).
% literals occurring in heads or in bodies
```

235

```
hlit(L)  :- rule_head(_,L).
% literals occurring in bodies
blit(L)  :- rule_body(_,L).
atom(L)  :- hlit(L).
atom(A)  :- blit(neg(A)), @functor(A) != "neg",
    @functor(A) != "k".
atom(A)  :- blit(k(A)), @functor(A) != "neg",
    @functor(A) != "k".
blit(L)  :- blit(neg(L)).
blit(L)  :- blit(k(L)).
% literals/atoms in subj. literals
klit(A)  :- blit(k(A)).
klit(L)  :- klit(neg(L)).
```

In the output of this ingredient, one obtains the set At_Π encoded as a collection of facts of the form atom(A). At this point, another SEARCH MODELS ingredient evaluates such collection together with the simple ASP program:

```
{guess_true(A)}  :- atom(A), hlit(A).
```

to generate a sequence of answer sets, each of them containing a possible subset of At_Π. Then, a MERGE ingredient combines all these sets in a single set by distinguishing/indexing their elements by the predicate __atomset__ (i.e., each atom $\langle atom \rangle$ occurring in the i-th set is encoded in output by an atom of the form __atomset__$(i, \langle atom \rangle)$). Now a SEARCH MODELS ingredient processes such "indexed atoms" together with the ASP rules

```
numset(SetID)  :- __atomset__(SetID,_).
{selectset(SetID)}  :- numset(SetID).
numOfSetsInW(Count)  :- Count==#count{I:selectset(I)}.
setInW(SetID,A)  :- __atomset__(SetID,A),
    selectset(SetID).
```

The first rule simply collects all sets indices SetID introduced by the previous ingredient. The second rule generates all possible selections of such indices/sets (i.e., the possible epistemic interpretations). Each of these epistemic interpretations $\mathcal{W}_i$ appears in one answer set of the ingredient output, and its members (the candidate stable models of Π_i) are encoded by facts of the form setInW(SetID,A). This completes step 1 in the pipeline.

A further SEARCH MODELS ingredient evaluates the following program for each $\mathcal{W}_i$ generated by the previous ingredient and infers which literals $\mathcal{W}_i$ models (by simply counting the modelling sets/IDs and comparing their number with the cardinality of $\mathcal{W}_i$):

```
modeledByW(A)  :- atom(A), numOfSetsInW(N), N>0,
    N == #count{I,A : setInW(I,guess_true(A)),
    selectset(I)}.
modeledByW(neg(A))  :- atom(A),
    0 == #count{I,A : setInW(I,guess_true(A)),
    selectset(I)}.
```

At this point the reduct of Π can be computed, w.r.t. each of the epistemic interpretation $\mathcal{W}_i$. In the case of semantics G94, this can be achieved by a SEARCH MODELS ingredient processing the following program (for each of the $\mathcal{W}_i$s in its input sequence):

```
red_blit(k(L),true)  :- blit(k(L)), modeledByW(L).
red_blit(k(L),false)  :- blit(k(L)), not modeledByW(L).
red_blit(neg(L),neg(L))  :- blit(neg(L)),
    @functor(L) != "k".
red_blit(L,L)  :- blit(L), @functor(L) != "neg",
    @functor(L) != "k".
red_blit(neg(k(L)),false)  :- blit(neg(k(L))),
    modeledByW(L).
red_blit(neg(k(L)),true)  :- blit(neg(k(L))),
    not modeledByW(L).
% reduced rules head and body literals:
red_rule_head(rule(H,B), @argument(H,I))  :- rule(H,B),
    I = 1..@arity(H).
red_rule_body(rule(H,B), R)  :- rule_body(rule(H,B), L),
    red_blit(L,R).
```

The first six rules assign to each literal in the input program Π a "substitute" (either the literal itself or one of the truth values false and true), according to Def. 4.1. Then, depending on which subjective literals are modeled by $\mathcal{W}_i$, facts of the form red_rule_head($\langle rule \rangle, \langle lit \rangle$) and red_rule_body($\langle rule \rangle, \langle lit \rangle$) are derived, representing the rules of the reduced program. Each reduced program is represented in a different set of atoms in the output sequence of the ingredient. This completes step 2 of the pipeline. Each set so obtained is joined to the program below and the answer sets are computed. Each element of each SMs_i

(for all i) is encoded in a distinct set of atoms in the output sequence of the ingredient by a collection of facts true($\langle atom \rangle$).

```
% detect falsified reduced rules bodies:
red_body_false(R)  :- red_rule_body(R,false).
% infer true literals w.r.t.\ reduced rules
true(L) : red_rule_head(rule(H,B),L) :- rule(H,B),
    not red_body_false(rule(H,B));
  true(N):red_rule_body(rule(H,B),N),
      @functor(N) != "neg", @functor(N) != "true";
  not true(M):red_rule_body(rule(H,B),neg(M)),
      @functor(M) != "neg", @functor(M) != "false";
  not not true(M):red_rule_body(rule(H,B),neg(neg(M))),
      @functor(M) != "false".
```

Few ingredients are used to gather in a single set the collection SMs_i, for each i. This operation outputs the sequence of SMs_is and completes step 3 in the pipeline.

Finally, step 5 of the pipeline is performed (note that step 4 is not needed for G94). For brevity, we omit the details: the remaining part of the recipe compares each SMs_i with $\mathcal{W}_i$ filtering out those that do not match. The output is then processed for better readability, and the world views of Π are listed by facts of the form worldView_SMid_Atom($\langle SMid \rangle, \langle atom \rangle$). For instance, this is the output for the ELP program Π_1 of Section 3:

```
worldView_SMid_Atom(101,a).
worldView_SMid_Atom(101,e).
worldView_SMid_Atom(102,b).
worldView_SMid_Atom(102,e).
§
worldView_SMid_Atom(157,a).
worldView_SMid_Atom(157,f).
worldView_SMid_Atom(158,b).
worldView_SMid_Atom(158,f).
```

The two sets in the output sequence (separated by §) represent two world views made of two answer sets (identified by different numeric IDs), each composed of two atoms.

We may observe that the pipeline we described and the implemented in ASP Chef can be easily exploited to mechanise any reduct-based semantics for ELP (hence the "fast prototyping"). It simply suffices to modify a single ingredient: namely, the one that implements the module M_{red} of the pipeline and, for each $\mathcal{W}_i$, computes the reduct program Π_i. This is equivalent to providing adequate alternative definitions of the predicates `red_blit/2` and (possibly) `red_rule_body/2` seen before, according to the specific notion of reduct at hand. The rest of the recipe remains unchanged.

The following is a possible encoding of the G11-reduct, which involves a change in the definition of `red_blit/2` only:

```
red_blit(k(L),false) :- blit(k(L)), not modeledByW(L).
red_blit(neg(k(L)),false) :- blit(neg(k(L))),
    modeledByW(L).
red_blit(neg(k(L)),true) :- blit(neg(k(L))),
    not modeledByW(@argument(@argument(L,1),1)).
red_blit(k(L),L) :- blit(k(L)), modeledByW(L).
red_blit(neg(L),neg(L)) :- blit(neg(L)),
    @functor(L) != "k".
red_blit(L,L) :- blit(L), @functor(L) != "neg",
    @functor(L) != "k".
```

Similarly, one can customise the solver for the CF24F semantics by using the following encoding (where `knowP(L)` and `knowN(L)/1` denote knowledge atoms $\mathbf{K}L$ and $\mathbf{K}notL$, resp.):

```
red_blit(k(neg(L)),knowN(L)) :- blit(k(neg(L))).
red_blit(k(L),knowP(L)) :- blit(k(L)),
    @functor(L) != "neg".
red_blit(neg(L),neg(L)) :- blit(neg(L)),
    @functor(L) != "k".
red_blit(L,L) :- blit(L), @functor(L) != "neg",
    @functor(L) != "k".
red_blit(neg(k(L)),neg(knowP(L))) :- blit(neg(k(L))),
    red_blit(k(L),knowP(L)).
red_blit(neg(k(L)),neg(knowN(L))) :- blit(neg(k(L))),
    red_blit(k(L),knowN(L)).
red_hlit(A,knowP(A)) :- hlit(A), modeledByW(A).
red_hlit(A,A) :- hlit(A), not modeledByW(A).
nonunit(R) :- rule_body(R, L), L!=true.
changehead(rule(H,B)) :- nonunit(rule(H,B)),
    I=1..@arity(H), L=@argument(H,I),
    @functor(L) != "neg", modeledByW(L).
red_rule_head(rule(H,B), knowP(L)) :-
    changehead(rule(H,B)), I=1..@arity(H),
```

239

```
      L=@argument(H,I), @functor(L) !="neg", modeledByW(L).
red_rule_head(rule(H,B), @argument(H,I)) :- rule(H,B),
      I=1..@arity(H), not changehead(rule(H,B)).
red_rule_body(rule(H,B), R) :- rule_body(rule(H,B), L),
      red_blit(L,R).
rule(head(L),body(knowP(L))) :- modeledByW(L),
      @functor(L) !="neg".
red_rule_head(rule(head(L),body(knowP(L))),L) :-
      modeledByW(L), @functor(L) !="neg".
red_rule_body(rule(head(L),body(knowP(L))),knowP(L)) :-
      modeledByW(L), @functor(L) !="neg".
rule(head(knowN(L)),body(neg(L))) :- modeledByW(neg(L)).
red_rule_head(rule(head(knowN(L)),body(neg(L))),
      knowN(L)) :- modeledByW(neg(L)).
red_rule_body(rule(head(knowN(L)),body(neg(L))),
      neg(L)) :- modeledByW(neg(L)).
```

As before, the predicate `red_blit/2` describes how body literals have to be rewritten to implement Def. 4.4. Similarly, `red_hlit/2` concerns atoms in rule heads. Note, in particular, that the substitutes for the subjective body literals `k(L)` and `k(neg((L))` are the knowledge atoms `knowP(L)` and `knowN(L)`, resp. The auxiliary predicates `nonunit/1` and `changehead/1` identify the unit rules and each rule whose head needs to be modified to obtain the CF24F-adaptation of the program (cf., (i) in Def. 4.4). For such rules, the first two clauses defining `red_rule_head/2` determine the head literals of the rewritten rules. The first clause of `red_rule_body/2` defines their bodies. The last six clauses extend the program by adding the new ASP rules $G \leftarrow \mathbf{K}G$ and $\mathbf{K}notG \leftarrow not\, G$ whenever one of the conditions $\mathcal{W} \models \mathbf{K}G$ and $\mathcal{W} \models \mathbf{K}not\, G$ holds, resp.

6   Querying World Views

In this section, we resort to the RAS semantics (Resource-based Answer Set Semantics, cf. [14]) for formalizing a new characterisation of world views, that allows one to query them, without actually computing them in advance.

240

6.1 Resource-based Answer Set Semantics

The Resource-based Answer Set semantics (RAS), as presented in [14], is characterised by its ability to provide answer sets for every program, thereby ensuring that no programs are inconsistent under this semantics. The foundational principles of the RAS semantics are as follows: (i) atoms belonging to a RAS answer set are either definitely true (i.e., a RAS answer set includes all atoms that are true w.r.t. the well-founded semantics[1]), or have been rationally (though defeasibly) assumed to hold; (ii) atoms not included in a RAS answer set are either definitely false (i.e., false w.r.t. the well-founded semantics), or have been rationally assumed not to hold to draw some conclusion or no judgment about them has been devised because any such assessment would lead to a contradiction. For instance, the program $\{p \leftarrow not\, p.\}$ has an empty RAS answer set because, in our view, a rational agent cannot believe both $not\, p$ and p, so no rational judgment on p being true can be given; differently from the case of AS semantics, this is not seen as a reason not to provide semantics. Therefore, an RAS answer set includes atoms either proved or assumed to be true, does not contain atoms either proved or assumed to be false, or atoms whose truth value cannot be assessed.

To formally explain the difference between AS and RAS we may resort to a modal logic formulation. For AS, as discussed in [37] an answer set programming rule can be transposed, to express its logical meaning, into its "modal image", where $L\, A$ is intended as "A is believed" under any modal logic contained in **S5**:

$$L\, A_1 \wedge \cdots \wedge L\, A_n \wedge L\, \neg\, L\, B_1 \wedge \cdots \wedge L\, \neg L\, B_m \supset L\, A \quad (Ae0)$$

In RAS, as discussed in [14], the logical meaning of each rule ρ is expressed by the following couple of modal rules, that form its modal

---

[1]The well-founded semantics [36] provides to every program a unique three-valued model $\langle W^+, W^- \rangle$, where atoms in W^+ are *true*, those in W^- are *false*, and all the others are *undefined*. All atoms in W^+ are true in every answer set under the AS semantics, and all atoms in W^- are false; the AS semantics in fact assigns, for consistent programs, truth values to the undefined atoms, and so does RAS, though in a slightly different way.

image:

$$L\,A_1 \wedge \cdots \wedge L\,A_n \wedge L\,\neg\,L\,B_1 \wedge \cdots \wedge L\,\neg L\,B_m \supset L\,\dot{A} \qquad (Ae1)$$
$$L\,\dot{A} \wedge \neg\,L\neg L\,A \supset L\,A \qquad\qquad\qquad\qquad\qquad (Ae2)$$

Rule $(Ae1)$ modifies $(Ae0)$ in the sense that, based on the same premises, one concludes $L\dot{A}$, which means that one believes *to be enabled* to prove A. Rule $(Ae2)$ states that $L\,A$ is derived only if $L\,\dot{A}$ holds, and one does not believe not to believe A. Thus, in the case for instance of the unary odd cycle $p \leftarrow not\,p$ which makes, under AS, an ASP program inconsistent, by $(Ae1)$ one from $not\,p$, that in terms of the modal image is expressed as $L\,\neg\,L\,p$, can derive $L\dot{p}$, i.e., to be enabled to prove p. However, one cannot do so, as $L\,\neg\,L\,p$ precisely accounts to believing not to believe p. Consequently, rule $(Ae2)$ cannot be applied, and p therefore is false without raising inconsistencies.

Each AS answer set is also a RAS answer set, but a RAS answer set is not necessarily a classical model of the program. RAS answer sets are, in fact, all the "Maximal Consistently Supported" (MCS) sets of atoms that a given program Π admits. That is, for each RAS answer set M of Π: every atom A in M is *consistently supported* by a set S of rules of Π, where each rule in S is supported in M, A is the head of exactly one rule in S, and each atom (positive literal) in the body of rules in S is different from A, and is in turn consistently supported; moreover, M is maximal, meaning that there exists no M' with the same properties such that $M \subset M'$. E.g., for the program $\{a \leftarrow not\,b.\ \ b \leftarrow not\,c.\ \ c \leftarrow not\,a.\}$, the sets of atoms $\{a\}$, $\{b\}$ and $\{c\}$ are the only MCSs and thus they are the RAS answer sets of this program. In fact, each of the composing atoms a, b, c is consistently supported by the set S consisting of the single rule of which that atom is the head and the three sets are maximal, as all their supersets (among which are the classical models) are not consistently supported. We made the examples of unary and ternary odd cycles because what makes programs inconsistent under AS are indeed (direct or indirect) odd loops, i.e., when an atom depends upon its own negation through an odd number of negative dependencies.

RAS answer sets of program Π can be computed via an operator GL_Π which is analogous to the traditional GL-Operator GL_Π (recalled in Section 2) though it takes as input those interpretations I such that for every $A \in I$ there exists rule ρ in Π with head A, and is based upon a modified definition of reduct (where, with respect to the traditional GL_Π, step 2 is not performed), and a modified immediate consequence operator T_Π, that computes consistently supported sets of atoms by discarding those atoms for which all possible derivations depend on their own negation. As discussed before, such atoms are excluded from any RAS answer set because their truth value cannot be assessed. The set $M = \hat{\Gamma}_\Pi(I)$ is a RAS answer set iff $M \subseteq I$ and M is maximal, i.e., there is no proper subset I_1 of I that determines $M' \supset M$.

For programs that are consistent under the AS semantics, the traditional answer sets are included among the RAS answer sets, with the latter potentially being more numerous. For examle, the program $\{a \leftarrow \ not\, b.\ b \leftarrow not\, a.\ p \leftarrow \ not\, p,\, a.\}$ has a unique answer set $\{b\}$ under AS and the answer sets $\{a\}$ and $\{b\}$ under RAS. For AS, the falsity of a is required in order to bypass the odd cycle by falsifying p. For RAS this is no longer required, as p is deemed to be false in any answer set because it depends on its own negation. Odd cycles are, in fact, the only possible source of difference between RAS and AS on programs which are consistent under AS. For programs that are either 'call-consistent' (i.e., they do not involve odd cycles) or that fulfil straightforward syntactic sufficient conditions (discussed in [14], and based on the considerations proposed in [38]) concerning odd cycles, the answer sets returned by the two semantics are the same. Hence, RAS can be seen as a variant of AS.

Differently from AS, RAS enjoys the properties that, ideally, every non-monotonic formalism should enjoy [39]: (i) cumulativity, i.e., the possibility of asserting lemmas while keeping the same answer sets; (ii) relevance, i.e., the fact that the truth value of each atom is determined by the subprogram consisting of the *relevant rules*, which are those upon which the atom depends (directly or indirectly, positively or negatively). AS does not enjoy such properties because some atoms may be forced to assume certain truth values in order to prevent inconsistencies; this is no longer the case for RAS. An advantage of the property of relevance is to make top-down query-answering possible [15] in Prolog-style, i.e., without computing the answer sets in advance but rather via an enhanced resolution procedure. This is because in RAS, only the relevant rules have a role in proving/disproving any atom A. Instead, under AS a query $?A$ might *locally* succeed, but still, for the lack of relevance, the overall program may not have answer sets including A. Under RAS, a query $?A$ w.r.t. ground program Π asks whether A is true in (belongs to) some answer set of Π. Query $not\ A$ asks whether $not\ A$ is true in some answer set of Π, which implies that there exists some answer set to which A *does not* belong. Series of queries can be, upon user's choice: (i) *contextual*, i.e., query $?A, B$ asks whether A is true (belongs to) some answer set, and B is true in (belongs to) some of those; (ii) independent.

RAS query-answering can be performed via RAS-XSB-resolution, described in [15]. It is meant to be implemented on top of XSB-resolution [40, 41], which is an efficient, fully described and implemented procedure, correct and complete w.r.t. the well-founded semantics. Features of XSB-resolution that are crucial for the implementation of RAS-XSB-resolution are negative cycles detection and the tabling mechanism, that associates to program Π a table $\mathcal{T}(\Pi)$, which is initialised before posing queries. Such table contains information about true and false atoms useful for both the present and the subsequent queries (if they are in conjunction).

The principle of functioning for table $\mathcal{T}(\Pi)$ under RAS-XSB-resolution are the following. During a proof:

(a) the negation of any atom which is not a program fact is available unless this atom has been proved;

(b) the negation of an atom which has been proved becomes unavailable, and the atom is asserted as true; so,

(c) the negation of an atom which cannot be proved remains always available.

Since RAS-XSB-resolution is a top-down proof procedure, modifications to the table might be undone and redone differently upon backtracking. The present (very initial prototype) implementation exploits XSB (or, more precisely, its basic version XOLDTNF), as a "plugin" for definite success and failure, where new cases are added to manage atoms with truth value *undefined* under XSB. RAS-XSB-resolution is correct and complete w.r.t. resource-based answer set semantics, in the sense that, given a program Π, a query $?A$ succeeds under RAS-XSB-resolution with an initialised $\mathcal{T}(\Pi)$ iff there exists some resource-based answer set M for Π where $A \in M$. The result extends to sequences of queries.

An implication of making odd cycles consistent under RAS is that constraints must be explicitly defined as such, since they can no longer be reinterpreted as unary odd cycles. These constraints need to be incorporated into the program as an additional layer. However, since RAS answer sets computation is a variation of the traditional answer set computation, constraints can still be verified during the computation process, as solvers typically do. Henceforth, when the context is clear, we will refer to RAS answer sets simply as 'answer sets'.

6.2 Characterizing and Querying World Views

This subsection discusses how to exploit RAS to characterise and query world views without fully computing them. First, let us consider a program Π that does not include subjective literals. Consequently, its single world view will correspond directly with the collection of its answer sets. As observed in [27], via RAS-XSB-resolution, it is possible to implement epistemic queries, among which the following (where A is an atom):

- Query $?\,\mathbf{not}\,A$ (where $\mathbf{not}$ is the epistemic operator introduced in [4]) asks whether A is false w.r.t. some answer set of Π, and therefore succeeds if $not\,A$ is true in some of them. This can be implemented via RAS query $?not\,A$.

- Query $?\,\mathbf{M}A$, or equivalently $?\,\mathbf{not}\,not\,A$ asks whether A is possible, i.e., $not\,A$ is false in some answer set, and therefore succeeds if A is true in some of them. This can be implemented via RAS query $?A$.

- Query $?\mathbf{K}A$, or, equivalently, $?\textit{not}\,\mathbf{not}\,A$, asks whether it is not true that A is false w.r.t. some answer set of Π, i.e., that A is true in all of them. This can be implemented via RAS query $?A, \textit{not}\,A(fail)$, where this overall query succeeds if $?A$ succeeds, i.e., A belongs to some answer set of Π, whereas $?\,\textit{not}\,A$ fails, so A is not false in any of them.

- Query $?\textit{not}\,\mathbf{not}\,\textit{not}\,A$ asks whether A is false in every answer set, meaning $\mathbf{K}\,\textit{not}\,A$, i.e., $\textit{not}\,\mathbf{M}A$. This can be implemented via RAS query $?A(fail)$ that succeeds if A fails, i.e., exactly whenever there is no answer set where A is true.

We may introduce a new operator **NOT** as a shorthand for $\mathbf{K}\,\textit{not}\,A$.

We will now proceed to consider ELPs proper, i.e., programs actually involving subjective literals, by modifying and extending what was presented in [27]. We start by defining a *guess*, which is a set of subjective literals occurring in a given program that are assumed to be true (where all the others are supposed to be false).

Definition 6.1. *Let a guess* Φ *be any subset of the subjective literals* $\mathbf{K}A$ *and* $\mathbf{K}\textit{not}\,A$ *occurring in a given program* Π.

An epistemic interpretation is said to fulfil a guess if the truth values for the subjective literals w.r.t. it are in accordance to the guess.

Definition 6.2. *An epistemic interpretation* wv *for program* Π *fulfils a guess* Φ *if it entails all subjective literals in* Φ *and does not entail all the other subjective literals occurring in* Π.

An epistemic interpretation that fulfils a guess Φ may or may not be a (candidate) world view, according to a given semantics $\mathcal{S}$. In the former case, Φ is said to be a *valid guess*. In the case of reduct-based semantics, the validity of a guess can be checked as follows.

Theorem 6.1 (RASCGK test: RAS Candidate Guess check. Reformulation and extension of Theorem 6.1 of [27]). *Given guess* Φ

w.r.t. program Π, *whether* Φ *is a valid guess w.r.t. semantics* $\mathcal{S}$ *can be checked as follows.*

1. *Derive reduct* Π^Φ *as said in the definition of* $\mathcal{S}$, *but with a fundamental variation. Whenever the definition of the reduct dictates* $\mathcal{W} \models L$ *or* $\mathcal{W} \models \mathbf{K}L$ *interpret this as* $\mathbf{K}L \in \Phi$; *whenever it dictates* $\mathcal{W} \models not\, L$ *or* $\mathcal{W} \models \mathbf{K}not\, L$ *interpret this as* $\mathbf{K}not\, L \in \Phi$. *Note that the reduct is a plain ASP program.*

2. *Given* Π^Φ: *for every subjective literal* $\mathbf{K}A$ *occurring in* Π *pose the query* $?A, not\, A(fail)$, *which as said succeeds if* $?A$ *succeeds, i.e.,* A *belongs to some answer set of* Π, *whereas* $?not\, A$ *fails, so* A *is not false in any of them; for every subjective literal* $\mathbf{K}not\, A$ *occurring in* Π, *pose the query* $?A(fail)$ *that succeeds if* A *fails, i.e., exactly whenever there is no answer set where* A *is true.*

A guess Φ *is valid if all queries concerning subjective literals in* Φ *succeed, while instead, all queries concerning the remaining subjective literals fail.*

Proof. Straightforward, given the above observations about the meaning of the reduct and of the queries. $\qquad\square$

To check the CF24F+S16C criterion, one has to verify that Φ passes the RASCGK test whereas no subset $\Phi' \subset \Phi$ where Φ' contains less literals of the form $\mathbf{K}A$ does.

We now proceed to illustrate how to query world views without fully computing them, via the method devised in [27]. The principle of functioning for table $\mathcal{T}(\Pi)$ under RAS-XSB-resolution extends those adopted in XSB and recalled earlier in Section 6.1.

In order to query the program under a world view determined by a certain valid guess Φ, the table initialisation is customised accordingly, by setting to true all and only the subjective literals occurring in the valid guess under consideration.

Definition 6.3 (Guess-tailored table initialisation). *Given program* Π *and a valid guess* Φ, *guess-tailored table initialisation will be performed in addition to normal RAS-XSB initialisation in the following* *way. Each subjective literal occurring in* Φ *will be set to true, and all the other subjective literals occurring in* Π *will be set to false.*

Definition 6.4 (G-RAS-XSB-Resolution). *The variant of RAS-XSB-resolution where, given program* Π *and valid guess* Φ*, guess-tailored table initialisation is applied is called G-RAS-XSB-resolution (tailored to* Φ*).*

Given program Π and a valid guess Φ, G-RAS-XSB-resolution tailored to Φ has been proved correct and complete w.r.t. the world view $\mathcal{W}$ that Φ fulfils.

Moreover, via simple extensions, one can query the whole set of world views, for instance, as follows. Query ? $\mathbf{K}^W A$ asks whether A is true in all world views of Π. Query ? $\mathbf{not}^W A$ asks whether A is false w.r.t. some world view of Π, and therefore if *not* A is true in some of them. Query ? $\mathbf{NOT}^W A$ asks whether A is false in every answer set of every world view.

Below is an example of application of the new operators, concerning an elaboration of the seminal example by Bowen and Kowalski [42] about establishing one guilty only if provably so:

$$
\begin{aligned}
innocent(X) &\leftarrow suspect(X), provably\_innocent(X). \\
innocent(X) &\leftarrow suspect(X), presumed\_innocent(X). \\
provably\_innocent(X) &\leftarrow \mathbf{NOT}^W guilty(X). \\
presumed\_innocent(X) &\leftarrow \mathbf{not}^W guilty(X).
\end{aligned}
$$

In this formulation, one is *presumed_innocent* if a world view exists where (s)he is not guilty. However, one is provably innocent if world views do not exist where (s)he is guilty.

7 Conclusions

The pursuit of identifying the most suitable semantics for Epistemic Logic Programs persists, and the exploration of new semantic avenues is crucial, extending beyond theoretical considerations to practical experimentation.

The proposed rapid prototyping method for creating solvers for reduct-based semantic frameworks can be a valid aid. We demonstrated the efficacy of this method with the G94 semantics (implicitly extending to FAAEL) and the newly proposed CF24F semantics. We extensively validated our tool using existing literature examples and additional ones. Notably, our approach is highly adaptable to other reduct-based methodologies owing to its modular definition and implementation within ASP Chef, as well as the flexibility of this system. A current limitation lies in the limited scalability of the implementation, preventing its application to larger programs. Consequently, future endeavours will focus on implementing and testing alternative semantics, even non-reduct-based, while enhancing the efficiency of our implementation.

A possible way of proceeding could rely on the fact that each world view is uniquely characterised by the set of subjective literals it entails [35]. Hence, instead of generating all possible epistemic interpretations and then filtering out those that are actually world views, one can guess the sets of (entailed) subjective literals. Intuitively, the process should generate fewer guesses of smaller size, simply because the objective parts of the world views are left implicit. The drawback of this approach is that the pipeline described previously must be modified in order to process epistemic interpretations implicitly represented by subjective guesses. Furthermore, to explicitly obtain the world views, it will be necessary to add a module that completes with the objective parts each of the guesses that turns out to encode a real world view. We leave the design of this variant of the pipeline as future work.

An objection that can be made is that, instead of adopting the proposed method, one might modify and extend the implementation of some of the existing ELP solvers (see Appendix B for a list). However, understanding and then modifying the code would, in our opinion, be more difficult and time-consuming, and less reliable as to the results.

The possibility of specifying world views of a given program in terms of guessing the subjective literals that they entail and of being able to query (top-down, in Prolog style) these world views, also by means of advanced operators, increases the possibility of fast prototyping and experimenting with new reduct-based semantics. In fact, guess-checking and querying are realised on top of an existing query-answering procedure and only require computing the reduct in a slightly different way and enacting suitable table initialisation. The limitation is that the present implementation of RAS-XSB-resolution is prototypical but could be easily improved.

In conclusion, we believe we have offered both conceptual and practical tools that are well-suited for investigating existing and novel semantics of Epistemic Logic Programs. We hope these tools will be beneficial to researchers and users alike, facilitating deeper exploration and understanding in this area.

References

[1] M. Gelfond and H. Przymusinska, "Definitions in epistemic specifications," in *Proc. of the 1st Intl. Workshop on Logic Programming and Non-monotonic Reasoning* (A. Nerode, V. W. Marek, and V. S. Subrahmanian, eds.), pp. 245–259, The MIT Press, 1991.

[2] M. Gelfond, "Logic programming and reasoning with incomplete information," *Ann. Math. Artif. Intell.*, vol. 12, no. 1-2, pp. 89–116, 1994.

[3] M. Gelfond and V. Lifschitz, "The stable model semantics for logic programming," in *Proc. of the 5th Intl. Conf. and Symp. on Logic Programming* (R. Kowalski and K. Bowen, eds.), pp. 1070–1080, MIT Press, 1988.

[4] Y. Shen and T. Eiter, "Evaluating epistemic negation in answer set programming," *Artificial Intelligence*, vol. 237, pp. 115–135, 2016.

[5] M. Gelfond, "New semantics for epistemic specifications," in *Proc. of LPNMR'11* (J. P. Delgrande and W. Faber, eds.), vol. 6645 of *LNCS*, pp. 260–265, Springer, 2011.

[6] M. Truszczyński, "Revisiting epistemic specifications," in *Logic Programming, Knowledge Representation, and Nonmonotonic Reasoning* (M. Balduccini and T. C. Son, eds.), vol. 6565 of *LNCS*, pp. 315–333, Springer, 2011.

[7] L. Fariñas del Cerro, A. Herzig, and E. I. Su, "Epistemic equilibrium logic," in *Proceedings of the Twenty-Fourth International Joint*

Conference on Artificial Intelligence, IJCAI 2015 (Q. Yang and M. J. Wooldridge, eds.), pp. 2964–2970, AAAI Press, 2015.

[8] P. T. Kahl and A. P. Leclerc, "Epistemic logic programs with world view constraints," in *Tech. Comm. of ICLP 2018* (A. Dal Palù, P. Tarau, N. Saeedloei, and P. Fodor, eds.), vol. 64 of *OASIcs*, pp. 1:1–1:17, Schloss Dagstuhl, 2018.

[9] E. I. Su, "Epistemic answer set programming," in *Proc. of JELIA'19* (F. Calimeri, N. Leone, and M. Manna, eds.), vol. 11468 of *LNCS*, pp. 608–626, Springer, 2019.

[10] P. Cabalar, J. Fandinno, and L. Fariñas del Cerro, "Autoepistemic answer set programming," *Artif. Intell.*, vol. 289, p. 103382, 2020.

[11] S. Costantini and A. Formisano, "Fast prototyping of a solver for reduct-based ELP semantics," in *Proc. of CILC'23* (A. Dovier and A. Formisano, eds.), vol. 3428 of *CEUR Workshop Proceedings*, CEUR-WS.org, 2023.

[12] M. Alviano, D. Cirimele, and L. A. Rodriguez Reiners, "Introducing ASP recipes and ASP Chef," in *Proceedings of the ICLP'23 Workshops* (J. Arias, S. Batsakis, W. Faber, G. Gupta, F. Pacenza, E. Papadakis, L. Robaldo, K. Rückschloß, E. Salazar, Z. G. Saribatur, I. Tachmazidis, F. Weitkämper, and A. Z. Wyner, eds.), vol. 3437 of *CEUR Workshop Proceedings*, CEUR-WS.org, 2023.

[13] S. Costantini and A. Formisano, "RASP and ASP as a fragment of linear logic," *Journal of Applied Non-Classical Logics*, vol. 23, no. 1-2, pp. 49–74, 2013. DOI:10.1080/11663081.2013.798997.

[14] S. Costantini and A. Formisano, "Negation as a resource: a novel view on answer set semantics," *Fundamenta Informaticae*, vol. 140, no. 3-4, pp. 279–305, 2015. DOI:10.3233/FI-2015-1255.

[15] S. Costantini and A. Formisano, "Query answering in resource-based answer set semantics," *Theory and Practice of Logic Programming*, vol. 16, no. 5-6, pp. 619–635, 2016. DOI:10.1017/S1471068416000478.

[16] F. Cantucci and R. Falcone, "A computational model for cognitive human-robot interaction: An approach based on theory of delegation," in *Proceedings of the 20th Workshop "From Objects to Agents", Parma, Italy, June 26th-28th, 2019* (F. Bergenti and S. Monica, eds.), vol. 2404 of *CEUR Workshop Proceedings*, pp. 127–133, CEUR-WS.org, 2019.

[17] A. Sapienza and R. Falcone, "Evaluating agents' trustworthiness within virtual societies in case of no direct experience," *Cogn. Syst. Res.*, vol. 64, pp. 164–173, 2020.

[18] F. Cantucci and R. Falcone, "A cognitive approach to model intelligent collaboration in human-robot interaction," in *Proceedings of the 24th Workshop "From Objects to Agents", Roma, Italy, November 6-8, 2023* (R. Falcone, C. Castelfranchi, A. Sapienza, and F. Cantucci, eds.), vol. 3579 of *CEUR Workshop Proceedings*, pp. 138–150, CEUR-WS.org, 2023.

[19] G. Brewka, T. Eiter, and M. T. (eds.), "Answer set programming: Special issue," *AI Magazine*, vol. 37, no. 3, 2016.

[20] E. Erdem, M. Gelfond, and N. Leone, "Applications of answer set programming," *AI Mag.*, vol. 37, no. 3, pp. 53–68, 2016.

[21] Y. Lierler and M. Truszczyński, "On abstract modular inference systems and solvers," *Artif. Intell.*, vol. 236, pp. 65–89, 2016.

[22] M. Gebser, R. Kaminski, B. Kaufmann, and T. Schaub, *Answer Set Solving in Practice*. Synthesis Lectures on Artificial Intelligence and Machine Learning, Morgan & Claypool Publishers, 2012.

[23] B. Kaufmann, N. Leone, S. Perri, and T. Schaub, "Grounding and solving in answer set programming," *AI Mag.*, vol. 37, no. 3, pp. 25–32, 2016.

[24] ASP solvers, 2024. Clingo:`potassco.sourceforge.net`; Cmodels:`www.cs.utexas.edu/users/tag/cmodels`; DLV:`www.dlvsystem.it`; WASP:`alviano.github.io/wasp`.

[25] V. Lifschitz, *Answer Set Programming*. Springer, 2019.

[26] J. W. Lloyd, *Foundations of Logic Programming*. Springer-Verlag, 1987.

[27] S. Costantini, "About epistemic negation and world views in epistemic logic programs," *Theory Pract. Log. Program.*, vol. 19, no. 5-6, pp. 790–807, 2019.

[28] P. Cabalar, J. Fandinno, and L. Fariñas del Cerro, "Splitting epistemic logic programs," *Theory Pract. Log. Program.*, vol. 21, no. 3, pp. 296–316, 2021.

[29] P. Kahl, R. Watson, E. Balai, M. Gelfond, and Y. Zhang, "The language of epistemic specifications (refined) including a prototype solver," *J. Log. Comp.*, vol. 30, no. 4, pp. 953–989, 2015.

[30] E. I. Su, L. Fariñas del Cerro, and A. Herzig, "Autoepistemic equilibrium logic and epistemic specifications," *Artif. Intell.*, vol. 282, p. 103249, 2020.

[31] D. Pearce and A. Valverde, "Synonymous theories in answer set programming and equilibrium logic," *Proc. of ECAI'04*, pp. 388–390, 2004.

[32] E. I. Su, "Refining the semantics of epistemic specifications," in *Techni-*

cal Communications of ICLP'21 (A. Formisano, Y. A. Liu, B. Bogaerts, A. Brik, V. Dahl, C. Dodaro, P. Fodor, G. L. Pozzato, J. Vennekens, and N. Zhou, eds.), vol. 345 of *EPTCS*, pp. 113–126, 2021.

[33] S. Costantini and A. Formisano, "Epistemic logic programs: a novel perspective and some extensions," in *Proceedings of the ICLP'22 Workshops* (J. Arias, R. Calegari, L. Dickens, W. Faber, J. Fandinno, G. Gupta, M. Hecher, D. Inclezan, E. LeBlanc, M. Morak, E. Salazar, and J. Zangari, eds.), vol. 3193 of *CEUR Workshop Proceedings*, CEUR-WS.org, 2022.

[34] M. Hecher, M. Morak, and S. Woltran, "Structural decompositions of epistemic logic programs," in *Proc. of AAAI'20, IAAI'20, EAAI'20*, pp. 2830–2837, AAAI Press, 2020.

[35] M. Morak, "Epistemic logic programs: A different world view," in *Technical Communications of ICLP'19* (B. Bogaerts, E. Erdem, P. Fodor, A. Formisano, G. Ianni, D. Inclezan, G. Vidal, A. Villanueva, M. D. Vos, and F. Yang, eds.), vol. 306 of *EPTCS*, pp. 52–64, 2019.

[36] A. Van Gelder, K. A. Ross, and J. S. Schlipf, "The well-founded semantics for general logic programs," *Journal of the ACM*, vol. 38, no. 3, pp. 620–650, 1991.

[37] V. W. Marek and M. Truszczyński, "Reflective autoepistemic logic and logic programming," in *LPNMR, Logic Programming and Nonmonotonic Reasoning, Proc. of the Second Intl. Workshop* (A. Nerode and L. M. Pereira, eds.), pp. 115–131, The MIT Press, 1993.

[38] S. Costantini, "On the existence of stable models of non-stratified logic programs," *Theory and Practice of Logic Programming*, vol. 6, no. 1-2, 2006.

[39] J. Dix, "A classification theory of semantics of normal logic programs I-II.," *Fundamenta Informaticae*, vol. 22, no. 3, pp. 227–255 and 257–288, 1995.

[40] T. Swift and D. S. Warren, "XSB: Extending prolog with tabled logic programming," *Theory and Practice of Logic Programming*, vol. 12, no. 1-2, pp. 157–187, 2012.

[41] W. Chen and D. S. Warren, "A goal-oriented approach to computing the well-founded semantics," *Journal of Logic Programming*, vol. 17, no. 2/3&4, pp. 279–300, 1993.

[42] K. Bowen and R. A. Kowalski, "Amalgamating language and metalanguage in logic programming, in: Logic programming," in *Logic Programming* (K. Clark and S. A. Tarnlund, eds.), pp. 153–173, Academic

Press, 1982.

A   Semantic Results for Interesting ELPs

In Tables 1, 2, and 3, a summary is reported of how the semantics presented in this chapter behave on some examples which are considered to be significant of situations that can be found in practical programming. (We integrated the results that appeared in the literature cited, mainly those in [10], with those relating to CF24F.)

Program	World views
$a \vee b$	$[\{a\}, \{b\}]$
$a \vee b$ $a \leftarrow \mathbf{K}b$	$[\{a\}, \{b\}]$
$a \vee b$ $a \leftarrow not\,\mathbf{K}b$	$[\{a\}]$
$a \vee b$ $c \leftarrow not\,\mathbf{K}b$	$[\{a, c\}, \{b, c\}]$
$a \leftarrow not\,\mathbf{K}b$ $b \leftarrow not\,\mathbf{K}a$	$[\{a\}], [\{b\}]$
$a \leftarrow not\,\mathbf{K}not\,a$ $a \leftarrow not\,\mathbf{K}a$	$[\{a\}]$

Table 1: Examples where semantics G94, G11, K15, F15, S16, FAEEL, and CF24F agree.

B   Available ELP Solvers

Table 4 shows, to the best of our knowledge, a list of available solvers for ELP.

Program	G94/G11/FAEEL	K15/F15/S16/CF24F
$a \leftarrow not \, \mathbf{K} not \, a$	$[\emptyset], [\{a\}]$	$[\{a\}]$
$a \vee b$ $a \leftarrow not \, \mathbf{K} not \, b$	none	$[\{a\}]$
$a \vee b$ $a \leftarrow \mathbf{K} not \, b$	$[\{a\}], [\{a\}, \{b\}]$	$[\{a\}, \{b\}]$
$a \leftarrow b$ $b \leftarrow not \, \mathbf{K} not \, a$	$[\emptyset], [\{a, b\}]$	$[\{a, b\}]$
$a \leftarrow not \, \mathbf{K} not \, b$ $b \leftarrow not \, \mathbf{K} not \, a$	$[\emptyset], [\{a, b\}]$	$[\{a, b\}]$

Table 2: Examples where semantics G94/G11/FAEEL differ from K15/F15/S16/CF24F.

Program	World views		
	G94	G11/FAEEL	K15	F15/S16/CF24F
$a \leftarrow not \, \mathbf{K} not \, b \wedge not \, b$ $b \leftarrow not \, \mathbf{K} not \, a \wedge not \, a$	$[\emptyset], [\{a\}, \{b\}]$		$[\{a\}, \{b\}]$
$a \leftarrow \mathbf{K} a$	$[\emptyset], [\{a\}]$		$[\emptyset]$
$a \leftarrow \mathbf{K} a$ $a \leftarrow not \, \mathbf{K} a$	$[\{a\}]$		none

Table 3: Examples showing differences among several semantics.

Solver	Year	Semantics	Underlying ASP-solver	Impl. language	Availability
ELMO	1994	G94	dlv	Prolog	n/a
sismodels	1994	G94	claspD	C++	n/a
Wviews	2007	G94	clingo	C++	Windows binary
Esmodels	2013	G11	clingo	(unknown)	Windows binary
ELPS	2014	K15	clingo	Java	source+binary
GISolver	2015	K15	clingo	(unknown)	Windows binary
ELPsolve	2016	K15/S16	clingo	C++	binary only
Wviews2	2017	G94	clingo	Python	Windows binary
EP-ASP	2017	K15/S16	clingo	Python+ASP	Windows binary
PelpSolver	2017	S16	clingo	Java	Windows binary
ELPsolve2	2017	S16	clingo	C++	not public release
EHEX	2018	S16	clingo	Python	source
selp	2018	S16	clingo	Python	source
eclingo	2020	G94	clingo	Python	source

Table 4: List of some available solvers for ELP.

Should I Trust?
On the Epistemic Foundation of
Arguments

Emiliano Lorini

IRIT, CNRS, Toulouse University, France

Emiliano.Lorini@irit.fr

Abstract

We present an epistemic analysis of the concept of argument that lays the foundation for a theory of the epistemic supports of trust. To formalize the concept of an argument, we use a semantics for epistemic logic relying on belief bases. It allows us to naturally represent the reasons for an agent to believe something as well as the cause of the possible inconsistency of the agent's beliefs.

1 Introduction

According to Castelfranchi & Falcone's (C&F's) theory [1], trust is a cognitive pattern involving:

1. a goal φ of the truster, and

2. three beliefs of the truster:

   (a) the truster's belief that the trustee is capable to do a certain action A,

   (b) the truster's belief that if the trustee does action A, φ will come true as consequence,

   (c) the truster's belief that the trustee intends to do action A.

Under the assumption that the truster's beliefs satisfy the deductive rationality principles of the basic modal logic K [2][1] and that "an agent is going to do a certain action if and only if it intends and is capable to do it", the beliefs 2a, 2b and 2c are together reducible to the truster's belief that the trustee is going to do action A and the fact φ will come true as a consequence. Using the terminology of the logics STIT and BIAT, the logics of "seeing to it that" [3] and "bringing it about that" [4], the latter can be rephrased as the truster's belief that the trustee sees to it that φ by doing action A. In other words, under the previous assumptions, the notion of trust can be simplified as follows:

> Agent i trusts agent j to help it achieve its goal φ through action A if i wants φ to be true and believes that j sees to it that φ by doing action A.

So, according to C&F's theory, for an agent i to trust another j, i has to form a belief about j's disposition to do an action whose execution guarantees the achievement of i's goal. For example, Anna may trust her husband Bob to make the bike usable again since she believes that Bob is going to repair the bike tire so that she could use the bike in the future.

Most logic-based formalizations of C&F's theory and, more generally, of the cognitive structure of trust are based on epistemic logic and on its standard semantics relying on the notion of Kripke model [5, 6, 7, 8, 9]. A Kripke model is a multi-relational structure over a set of possible worlds by means of which the modeler can naturally represent the agents' beliefs and uncertainty about the environment as well as the agents' higher-order beliefs and uncertainty about other agents' beliefs and uncertainty. Nonetheless, one limitation of the Kripke semantics is that it does not explicitly represent the justifications (or arguments in support) of the agents' beliefs nor the cause of their possible inconsistency. Therefore, using Kripke models it is not possible to represent the reason why an agent should trust another

---

[1]These include the so-called axiom K of the form $\big(\Box_{truster}\varphi \wedge \Box_{truster}(\varphi \rightarrow \psi)\big) \rightarrow \Box_{truster}\psi$, and the so-called rule of necessitation $\frac{\varphi}{\Box_{truster}\varphi}$.

agent which could be counterbalanced by a reason not to trust. For example, the reason for Ann to believe that her husband Bob is going to fix her bike tire so that she can use the bike for going to work may consist in Ann believing that Bob believes that the bike tire is flat, and Ann believing that if Bob believes that the bike tire is flat, he will fix it. The previous reason could be counterbalanced by the reason to believe that Bob will not fix the bike tire since Bob believes that Ann does not wish to use the bike for going to work, and if Bob believes that Ann does not wish to use the bike, he will not fix the bike tire.

In this short paper we are going to provide a formal semantics for epistemic logic that allows us to represent the arguments and reasons in support of an agent's beliefs. This paves the way for the foundation of a logical theory of trust supports. The current proposal grounds on previous works devoted to redefining the semantics for epistemic using belief (or knowledge) bases, as an alternative to Kripke models [10, 11, 12]. The belief base semantics allows us to naturally distinguish an agent's explicit beliefs from its implicit beliefs. While an explicit belief corresponds to a piece of information in the agent's belief base, an implicit belief of the agent corresponds to an information that is deducible from the agent's explicit beliefs. Explicit beliefs can also be seen as the supports of an implicit belief and, in the case of trust, as the supports of the truster's trust in the trustee.

The paper is organized as follows. In Section 2, we present the belief base semantics. Then, in Section 3 we introduce an epistemic language that distinguishes the concepts of explicit belief, plain implicit belief, and focused implicit belief. Finally, in Section 4 we use the language to formalize the concept of argument and to analyze its epistemic properties.

2 Belief Base Semantics

Following [10], in this section we present a semantics for epistemic logic exploiting the notion of belief base. Unlike the standard Kripke semantics for epistemic logic in which possible worlds and epistemic alternatives are primitive, in the belief base semantics they are defined from the primitive concept of belief base.

Assume a countably infinite set of atomic propositions $Atm = \{p, q, \ldots\}$ and a finite set of agents $Agt = \{1, \ldots, n\}$. We define the language $\mathcal{L}_0$ for talking agents' explicit beliefs by the following grammar:

$$\mathcal{L}_0 \stackrel{\text{def}}{=} \quad \alpha \quad ::= \quad p \mid \neg\alpha \mid \alpha \wedge \alpha \mid \triangle_i\alpha,$$

where p ranges over Atm and i ranges over Agt. The formula $\triangle_i\alpha$ is read "agent i has the explicit belief that α".

The following notion of state is needed to provide a semantic interpretation of the formulas in the language $\mathcal{L}_0$. A state has two components: a propositional valuation representing the atomic facts that are true (resp. false) in the environment and one belief base for every agent in Agt made of formulas from $\mathcal{L}_0$. Thus, an agent can have at its disposal in its belief base not only propositional information (i.e., information about the environment) but also higher-order information about other agents' explicit beliefs.

Definition 1 (State). *A state is a tuple $S = ((B_i)_{i \in Agt}, V)$ where:*

- *B_i is a finite set of formulas from $\mathcal{L}_0$ representing agent i's belief base,*

- *V is a set of atomic propositions from Atm representing the actual environment.*

The set of all states is denoted by $\mathbf{S}$.

The following definition specifies the truth conditions for the formulas in the base language $\mathcal{L}_0$ relative to a state.

Definition 2 (Satisfaction relation). *Let $S = ((B_i)_{i \in Agt}, V) \in \mathbf{S}$. Then,*

$$S \models p \iff p \in V,$$
$$S \models \neg\alpha \iff S \not\models \alpha,$$
$$S \models \alpha_1 \wedge \alpha_2 \iff S \models \alpha_1 \text{ and } S \models \alpha_2,$$
$$S \models \triangle_i\alpha \iff \alpha \in B_i.$$

According to the previous semantic interpretation, p is actually true (i.e., $S \models p$) if p is a property of the actual environment (i.e., $p \in V$). Observe in particular the set-theoretic interpretation of the explicit belief operators in the previous definition: agent i has the explicit belief that α (i.e., $S \models \triangle_i \alpha$) if α is included in its belief base (i.e., $\alpha \in B_i$).

The following definition introduces the notion of epistemic alternative.

Definition 3 (Epistemic alternatives). *Let $i \in Agt$. Then, $\mathcal{R}_i$ is the binary relation on the set $\mathbf{S}$ such that, for all $S = ((B_i)_{i \in Agt}, V)$, $S' = ((B_i')_{i \in Agt}, V') \in \mathbf{S}$:*

$$S \mathcal{R}_i S' \text{ if and only if } \forall \alpha \in B_i : B' \models \alpha.$$

$S\mathcal{R}_i S'$ means that S' is an epistemic alternative for agent i at S, that is to say, S' is a state that at S agent i considers possible. The idea of the previous definition is that S' is an epistemic alternative for agent i at S if and only if, S' satisfies all facts that agent i explicitly believes at S.

The following definition introduces the notion of focus-based epistemic alternative, namely, an alternative that an agent considers possible when focusing its attention on a restricted set of formulas from $\mathcal{L}_0$.

Definition 4 (Focus-based epistemic alternatives). *Let $i \in Agt$ and $X \subseteq \mathcal{L}_0$ Then, $\mathcal{R}_i^X$ is the binary relation on the set $\mathbf{S}$ such that, for all $S = ((B_i)_{i \in Agt}, V)$,*
$S' = ((B_i')_{i \in Agt}, V') \in \mathbf{S}$:

$$S \mathcal{R}_i^X S' \text{ if and only if } \forall \alpha \in (B_i \cap X) : B' \models \alpha.$$

$S\mathcal{R}_i^X S'$ means that S' is an epistemic alternative for agent i at S under focus X, that is to say, S' is a state that at S agent i considers possible when focusing on the formulas in X. According to Definition 4, the latter means that S' satisfies all facts in the restriction of agent i's belief base at S to X.

The last notion we take into consideration is that of model. A model is a state supplemented with a set of states, called *context* or *universe*, that captures the agents' common ground, namely, the body of information of which the agents have common belief [13]. The context is not necessarily equal to the set of all states **S**, since there could be states in **S** which are incompatible with the agents' common belief of the "laws of the domain". For example, we might want to exclude from the context all states in which the propositions "1+1=2" and "1+1=3" are true concomitantly under the assumption that the agents have common belief of the basic principles of integer arithmetic.

Definition 5 (Model). *A model is a pair* (S, U), *where* $S \in \mathbf{S}$ *and* $U \subseteq \mathbf{S}$. *The class of models is denoted by* **M**.

Note that in Definition 5 we do not require $S \in U$ since we conceive the agents' common ground as their common belief and not as their common knowledge. So, the agents' common ground could be incorrect and not include the actual state.

3 Epistemic Language

We extend the language $\mathcal{L}_0$ of explicit beliefs with modal operators for plain implicit belief and focused implicit belief. We call $\mathcal{L}$ the resulting language and define it by the following grammar:

$$\mathcal{L} \overset{\text{def}}{=} \quad \varphi \quad ::= \quad \alpha \mid \neg\varphi \mid \varphi \wedge \varphi \mid \Box_i\varphi \mid \Box_i^X\varphi,$$

where α ranges over $\mathcal{L}_0$, i ranges over *Agt* and $X \subseteq \mathcal{L}_0$. The other Boolean constructions $\top$, $\bot$, $\vee$, $\rightarrow$ and $\leftrightarrow$ are defined from α, $\neg$ and $\wedge$ in the standard way. The formula $\Box_i\varphi$ has to be read "agent i implicitly believes that φ" or "agent i can deduce that φ from its explicit beliefs", while the formula $\Box_i^X\varphi$ has to be read "agent i implicitly believes that φ, when focusing on the formulas in X" or "agent i can deduce that φ from its explicit beliefs, when focusing on the formulas in X".

The following definition extends Definition 2 to the full language $\mathcal{L}$. Its formulas are interpreted with respect to a model (S, U), as defined in Definition 5. We omit the Boolean cases, as they are defined in the usual way.

Definition 6 (Satisfaction relation (cont.)). *Let* $(S, U) \in \mathbf{M}$. *Then:*

$$(S, U) \models \alpha \iff S \models \alpha,$$
$$(S, U) \models \Box_i \varphi \iff \forall S' \in U, \text{ if } S\mathcal{R}_i S' \text{ then } (B', U) \models \varphi,$$
$$(S, U) \models \Box_i^X \varphi \iff \forall S' \in U, \text{ if } S\mathcal{R}_i^X S' \text{ then } (B', U) \models \varphi.$$

According to the previous definition, agent i implicitly believes that φ if φ is true at all states in the context that i considers possible. Moreover, agent i implicitly believes that φ when focusing on X if φ is true at all states in the context that i considers possible when focusing on X.

For notational convenience, we write $S \models \varphi$ instead of $(S, \mathbf{S}) \models \varphi$. The model $(S, \mathbf{S})$ is a model with maximal uncertainty since it includes all possible states. This means that the agents have no shared information in their common ground. It is analogous to the notion of canonical model in modal logic [14].

Let $\varphi \in \mathcal{L}$. We say that φ is valid relative to the class $\mathbf{M}$, denoted by $\models_{\mathbf{M}} \varphi$, if for every $(S, U) \in \mathbf{M}$ we have $(S, U) \models \varphi$. We say that φ is satisfiable for the class $\mathbf{M}$ if $\neg\varphi$ is not valid for the class $\mathbf{M}$.

As the following validities highlight, the epistemic operators $\Box_i$ and $\Box_i^X$ are both normal modalities satisfying the basic principles of the modal logic K. For $\blacksquare \in \{\Box_i, \Box_i^X\}$:

$$\models (\blacksquare\varphi \wedge \blacksquare(\varphi \to \psi)) \to \blacksquare\psi \tag{1}$$
$$\text{if } \models \varphi \text{ then } \models \blacksquare\varphi \tag{2}$$

Moreover, focused implicit belief is stronger than plain implicit belief, and focused implicit belief is monotonic wrt to focus:

$$\models \Box_i^X \varphi \to \Box_i \varphi \tag{3}$$
$$\models \Box_i^X \varphi \to \Box_i^{X'} \varphi \text{ if } X \subseteq X' \tag{4}$$

263

4 Argument

After introducing semantics and language in the previous two sections, we can finally provide a definition of the notion of argument. We define it as an abbreviation. An argument can be seen as a 3-ary predicate. It involves an agent $i \in Agt$ who puts forward the argument, a finite set of formulas $X \subseteq \mathcal{L}_0$ which represent the support (or the content of the argument), and a formula $\varphi \in \mathcal{L}$ which represents the supported conclusion:

$$\mathsf{Arg}_i(X,\varphi) \stackrel{\text{def}}{=} \bigwedge_{\alpha \in X} \triangle_i \alpha \wedge \Box_i^X \varphi \wedge \neg \Box_i^X \bot \wedge \bigwedge_{X' \subset X} \neg \Box_i^{X'} \varphi.$$

The abbreviation $\mathsf{Arg}_i(X,\varphi)$ stands for "the information in X is for agent i an argument in the support of the conclusion that φ". This means that: i) every fact in X is explicitly believed to be true by agent i (i.e., $\bigwedge_{\alpha \in X} \triangle_i \alpha$), ii) agent i can deduce that φ when focusing on the information in X (i.e., $\Box_i^X \varphi$), iii) the information in X is consistent (i.e., $\neg \Box_i^X \bot$), and iv) agent i can no longer deduce that φ when neglecting some information in X (i.e., $\bigwedge_{X' \subset X : |X'| = |X| - 1} \neg \Box_i^{X'} \varphi$).

Example 1. *Let us consider the state* $S = ((B_{Ann}, B_{Bob}), V)$ *that represents the explicit beliefs of Ann and Bob in the situation we briefly described in the introduction:*

$$B_{Ann} = \{\triangle_{Bob} tf, \triangle_{Bob} tf \rightarrow br, \triangle_{Bob} \neg ab, \triangle_{Bob} \neg ab \rightarrow \neg br\},$$
$$B_{Bob} = \{tf, br\},$$
$$V = \{tf, br\}.$$

In particular, Ann explicitly believes

- *that Bob explicitly believes that the bike tire is flat (i.e., $\triangle_{Bob} tf$ in B_{Ann}),*

- *that if Bob explicitly believes that the bike tire is flat then he will fix it (i.e., $\triangle_{Bob} tf \rightarrow br$ in B_{Ann}),*

- *that Bob explicitly believes that Ann does not wish to use the bike for going to work (i.e., $\triangle_{Bob} \neg ab$ in B_{Ann}), and*

- *that if Bob explicitly believes that Ann does not wish to use the bike for going to work then he will not fix the bike tire (i.e., $\triangle_{Bob}\neg ab \rightarrow \neg br$ in B_{Ann}).*

Moreover, Bob explicitly believes

- *that the bike tire is flat (i.e., tf in B_{Bob}), and*

- *that he is going to fix it (i.e., br in B_{Bob}).*

Finally, the bike tire is actually flat and Bob is going to fix it (i.e., tf and br in V). It is straightforward to verify that:

$$S \models \mathsf{Arg}_{Ann}(\{\triangle_{Bob}tf, \triangle_{Bob}tf \rightarrow br\}, br) \wedge$$
$$\mathsf{Arg}_{Ann}(\{\triangle_{Bob}\neg ab, \triangle_{Bob}\neg ab \rightarrow \neg br\}, \neg br).$$

This means that Ann has both an argument is support of the fact that Bob is going to fix the bike and a counter-argument in support of the fact that Bob is not going to fix the bike.

As the following proposition highlights, in a model with maximal uncertainty and when the supported formula is propositional, the previous notion of argument has a precise deductive characterization which is in line with the theory of deductive arguments [15, 16].

Proposition 1. *Let φ be a propositional formula and let $S = ((B_i)_{i \in Agt}, V)$ be a state. Then,*

$$S \models \mathsf{Arg}_i(X, \varphi) \text{ iff } i) \ X \subseteq B_i,$$
$$ii) \ \varphi \in Cn(X),$$
$$iii) \ \bot \notin Cn(X),$$
$$iv) \ \forall X' \subset X, \varphi \notin Cn(X).$$

The following are further remarkable properties of the previous notion of argument:

$$\models \neg\mathsf{Arg}_i(X, \top) \tag{5}$$
$$\models \neg\mathsf{Arg}_i(X, \bot) \tag{6}$$
$$\models (\mathsf{Arg}_i(X, \varphi) \wedge \mathsf{Arg}_i(X, \psi)) \rightarrow \mathsf{Arg}_i(X, \varphi \wedge \psi) \tag{7}$$
$$\models \mathsf{Arg}_i(X, \varphi) \rightarrow \mathsf{Arg}_i(X, \varphi \vee \psi) \tag{8}$$

According to the validity (5), an agent cannot have an argument in support of a tautology, while according to the validity (6), it cannot have an argument in support of a contradiction. According to the validity (7), arguments aggregate under conjunction: if X is an argument in support of φ and X is an argument in support of ψ, then X is an argument in support of $\varphi \wedge \psi$. Validity (8) is a form of weakening: if X is an argument in support of φ then it is also an argument in support of $\varphi \vee \psi$. Note that weakening for conjunction, i.e., $\mathsf{Arg}_i(X,\varphi \wedge \psi) \rightarrow \mathsf{Arg}_i(X,\varphi)$, does not hold due to the negative condition $\bigwedge_{X' \subset X} \neg \Box_i^{X'} \varphi$ in the definition of argument.

5 Conclusion

We have provided an epistemic analysis of the notion of argument with the support of a formal semantics relying on belief bases. This is just the preliminary step towards a formal theory of trust supports, an aspect that is ignored altogether in existing logical formalizations of Castelfranchi & Falcone's theory.

References

[1] C. Castelfranchi and R. Falcone, "Social trust: A cognitive approach," in *Trust and Deception in Virtual Societies* (C. Castelfranchi and Y. H. Tan, eds.), pp. 55–90, Dordrecht: Kluwer Academic Publishers, 2001.

[2] R. Fagin, J. Halpern, Y. Moses, and M. Vardi, *Reasoning about Knowledge*. Cambridge: MIT Press, 1995.

[3] N. Belnap, M. Perloff, and M. Xu, *Facing the future: agents and choices in our indeterminist world*. Oxford, 2001.

[4] G. Governatori and A. Rotolo, "On the axiomatisation of Elgesem's logic of agency and ability," *Journal of Philosophical Logic*, vol. 34, no. 4, pp. 403–431, 2005.

[5] A. Herzig, E. Lorini, J. F. Hubner, and L. Vercouter, "A logic of trust and reputation," *Logic Journal of the IGPL*, vol. 18, pp. 214–244, 12 2009.

[6] R. Demolombe, "Reasoning about trust: A formal logical framework," in *Proceedings of the Second International Conference on Trust Management (iTrust 2004)*, vol. 2995 of *LNCS*, pp. 291–303, Springer, 2004.

[7] C. Liau, "Belief, information acquisition, and trust in multi-agent systems–a modal logic formulation," *Artificial Intelligence*, vol. 149, no. 1, pp. 31–60, 2003.

[8] F. Liu and E. Lorini, "Reasoning about belief, evidence and trust in a multi-agent setting," in *PRIMA 2017: Principles and Practice of Multi-Agent Systems - 20th International Conference, Nice, France, October 30 - November 3, 2017, Proceedings*, vol. 10621 of *Lecture Notes in Computer Science*, pp. 71–89, Springer, 2017.

[9] E. Lorini and R. Demolombe, "Trust and norms in the context of computer security: A logical formalization," in *Deontic Logic in Computer Science, 9th International Conference, DEON 2008, Luxembourg, Luxembourg, July 15-18, 2008. Proceedings*, vol. 5076 of *Lecture Notes in Computer Science*, pp. 50–64, Springer, 2008.

[10] E. Lorini, "Rethinking epistemic logic with belief bases," *Artificial Intelligence*, vol. 282, 2020.

[11] E. Lorini, "Exploiting belief bases for building rich epistemic structures," in *Proceedings of the Seventeenth Conference on Theoretical Aspects of Rationality and Knowledge (TARK 2019)*, vol. 297 of *EPTCS*, pp. 332–353, 2019.

[12] E. Lorini, "In praise of belief bases: Doing epistemic logic without possible worlds," in *Proceedings of the Thirty-Second AAAI Conference on Artificial Intelligence (AAAI-18)*, pp. 1915–1922, AAAI Press, 2018.

[13] R. Stalnaker, "Common ground," *Linguistics and philosophy*, vol. 25, no. 5/6, pp. 701–721, 2002.

[14] P. Blackburn, M. de Rijke, and Y. Venema, *Modal Logic*. Cambridge: Cambridge University Press, 2001.

[15] L. Amgoud and R. Demolombe, "An argumentation-based approach for reasoning about trust in information sources," *Argument and Computation*, vol. 5, no. 2-3, pp. 191–215, 2014.

[16] P. Besnard and A. Hunter, "A logic-based theory of deductive arguments," *Artificial Intelligence*, vol. 128, no. 1-2, pp. 203–235, 2001.

When Transparency Shapes Trust: A Signalling Approach

Aron Szekely

Collegio Carlo Alberto (CCA)
Piazza Arbarello 8, 10122, Torino, Italy
aron.szekely@carloalberto.org

Luca Tummolini

Institute of Cognitive Sciences and Technologies (ISTC)
National Research Council of Italy
Via Giandomenico Romagnosi 18A, 00196, Rome, Italy
luca.tummolini@istc.cnr.it

Abstract

Transparency is advocated as a key solution to trust problems in contexts spanning governments, charities, companies, and individuals. By increasing the amount of public information, transparency should increase trust. Empirical research, however, finds that this intuitive relationship between trust and transparency is less clear than often assumed. Here we highlight some ways in which such mandated transparency measures can misfire, helping to explain mixed findings, and propose that voluntary transparency signals are likelier to provide the desired effect. We draw on signalling theory to support this argument and distinguish different ways in which transparency can act as a signal.

LT acknoweldges the support by project PRIN WHIM (Prot. 2022LYRT8E) and PRIN 2022 PNRR NOJA (Prot. P2022YYRK3, CUP B53D23030380001).

1 Introduction

Fundamental human needs can be met in society only when cooperation on a large scale is possible. In particular, given their inescapable cognitive, material, and physical limitations, human beings depend on each other for most of their goals. Therefore their willingness to make themselves vulnerable by relying on each other —trust— is a necessary lubricant of social interaction which enable cooperation. Moreover, as the sociocognitive model of Castelfranchi and Falcone has helped us see, both at the interpersonal and at the societal level, trust is a fundamental dynamic phenomenon that changes over time as a function of many different factors, from intrapsychic cognitive dynamics to socially emergent effects [1]. Despite this complexity, however, there is a recurring solution to the problem of fostering trust when it is lacking across many different domains and scales: increase transparency.

As a policy measure, prescribing transparency means to require that candidate "trustees" provide information that may be relevant to assess their trustworthiness. Adoption of this approach is widespread, from Open Governance initiatives to the Open Science movement that is on the rise in these days. However whether mandated transparency can actually enhance trust remains unclear, and evidence of its effect is mixed. In this contribution we raise this challenge and propose that an important dimension is whether transparency is voluntarily chosen or not. We suggest that focusing on this often-neglected dimension can help re-orient policy interventions in new and more promising directions.

2 The limits of transparency as a top-down solution

The thaumaturgical benefits of transparency have been acknowledged for long. Already in in the early 1900s, Louis Brandeis wrote that "sunlight is said to be the best of disinfectants; electric light the most efficient policeman", and echoing this motto Barack Obama, stated that "restoring transparency is not only the surest way to achieve results, but also to earn back the trust in government" [2]. Indeed, the top response of Americans for how to improve trust in the federal government was "more transparency, less secrecy" in a 2019 survey of the Pew Research Center. Thus, in defiance of the underlying complexity of trust as a social and collective phenomenon that increasing transparency measures is sufficient to boost trust is in fact a widespread assumption (see e.g. [3, 4]).

The intuitive mechanism for how transparency should increase trust is two-sided: one for those revealing the information and the other for those receiving this information. By revealing their inner workings and making observable previously hidden properties, people and institutions should have incentives to "behave well", reducing inefficiency, corruption, and improving decision making. Observers, in turn, should react to the consequent increase in trustworthiness and efficiency by increasing their trusting and assessment of quality, leading to greater cooperation in the aggregate (see also [3]).

Such transparency measures can take many forms: requiring that governmental institutions and public companies reveal how they spend their revenues, introducing TV cameras in courtrooms—as the UK did for sentencing in Crown Courts in 2022, or televising political debates (as the USA does for the US House of Representative), mandating the use of body cameras on police officers, and publishing citizens' tax reports online, like in Norway. There can also be physical ways of increasing transparency: open kitchens in restaurants allow customers to see their food being prepared and presented, open offices make it easy for the supervisors to monitor their workers, and Jeremy Benthem's (in)famous Panopticon was designed as a prison that enforced full transparency on inmates by giving them no privacy—in cruel contrast guards would have full privacy.

While each implementation has its own particular features, they are united by top-down, typically legal, physical, or rule-based requirements that force their targets to reveal information that they otherwise would want to keep private. That is, these are all involuntary transparency measures that aim to improve trust by overriding the underlying incentives of the individuals and institutions involved

271

to conceal information. They try to solve the problem of trust by increasing surveillance and control and thus alleviating the very need to place trust in the first place.

But this leaves open an important, and underappreciated issue: when transparency measures are implemented in such top-down ways, many agents may still have incentives to hide the true state of affairs from their observers. This implies that in absence of the external compulsion to reveal information, these actors would prefer to keep such information hidden. Consequently, when this condition holds, they are incentivised to take the bare minimum needed to comply with the requirements, or, only technically comply with them while ignoring their spirit.

Examples illustrating this abound. Consider the terms and conditions that companies and websites set out whenever we install their software or use their service. While explicit and technically fully transparent, they are so long and written in a hard-to-understand way, that in practice, they place an unrealistic burden on the readers making them anything but transparent. Indeed, one estimate for how long it would take Americans to read the privacy policies of every website that they visit is 244 hours annually (with lower and upper bounds of 181 and 304 hours per year)[5] —clearly an unfeasible expectation. When scientific journals require that data for an article are made freely available, researchers may technically share the data, but do so in a highly restricted, "curated", way that does not allow data exploration any further than what is already presented in an article.

Onora O'Neill puts this point differently, writing that "It is easy to place information in the public domain, but hard to ensure that information is accessible to those whom it might be valuable, intelligible to them if they find it, or assessable by them if they find and understand it" [6]. Moreover, given the incentives that mandatory transparency measures create, not only are accessibility, intelligibility, and assessability hard to ensure, but these measures instead actually promote the very conditions for individuals and companies to make information as *in*accessible, *un*intelligible, and *non*-assessable as they can get away with.

Of course, this does not imply that imposing transparency measures will lead to worse outcomes than would be the case without them. But it does imply that there is an important countervailing pressure —*incentives from the interactions between senders and observers*— that may mitigate, partially or entirely, the hoped-for benefits.

There are other ways too of seeing this tension between externally imposed requirements and decision maker's endogenous incentives to maintain secrecy. Anticipating that their actions will be observed, institutions are incentivised to "massage the truth. Public reports may underplay sensitive information; head teachers and employers may write blandly uninformative reports and references; evasive and uninformative statements may substitute for truth-telling" [7]. These are all responses by the relevant institutions and individuals to avoid the downsides of transparency in response to its top-down imposition. Key decisions in politics may end up being guided more by how they look in the public eye and less by how much they actually serve the public interest [8].

We can also look more broadly at the literature on cooperation and social preferences —individuals' motivations for considering the outcomes of others— and how this is affected by institutional policies. In a wide-ranging and influential article, Bowles and Polanyia-Reyes [9] propose a conceptual framework for understanding how institutions can drive out social preferences and then test this framework using available experimental evidence from behavioural games. Our view is that the concerns that they identify, and general empirical result, also plausibly applies to institutional transparency measures.

Specifically, Bowles and Polanyia-Reyes propose three direct ways in which policy interventions can undermine, or "crowd out", prosocial motives: by implementing sanctions, or threatening sanctions, institutions (i) reveal their poor beliefs about the motives of the decision makers that they affect, (ii) negatively shape the "framing" of a situation —how individuals understand the context of the decision— potentially triggering expectations that self-interested behaviour is appropriate or at least alleviate social norms promoting

273

prosociality, and (iii) activating control aversion —the sense that people's self-determination is being overruled— leading to "rebellious" actions aimed to alleviate this. Based on extensive data of more than "twenty-six thousand subjects from thirty-six countries, playing Dictator, Trust, Ultimatum, Public Goods, Third Party Punishment, Common Pool Resource, Gift Exchange, and other principal-agent games" (p. 375) they find broad support for their framework. Indeed, they estimate that there is at least some crowding out in the majority (69%) of the studies in their survey.

Rephrased from the perspective of trust and transparency, imposing involuntary transparency measures on individuals can reveal to the subjects that others believe that they are untrustworthy and therefore need to be put under scrutiny and that they are operating in a context in which instrumental, and not trusting and trustworthy, behaviour is appropriate. Additionally, based on the external constraints, agents may feel that their actions, and possible choices, are being overly restricted thereby leading to deliberately counterproductive responses. Mandated transparency in fact can create the conditions for control —monitoring and intervention over the decisions and actions of the agents, which can in fact be antagonistic of (strict) trust itself [10].

Consistent with this hypothesis, reviews of the empirical research highlight just how unclear the effects of transparency measures are. As one review of experimental evidence highlights that "For every example where transparency seemed to produce more accountable and effective governance there is another where transparency either had no effect or produced a backlash that further insulated public officials from accountability to citizens" [3]. And, while they do find that transparency had an overall positive effect in 11 out of 16 (69%) available studies, the remaining 31% had neutral or backfire effects. Another, a 25 year review of 187 studies from across disciplines identifies two of the key takeaways as: "government transparency is no cure-all and does not always have positive outcomes" and that transparency can "work under some conditions but not under others" [11].

So what can be done to promote trust if the consequences of involuntary transparency are so ambiguous? One answer comes from the literature cited above. As argued there, it is essential to understand the specific context in which the transparency measures are implemented. And, while not our focus here, there are promising attempts to systematically identify these factors in a more comprehensive approach (e.g. [3]). One general point also emerges from the social preference literature: with involuntary measures interpretation is crucial. That is, the way in which decision makers interpret and understand the policy approach is essential [3, 9, 12].

Another answer, which is our focus here, is to move away from mandated transparency measures and turn to policies that promote voluntary transparency. Since mandated transparency has so many potential pitfalls, it may be better to work indirectly and create the conditions that allow trust-enhancing actions to be voluntarily taken. That is, adjusting or changing the environment in such a way as to align incentives so that individuals prefer to disclose information about themselves rather than to leave it hidden. They should prefer, even in absence of legal or rule-based pressures, to reveal relevant private information. A consequence of this focus, on revealing private information, is that it moves a concern with aggregate trust in a system to *well-placed* trust: trusting the trustworthy and not trusting the untrustworthy [6].

Here we propose that signalling theory is an essential framework that can help advance our understanding of these points. Specifically, it identifies the conditions under which agents can be incentivized to reveal private information about themselves to communicate with others without external pressure.

3   Signalling theory and transparency

In order to place their trust appropriately, observers face a crucial problem: how can they separate between trustworthy and untrustworthy others? Efficient and inefficient institutions? Or knowledgeable and less knowledgeable people? Conversely, trustworthy, efficient, and knowledgeable decision makers face the opposite problem: in what way can they "stand out" from the crowd and reliably separate themselves from untrustworthy, inefficient, and less knowledgeable agents? Signalling theory, formally a branch of game theory, provides answers to these questions [13, 14, 15]. We provide a brief overview of the theory here, focusing on the aspects most relevant for transparency. For a fuller exposition, see especially [16, 17].

The theory deals with situations in which -like often encountered in the real world- there is the possibility for the incentives of the participants to diverge and there is informational asymmetry such that one participant knows more about a state of affairs than the other. Consider the classic case of employers and a job seeker to illustrate the two assumptions. First, the potential for incentive conflict: an employer only wants to hire applicants who have the appropriate skills for the position, and, while many job seekers apply for a job only if they are suited for it, some would also happily receive a contract even if they are only a poor fit. Second, informational asymmetry: job seekers know more about their own skills and characteristics than the potential employers do.

In such conditions, signalling theory addresses the questions that both "senders", agents aiming to change the beliefs of observers about properties of theirs, and "receivers", those perceiving the actions of senders, should ask themselves. What signals can the former send to change the beliefs of the latter and what signals should the latter consider when updating their beliefs?

Despite the apparent difficulty in communicating in settings in which there is the possibility of deception, a core contribution of signalling theory is that it demonstrates that reliable information transmission (i.e. the signal accurately represents a state of affairs) is possible when two conditions are met: the Can condition and the Cannot condition [16]. The former requires that senders with the relevant properties, "high types" in the theory's terminology, must be able to afford a specific signalling action, call this s_h. While the latter requires that senders without the relevant properties, "low types", should not be able to afford the same signalling action as high types and only a lower level of this represented by s_l (there is an edge case in which some low types can afford s_h; this is known as the "Can

Just condition" and the proportion sending s_h depends up on the exact costs and benefits). If there is some action, a signal that satisfies both the Can and Cannot conditions, then credible signalling, in which some individuals and institutions voluntarily reveal private information about themselves, and observers interpret them correctly, is possible.

The Can and Cannot conditions can be met in three fundamental ways: sending the signal is associated with differential costs, differential benefits, or, it simply is a signal that cannot be faked.

The most well-known way relies on the first option and proposes that the cost to taking a signal for both high and low types differ. For instance, the cost to a well-trained athlete to running a marathon is much lower than it is for an untrained "average Joe". If this cost is sufficiently different-cheap enough for a high type and costly enough for a low type, then there is a signal that the high types can take to separate themselves. This allows observers to distinguish between the two types and reliable communication can occur. In biology, a version of this argument is known as the "handicap principle" [15].

In the second possibility, instead, costs remain the same between the two types but the benefits may differ. That is, a high type gains more from being treated accordingly than a low type does. In [18], Eric Posner argues that a vast range of behaviours can be understood as signals taken by individuals with differing discount rates, with high types valuing the future more than low types. The former have more to gain than the latter from long term cooperation, so they benefit more, allowing reliable signalling to occur.

Finally, a signal may be reliable because the signal is linked to the hidden feature to be signalled in such a way that cannot be in fact produced by those who lack the relevant properties. In biology these signals have been called "indices" [19] and have been considered reliable because of physiological constraints on signal production. The croaks made by toads during mating season are a classic example. Since the vocal apparatus of toads depends on their size, females can reliably assess the size of the male by detecting how deep the croak is. But in presence of such "verifiable signals" the problem is to explain why those who can only signals about their lower quality may be inclined to do so. A number of well-known results in information economics have established that in fact voluntary disclosure is in fact optimal if observers learn to interpret silence as a sign that one is the worst type. Under this assumption, disclosure is in the interests of all who prefer to distinguish themselves from the worst ones and the unravelling process is inevitable [20].

4 Promoting transparency with signalling

How can all of this help with transparency and trust? Fundamentally, it tells us that policy makers and institutions can promote genuine transparency by focusing on the environment in which interactions take place, instead of imposing external constraints. Specifically, we propose that this can happen in four ways.

First, and most directly, policy makers can ensure that both the Can and Cannot conditions are met for existing actions by increasing the cost or benefit difference between the types. Educational certificates, from university, secondary school, or technical schools, are simple examples of actions that have different costs on people who differ in terms of motivations, ambitions, and capabilities (unfortunately wealth also!). Yet if an educational certificate is too easy for all applicants to attain, or too hard for all applicants, then reliable signalling cannot occur. In the case of the former, the difficulty of passing, or obtaining a certain grade, can be increased. While in the case of the latter, they should be loosened. eBay, the online marketplace for second hand goods, introduced an "Authenticity Guarantee" for high-end watches; in this process, the seller of the watch needs to first send the watch to an independent authenticator who inspects it for indications of forgery, and, if it passes, it is then forwarded on to the buyer with a certificate of authenticity. By including this step automatically, eBay increases the cost to a would-be faker while only slightly burdening genuine sellers.

Second, enlarge the set of signalling actions that agents can take by introducing new potential actions. If there is no existing signal that satisfies both the Can and Cannot conditions, reliable signalling cannot take place. So, the development of new signalling possibilities can be promoted that do actually satisfy both conditions. One example of this is from The Royal Society for the Prevention of Cruelty to Animals (RSPCA) who introduced the possibility to farmers to label the packing of their animal produce as being "RSPCA Assured". This certification is given to farms that fulfil specific criteria for treating their animals humanely and was introduced to help consumers differentiate between unethical factory farming practices and those farmers that meet humane farming practices. By introducing this action, they allow farmers to credibly signal how they treat their animals to consumers. There are technological examples too of enlarging the set of possible actions. For instance, in the last years it has become increasingly clear that the methodological practices and norms that used to be standard in many disciplines like psychology and medicine are not able to ensure replicability of many published results questioning the reliability of these findings. In response to this crisis, technological organizations like the Center for Open Science (COS) were funded with the aim of creating new platforms to facilitate adherence to better practices. Thanks to these platforms it is now possible, for instance, to deposit hypotheses, experimental design and analysis plans before conducting experimental studies, which is considered essential to ensure the validity of confirmatory research. Crucially, by providing the infrastructure itself, these platforms also create a signalling system to send verifiable signals of scientists' intentions that before was simply non-existent, and with it the possibility of inducing an unravelling dynamics in which the transparent sharing of scientists' intentions may finally become the norm [21].

Third, actions often have, or can have, multiple meanings and policy makers can highlight one or the other to shape people's actions. Some of the meanings can be intentional while others are unintentional, but, in both cases they provide levers for increasing or reducing a signalling action. Fur coats, made from the pelts of animals including fox or mink, were stylish displays of wealth in the early to mid 1900s while today, after campaigns by activists most prominently from

PETA, they are often seen as vulgar and brutish. A less known but instructive example comes from the memoirs of Casanova —astutely identified and highlighted by Gambetta [22] albeit to illustrate a different point. Casanova recounts that in the late 1700s, when he was visiting Spain, one of the main controversies was about male fashion: specifically that of wearing "codpieces" or "braguettes"—an accessory that accentuates the bulge of men's crotches. Despite the church's best efforts to stop men wearing them, with priests and monks preaching against their use, the fashion was only stopped when an "edict was published and affixed to the doors of all the churches, in which it was declared that breeches with braguettes were only to be worn by the public hangmen. Then the fashion passed away; for no one cared to pass for the public executioner" (quoted in [22]). By associating the meaning of braguettes with hangmen, they changed how senders and receivers interpreted the action and thus reduced its use without imposing external fines. Both of the above examples illustrate broad lessons: once the Can and Cannot conditions are fulfilled, the meaning of specific acts can be changed through association, advertising, games, and or even education. Ultimately these all rely on reshaping the meaning of an action by coordinating the beliefs of senders and receivers onto one of multiple possibilities.

Finally, all of us have limits on our cognitive capacities. One aspect of this is that we can find it difficult to think in counterfactual terms and deviate from some status quo upon which we have settled. In signalling theory terms, even when both the Can and Cannot conditions are fulfilled, in addition to the "separating equilibrium", in which high and low type senders take different actions, the so-called "pooling equilibrium" also exists. In this latter kind of equilibrium, both high and low type senders emit the same signal and thus receivers cannot differentiate between them. Some systems may be stuck in a pooling equilibrium even though, if high type senders were to take a different action and receivers had beliefs that would treat those deviators as high type senders, then they could move to the separating equilibrium. Part of this failure to move can be a difficulty of counterfactual reasoning. But, it can also be due to the difficulty of coordinating senders and receivers; like in the case of changing the meaning of signal, education, games, marketing approaches, can all help overcome this.

5 Conclusions

So when does transparency shape trust? Here we have started from the observation that, notwithstanding its popularity, the widespread inclination to counteract trust crises by mandating transparency has often misfired in practice. We have suggested that an important reason for such limited effects is that releasing information into the public sphere does not by itself ensure its quality. Indeed, if the primary problem of trust is discriminating the trustworthy from the rest, the secondary problem is being able to trust the "signs" that may be used to infer such trustworthiness [16]. However, mandating transparency, while aimed to provide evidence that is relevant to assess a target, often produces signs that do little to separate the wheat from the chaff, and may even end up blurring the distinction.

But where an outpouring of signs may fail, enabling reliable signals may instead succeed.

With the help of signalling theory, we have suggested that if policy interventions aim to create the conditions for a reliable signalling system, signals of trustworthiness can be voluntarily emitted by those who desire to be trusted. In sum, while mandating transparency is not sufficient to motivate individuals to pay the costs of generating intelligible signs, an indirect approach that creates the appropriate conditions for reliable communication between the parties may be up to the task.

Perhaps the truth is in the middle, with an indirect strategy being better to establish a virtuous practice and a direct one more apt to maintain it once emerged. Whatever it is, we believe that, by combining the sociocognitive model of trust with the theory of reliable signalling, we finally have the tools to discover it.

References

[1] Castelfranchi, C., & Falcone, R. (2010). Trust theory: A socio-cognitive and computational model. John Wiley & Sons.

[2] Obama, B. (2009). Transparency and Open Government, memorandum for the heads of executive departments and agencies. .

[3] Kosack, S., & Fung, A. (2014). Does transparency improve governance?. Annual Review of Political Science, 17, 65-87.

[4] Zamir, E., & Engel, C. (2021). Sunlight Is the Best Disinfectant-Or Is It? Anonymity as a Means to Enhance Impartiality. Arizona Law Review, 63, 1063.

[5] McDonald, A.M. & Cranor, L.F. (2008) The Cost of Reading Privacy Policies. I/S: A Journal of Law and Policy for the Information Society, 4(3), 543-568.

[6] O'Neill, O. (2018). Linking trust to trustworthiness. International Journal of Philosophical Studies, 26(2), 293-300.

[7] O'Neill, O. (2002). A question of trust: The BBC Reith Lectures 2002. Cambridge University Press.

[8] Krastev, I. (2001). Think tanks: Making and faking influence. Southeast European and Black Sea Studies, 1(2), 17-38.

[9] Bowles, S., & Polania-Reyes, S. (2012). Economic incentives and social preferences: substitutes or complements? Journal of Economic Literature, 50(2), 368-425.

[10] Falcone, R., & Castelfranchi, C. (2002). Issues of trust and control on agent autonomy. Connection Science, 14(4), 249-263.

[11] Cucciniello, M., Porumbescu, G. A., & Grimmelikhuijsen, S. (2017). 25 Years of Transparency Research: Evidence and Future Directions. Public Administration Review, 77(1), 32-44.

[12] Gneezy, U., & Rustichini, A. (2000). A fine is a price. The Journal of Legal Studies, 29(1), 1-17.

[13] Grafen, A. (1990). Biological signals as handicaps. Journal of Theoretical Biology, 144(4), 517-546.

[14] Spence, M. (1974). Market Signaling: Informational Transfer in Hiring and Related Screening Processes. Cambridge, MA: Harvard University Press.

[15] Zahavi, A. (1975). Mate selection-A selection for a handicap. Journal of Theoretical Biology, 53(1), 205-214.

[16] Bacharach, M., & Gambetta, D. (2001). Trust in signs. In K. S. Cook

(Ed.), Trust in society (pp. 148-184). New York: Russell Sage Foundation

[17] Gambetta, D. (2009). Why prisoners fight (and signal). In Codes of the Underworld: How Criminals Communicate. Princeton, NJ: Princeton University Press.

[18] Posner, E. (2002). Law and Social Norms. Harvard University Press.

[19] Maynard Smith, J., & Harper, D.G.C. (2003). Animal Signals. Oxford: Oxford University Press.

[20] Frank, R. (1988). Passions within Reason: The Strategic Role of Emotions. New York: Norton.

[21] Nosek, B. A., & Lindsay, D. S. (2018). Preregistration becoming the norm in psychological science. APS observer, 31.

[22] Gambetta, D. (2009). Signaling. In P. Hedström & P. Bearman (Eds.), The Oxford Handbook of Analytical Sociology (pp. 169-194). Oxford: Oxford University Press.

Coordination and Trust in MAS towards Intelligent Socio-technical Systems

Andrea Omicini

Alma Mater Studiorum—Università di Bologna

andrea.omicini@unibo.it

Alessandro Ricci

Alma Mater Studiorum—Università di Bologna

a.ricci@unibo.it

Abstract

The discipline of intelligent systems engineering is slowly moving towards its presumable final destination – that is, agent-oriented software engineering –, whereas multi-agent systems emerge as the apparent source for abstractions, technologies, and methods for intelligent socio-technical systems. There, fundamental notions such as *trust* and *coordination* play a key role in the modelling and engineering of multi-agent systems: yet, even though they both insist on the same conceptual space – the space of agent and MAS *interaction* – the intrinsic nature of their relationship has not been explored deeply enough till now.

We try and provide a brief look at the two concepts, as they developed in the literature in the last decades, then focus on some possible dimensions around which their understanding and definition could possibly be framed in a comprehensive and coherent way; finally, we discuss a case study of a blockchain-based infrastructure providing for *trustworthy coordination*.

This paper was supported by the "ENGINES – ENGineering INtElligent Systems around intelligent agent technologies" project funded by the Italian MUR program "PRIN 2022" under grant number 20229ZXBZM.

1 Introduction

When looking back at the many works on both *trust* and *coordination* in the field of multi-agent systems (MAS), where the two notions mostly dwell and strive, clarity might not be the first thing to come to mind. On the one hand, largely diverse understanding of both terms have been widely used in the technical literature over the decades, in spite of the many efforts – e.g., [1, 2] for trust, [3, 4] for coordination, among the many others – to provide a common general definition, upon which computer scientists and engineers could rely for the modelling and design of their (intelligent) systems. On the other hand, both notions are fundamental, even essential, for the engineering of intelligent systems, in particular when *socio-technical system* (STS) henceforth) are concerned [5].

Three things, however, clearly emerge from a preliminary cursory examination of the literature. First, both trust and coordination are first of all non-technical terms, with a widespread acceptation, commonly used with relevant meaning in the context of human social systems. Then, they both have operational acceptations in the MAS field, which are essential not just for agent-based systems, but more generally for intelligent STS. Finally, whatever acceptations of the terms one might adopt, they are largely related notions, since they both deal with the dimension of *interaction* within MAS.

Accordingly, in the following we explore some of the intersections between the notions of trust and coordination in MAS and STS, pointing out some fundamental dimensions that could help illustrate their mutual strict relationship.

2 Trust and Coordination

2.1 The Many Ways of Coordination

Many are the viewpoints one could use to look at the issue of coordination, even when focusing on artificial systems only. One of the most general definitions originates from organisational science, where coordination is defined as the *management of dependencies between organisational activities* [6]: for instance, one workflow in an organisation may involve a secretary writing a letter, an official signing it, and another employee sending it to its final destination—each activity depending on each other for the success of the whole coordinated process. This notion straightforward generalises to coordination in MAS, where *agents* are the subjects whose activities need to be coordinated – the *coordinables* in [7] –, encapsulating the entities for which dependencies arise—like goals, actions, and plans. According to [8], the process of coordination involves two major tasks: *(i)* a detection of dependencies, *(ii)*, a decision about which coordination action to be taken. A coordination *mechanism* shapes the way that agents perform these tasks [9].

Shifting from a *modelling* to a *design* perspective, coordination is probably best conceived as the endeavour of *managing the space of interaction* [10] within a MAS. The design stance on coordination basically amounts at translating dependency detection and decision tasks into suitable interaction patterns. Diverse approaches such as planning, negotiation, norms, reputation management, and the like, aims at suitably *shaping* the space of interaction within a MAS by either *directly* making assumptions on agent behaviours or by *indirectly* modifying the environment around agents themselves.

Looking at the MAS interaction space, two distinct viewpoints can be taken: the individual agent one, and the external observer one [11]. The first perspective is what is called *subjective viewpoint* over interaction, where an agent looks at the MAS overall dynamics (other agents and environment), aiming at understanding and predicting the relevant behaviours there, so as to be able to coordinate its own behaviour with the rest of the system towards the achievement of its own goals. The second perspective is what is called *objective viewpoint*, where the space of interaction is basically seen as composed by the observable behaviour of all MAS agents and environment, as well as by all their interaction histories [12]. The objective viewpoint is typically identified as the *engineering perspective*, adopted by MAS designers to map out both individual and social interaction protocols and patterns so as to drive the overall system behaviour towards the achievement of the overall system goal. As a simple example of subjective vs. objective coordination, it is enough to consider the simple setting of the well-known *blocks world* [13] with many agent involved, as described in [14]. There, the subjective viewpoint of each agent pushes each of them at devising out the individual plan that would most likely allow the agent to reach its own goal, whatever other agents would do; whereas, the objective viewpoint favours the multiplan that would most likely drive the MAS towards the achievement of its global goal—for instance, by maximising the overall outcome for all planning agents in the MAS.

In particular, the ability to act on the MAS interaction space without affecting agents is a premise to any form of objective coordination. Thus, objective coordination first of all deals with MAS *environment*, and depends on the availability of suitable models of the agent environment, as well as on their proper embodiment within MAS *infrastructures* [15]. There, objective coordination takes the form of *coordination abstractions*, provided as run-time *coordination services* by the agent infrastructure [16].

2.2 The Many Facets of Trust

The common acceptation of the term *trust* is basically too wide to allow for a definition that could be at the same time comprehensive at the conceptual level and operable at the technical level. In [17], Castelfranchi e Falcone design a general framework accounting for the layered notion of trust, encompassing most (if not all) the many different acceptations of the term proposed and adopted in the literature. Among the many other results, the authors point out how "it is crucial to distinguish at least between two kinds and meanings of trust: (a) trust as psychological attitude of X towards Y relative to some possible desirable behavior or feature, (b) trust as the decision and the act of relying on, counting on, depending on Y." There, it is quite clear that a fundamental distinction has to be made between the inner, individual notion of agent trust, and its emergence in terms of the observable agent behaviour.

Even though, starting from [1], many definitions of trust mostly refer to the psychological viewpoint – which is also the nearest to the "folk acceptation" of the term –, a number of acceptations of "trust" either explicitly or implicitly refer to a "social" or "infrastructural" interpretation of the term. For instance, trust is quite a typically relevant notion in the security area. There, notions such as *trusted computing* [18] or *trusted third parties* [19] mostly refer to a general notion of *trusted infrastructure* or *trusted platform*, basically deferring trust to some external technological element (either hardware or software) rather than to software components (like agents).

More generally, while referring the reader to Chapter 12 of [17] for a comprehensive discussion of the many articulations of the relationship between trust and technology, the shared space where trust meets the technological infrastructure is especially relevant in the context of MAS, in particular in the field of electronic institutions [20], or *e-institutions*. There, the observable behaviours of autonomous agents are constrained by the rules enforced by the e-institution around which the global MAS behaviour is designed, and which is the implicit *locus* of trust for individual agents and MAS designers. Other approaches exist that deal with open MAS via non-centralised techniques, such as [21].

3 Trust vs. Coordination

Whereas almost any class or course on MAS or STS typically quotes trust and coordination as two fundamental and strictly related concepts, typically placing them in the areas of collaborative / cooperative approaches or *agreement technologies* [22], the number of scientific and technical works focussing specifically on both of them, and on their relationship, is relatively low. Works focussing on trust typically mention coordination as a sort of post-product of some form of trust between software or human components in a STS, whereas coordination approaches dealing with trust mostly deal with trust as a critical prerequisite of any agreement process [23]. The fact is that both notions are so wide and comprehensive that any attempt to systematically define their mutual relationship is doomed to failure when attempting to capture any sort of meaningful generality. On the other hand, a more fruitful approach could be undertook by drawing some significant parallels between the two notions, or, by devising out some shared criteria for their understanding—which is what we will do in the remainder of this section.

3.1 Dependence

Interdependency and coordination have been lasting topics in organisational research—and that has brought about the most widespread definition of coordination in MAS and STS as the the *management of dependencies* [6]. So, as we mentioned above, devising out the dependencies between agents – in terms of tasks, goals, actions, ... – is the conceptual and technical pre-condition for the coordination of both human and software agents.

In the dichotomy between modelling and design stances, the individual agent perception of dependence w.r.t. other agents is the premise to subjective coordination, whereas the overall network of dependencies among agents, and between individual agents and the whole system, accounts for the dimension of objective coordination .

The notion of dependence is clearly essential for trust, too. The notion of *dependence* and *trust networks* are discussed in [24] as the conceptual and technical representation of social dependence, whereas the role of dependence networks in the interaction between agents, with particular interest in the relationship between dependence and trust, are investigated in [25] There, different forms of *objective* and *subjective* dependence are defined and formalised, then used as the grounding for agent reasoning – in particular, in terms of negotiation power –, and for establishing operable trust relationships between agents. Even though coordination is not even mentioned in neither [24] nor [25], the same notions of dependence and trust networks could be easily seen as technical foundations for the whole coordination discourse in both MAS and STS.

Overall, the concept of dependence is at the very core of both the notions of trust and coordination. So, when dealing with the engineering of intelligent STS built out of autonomous components, modelling dependencies – e.g., in terms of dependence networks [25] – is definitely a necessary prerequisite for the management of both trust and coordination.[1]

3.2 Subjective vs. Objective Trust

As we mentioned above, the distinction between the subjective and the objective viewpoint is essential for understanding the many ways in which coordination articulates.[11]. On the one side, the subjective viewpoint of an autonomous agent takes on the system interaction space from the individual agent perspective so as to coordinate its own actions towards the achievement of its own goals—so it deals with the individual aspects of agent autonomy and intelligence. On the other side, the objective viewpoint oversees every interaction within a system, including both agent-to-agent and agent-to-environment interactions, and relates them to the global system behaviour—so it deals with the social aspects of system autonomy and intelligence through the system infrastructure (e.g., through the *coordination middleware* [15].

When adopting the design stance over intelligent STS, this distinction basically amounts at separating concerns: in short, when designing intelligent autonomous agents, engineers should adopt the individual subjective viewpoint over coordination; when designing intelligent STS overall, engineers should instead shift towards the global objective viewpoint, so as to shape the social interaction and the system infrastructure according to the system goal.

Whereas, to the best of our knowledge, a similar distinction has not explicitly emerged in the field of trust yet, it is however quite easy to set a line between the many works in the field mostly dealing with the modelling of trust within individual agents, and those focussing on the social or infrastructural aspects of trust. More specifically, most of

---

[1]In the literature, both *dependence* and *dependency* are widely adopted with no easy-to-discern rationale behind their use. In this paper, we choose to use the (uncountable) term "dependence" to label the conceptual relationship between distinct elements in a system, and "dependencies" to denote the diverse analytical forms that dependence can take within a system at either design or execution time.

the approaches dealing with the conceptualisation and formalisation of the notion of trust typically focus on subjective aspects—"X trusts Y" sorts of models. On the other hand, many approaches dealing with the technical and operable aspects of trust mostly focus on the technological infrastructure [18, 19], where trust-related issues can be faced at the system level.

It is then quite tempting to draw inspiration from coordination research, and observe that a suitably-shaped notion of *subjective trust* would easily help in modelling and engineering individual aspects of trust for autonomous agents, whereas a well-defined notion of *objective trust* would definitely provide a neat conceptual framework for modelling and engineering systemic aspects of trust within intelligent STS. Finally, this also suggests that middleware for MAS and STS should be designed so as to account for the objective viewpoint of both coordination and trust, handling social and environment issues in a coherent and comprehensive way.

4 Trustworthy Coordination: A Case Study

As a case study for our discussion, in the following we shortly illustrate the case of *trustworthy coordination*—that is, a coordination model implemented over a distributed infrastructure for trusted computing [18].

In the last years, *distributed ledger technologies* (DLT [26]) have emerged as the most reliable technology standard for trusted infrastructure for open distributed systems. In particular, *blockchain* technologies (BCT [26]) denote a heterogeneous number of technologies providing a novel way of dealing, essentially, with distributed asset tracking or identity management—whose most prominent use case is represented by cryptocurrencies. BCT provide secure decentralisation of both data and control in distributed systems, by effectively combining several results from cryptography, distributed consensus, game theory, and state machine replication. In particular, blockchain provides *untrusted parties* (such as autonomous agents in an open MAS) with a layer of trust enabling secure and trusted records along with secure and trusted transactions. Since trust, secured communications, and data consistency are critical issues for MAS engineering, BCT have gained traction as a straightforward way for injecting MAS with those features. Overall, as a middleware for MAS and STS, BCT currently represent one of the most prominent examples of infrastructure for objective trust.

BCT are also appealing from the more specific perspective of coordination [27] since they provide MAS with highly-desirable properties such as total ordering of events, data consistency, accountability of actions, identity management, and fault tolerance. This is why in [28] *smart contracts* [29] are exploited to implement *tuple-based* coordination [30] over Ethereum BCT [31]. More generally, in [27] the authors take a general look at the mainstream BCT implementations, and show how the general pattern of *coordination as a service* [16] can be implemented over BCT. The proposed computational model for smart contracts is there implemented on a custom blockchain called Tenderfone [32] that embodies both the conceptual and technical requirements for both objective coordination and objective trust.

5 Conclusion

In this work we take a fast look at two of the key issues in the modelling and engineering of intelligent STS—that is, *trust* and *coordination*. We try and overcome the fuzziness of the current definitions of the two notions, by focussing on their essential role in the management of *interaction* in complex intelligent systems, and on the common issues of *dependence* and *objective vs. subjective* viewpoints. Finally, we illustrate a case study where a coordination service is built over a trusted BCT so as to implement a *trustworthy coordination* infrastructure.

In the end, we expect that when AI research will rightly shift its focus back towards *interaction* as a core issue for intelligent system, both trust and coordination will be easily acknowledged as two of the key aspects in the area of intelligent systems engineering. Then, some general frameworks accounting for them both in a coherent and comprehensive way will be needed—expressing dependencies between components in a systematic way, and consistently capturing both subjective and objective viewpoints over trust and coordination.

Along that line, for instance, the A&A meta-model [33] could work as the source for an coherent notion of *trust artefact* as the core abstraction for objective trust provided by a suitably-shaped infrastructure—in the same way as *coordination artefacts* [34] work for objective coordination, and on the same line of negotiation and argumentation artefacts for MAS [35]. We are confident that future research will shed some light on that subject.

References

[1] Diego Gambetta, editor. *Trust: Making and Breaking Cooperative Relations*. Basil Blackwell, 1988.

[2] Cristiano Castelfranchi and Rino Falcone. Principles of trust for MAS. cognitive anatomy, social importance, and quantification. In *Proceedings of the 3rd International Conference on Multi Agent Systems (ICMAS 1998)*, pages 72–79. IEEE Computer Society, 1998.

[3] George A. Papadopoulos and Farhad Arbab. Coordination models and languages. In Marvin V. Zelkowitz, editor, *The Engineering of Large Systems*, volume 46 of *Advances in Computers*, pages 329–400. Academic Press, 1998.

[4] Andrea Omicini, Franco Zambonelli, Matthias Klusch, and Robert Tolksdorf, editors. *Coordination of Internet Agents*. Springer Berlin Heidelberg, March 2001.

[5] Brian Whitworth. Socio-technical systems. In Claude Ghaou, editor, *Encyclopedia of Human Computer Interaction*, pages 533–541. IGI Global, 2006.

[6] Thomas W. Malone and Kevin Crowston. The interdisciplinary study of coordination. *ACM Computing Surveys*, 26(1):87–119, 1994.

[7] Paolo Ciancarini. Coordination models and languages as software integrators. *ACM Computing Surveys*, 28(2):300–302, June 1996.

[8] Frank von Martial. *Co-ordinating Plans of Autonomous Agents*, volume 661 of *Lecture Notes on Artificial Intelligence*. Springer-Verlag, Berlin, 1992.

[9] Sascha Ossowski. *Co-ordination in Artificial Agent Societies. Social Structures and Its Implications for Autonomous Problem-Solving Agents*, volume 1535 of *Lecture Notes in Artificial Intelligence*. Springer-Verlag, 1999.

[10] Nadia Busi, Paolo Ciancarini, Roberto Gorrieri, and Gianluigi Zavattaro. Coordination models: A guided tour. In Omicini et al. [4], chapter 1, pages 6–24.

[11] Michael Schumacher. *Objective Coordination in Multi-Agent System Engineering. Design and Implementation*, volume 2039 of *LNCS*. Springer, April 2001.

[12] Peter Wegner. Why interaction is more powerful than algorithms. *Communications of the ACM*, 40(5):80–91, May 1997.

[13] Earl D. Sacerdoti. *A Structure for Plans and Behavior*, volume 3 of *Artificial Intelligence Series*. Elsevier, 1977.

[14] Andrea Omicini and Sascha Ossowski. Objective versus subjective coordination in the engineering of agent systems. In Matthias Klusch, Sonia Bergamaschi, Peter Edwards, and Paolo Petta, editors, *Intelligent Information Agents: An AgentLink Perspective*, volume 2586 of *Lecture Notes in Computer Science*, pages 179–202. Springer Berlin Heidelberg, 2003.

[15] Andrea Omicini, Alessandro Ricci, Mirko Viroli, Marco Cioffi, and Giovanni Rimassa. Multi-agent infrastructures for objective and subjective coordination. *Applied Artificial Intelligence*, 18(9–10):815–831, October–December 2004. Special Issue: Best papers from EUMAS 2003: The 1st European Workshop on Multi-agent Systems.

[16] Mirko Viroli and Andrea Omicini. Coordination as a service. *Fundamenta Informaticae*, 73(4):507–534, 2006. Special Issue: Best papers of FOCLASA 2002.

[17] Cristiano Castelfranchi and Rino Falcone. *Trust Theory: A Socio-Cognitive and Computational Model*. Wiley Publishing, 1st edition, 2010.

[18] Chris Mitchell, editor. *Trusted Computing*. The Institution of Engineering and Technology (IET), November 2005.

[19] Peter J. Skevington and Timothy P. Hart. Trusted third parties in electronic commerce. *BT Technology Journal*, 15(2):39–44, April 1997.

[20] Pablo Noriega and Carles Sierra. Electronic institutions: Future trends and challenges. In Matthias Klusch, Sascha Ossowski, and Onn Shehory, editors, *Cooperative Information Agents VI*, pages 14–17, Berlin,

Heidelberg, 2002. Springer Berlin Heidelberg.

[21] Yosi Mass and Onn Shehory. Distributed trust in open multi-agent systems. In Rino Falcone, Munindar Singh, and Yao-Hua Tan, editors, *Trust in Cyber-societies. Integrating the Human and Artificial Perspectives*, volume 2246 of *Lecture Notes in Artificial Intelligence*, pages 159–174. Springer Berlin Heidelberg, 2001.

[22] Sascha Ossowski, editor. *Agreement Technologies*, volume 8 of *Law, Governance and Technology Series*. Springer Netherlands, 2012.

[23] Sascha Ossowski. Coordination in multi-agent systems: Towards a technology of agreement. In Ralph Bergmann, Gabriela Lindemann, Stefan Kirn, and Michal Pĕchouček, editors, *Multiagent System Technologies*, volume 5244 of *Lecture Notes in Artificial Intelligence*, pages 2–12, Berlin, Heidelberg, 2008. Springer Berlin Heidelberg.

[24] Cristiano Castelfranchi and Rino Falcone. Grounding human machine interdependence through dependence and trust networks: Basic elements for extended sociality. *Frontiers in Physics*, 10:946095, 2022.

[25] Rino Falcone and Alessandro Sapienza. Dependence networks and trust in agents societies: Insights and practical implications. In Rino Falcone, Cristiano Castelfranchi, Alessandro Sapienza, and Filippo Cantucci, editors, *24th Workshop "From Objects to Agents" (WOA 2023)*, volume 3579, pages 106–122. CEUR-WS.org, November 2023.

[26] ISO/TC 307. Blockchain and distributed ledger technologies — vocabulary. ISO 22739:2020, International Organization for Standardization, 2020.

[27] Giovanni Ciatto, Stefano Mariani, Alfredo Maffi, and Andrea Omicini. Blockchain-based coordination: Assessing the expressive power of smart contracts. *Information*, 11(1):1–20, January 2020. Special Issue "Blockchain Technologies for Multi-Agent Systems".

[28] Giovanni Ciatto, Stefano Mariani, and Andrea Omicini. Blockchain for trustworthy coordination: A first study with Linda and Ethereum. In *2018 IEEE/WIC/ACM International Conference on Web Intelligence (WI)*, pages 696–703, December 2018.

[29] Loi Luu, Duc-Hiep Chu, Hrishi Olickel, Prateek Saxena, and Aquinas Hobor. Making smart contracts smarter. In *2016 ACM SIGSAC Conference on Computer and Communications Security (CCS'16)*, pages 254–269. ACM Press, 2016.

[30] Davide Rossi, Giacomo Cabri, and Enrico Denti. Tuple-based technologies for coordination. In Omicini et al. [4], chapter 4, pages 83–109.

[31] Gavin Wood. Ethereum: a secure decentralised generalised transaction ledger, 2014.

[32] Giovanni Ciatto, Alfredo Maffi, Stefano Mariani, and Andrea Omicini. Towards agent-oriented blockchains: Autonomous smart contracts. In Yves Demazeau, Eric Matson, Juan Manuel Corchado, and Fernando De la Prieta, editors, *Advances in Practical Applications of Survivable Agents and Multi-Agent Systems: The PAAMS Collection*, volume 11523 of *Lecture Notes in Computer Science*, pages 29–41. Springer International Publishing, June 2019.

[33] Andrea Omicini, Alessandro Ricci, and Mirko Viroli. Artifacts in the A&A meta-model for multi-agent systems. *Autonomous Agents and Multi-Agent Systems*, 17(3):432–456, December 2008. Special Issue on Foundations, Advanced Topics and Industrial Perspectives of Multi-Agent Systems.

[34] Andrea Omicini, Alessandro Ricci, and Mirko Viroli. Coordination artifacts as first-class abstractions for MAS engineering: State of the research. In Alessandro F. Garcia, Ricardo Choren, Carlos Lucena, Paolo Giorgini, Tom Holvoet, and Alexander Romanovsky, editors, *Software Engineering for Multi-Agent Systems IV: Research Issues and Practical Applications*, volume 3914 of *Lecture Notes in Computer Science*, pages 71–90. Springer Berlin Heidelberg, April 2006. Invited Paper.

[35] Enrico Oliva, Peter McBurney, Andrea Omicini, and Mirko Viroli. Argumentation and artifacts for negotiation support. *International Journal of Artificial Intelligence*, 4(S10):90–117, Spring 2010. Special Issue on Negotiation and Argumentation in Artificial Intelligence.

Reasoning about Trust in Information Transmitted by a Sequence of Agents

Robert Demolombe

No Affiliation

robert.demolombe@orange.fr

Abstract

When an agent receives information from another agent, who in turn has received it from a sequence of agents, the question the receiver asks himself is: "Can I trust the validity of this information?" The answer to this question depends on the knowledge he has about these agents.

More precisely, we distinguish three types of situation. The situation in which the receiver knows only the last agent who directly transmitted the information to him, the situation in which he knows all the agents in the sequence, and finally the situation in which he knows only one agent who serves as a referent, and who informs him about the other agents.

To model the reasoning that the receiver can make, we first present definitions of the different types of agent properties on which trust is based: validity, sincerity, competence, completeness, cooperativeness and vigilance. These notions are formalized in modal logic and we show, on examples, the reasoning that the receiving agent can do to conclude on the information he has received in different types of situations.

Finally, we analyze the case where trust is graded, and trust levels are qualitative. In this case, we present only a sketch of formalization.

Key Words. Agent communication, Trust, Modal Logic.

1 Introduction

When an agent x transmits information to an agent r (the receiver), and x tells r that x tells r through which sequence of agents it has received this information, it is useful for r to have information about the agents in the sequence, so as to know to what extent he can trust the information he has received.

The aim of this article is to analyze the types of reasoning that can be used to r can arrive at this or that conclusion concerning the information emitted by the first agent in the sequence.

Example: Ines told Robert that (John told her that (Kenzo told John that: "vaccine V cures cancer K")). Or, in more general notation, i told r that j told i, that k told j that ϕ.

In the example we have: i = Ines, j = John, k = Kenzo and ϕ = "the vaccine V cures cancer K".

If we accept certain assumptions about r's belief in i, we can deduce that: j has told i that (k has told j that ϕ).

If we accept certain additional assumptions about the belief in the agents, r can deduce that: k told j that ϕ.

And if we accept yet further assumptions about trust in agents, r can deduce that: ϕ.

There are also situations where r has not received this information from i, and where r can deduce: i has not told r that (j has told i that (k has told j that ϕ)).

In this case, if we accept other assumptions about the agents' beliefs, r can similarly deduce: $\neg\phi$.

This type of situation can be illustrated, with the previous example, when r has not received information from i expressing the fact that "vaccine V cures cancer K", and when r can deduce that it is false that "vaccine V cures cancer K".

We can imagine a similar example, where j is the newspaper The New Island Times, and k is the institution National Health System that informs about the truth of ϕ. This example shows that the agents transmitting information are not necessarily individuals, but can also be institutions.

The trust that these agents have in each other can have different structures. In this article, we consider the following three types of scenario, corresponding to different structures.

Type A scenarios. The receiver r knows only the agent x who transmitted the information, and it is his trust in x that enables him to conclude about ϕ. In this scenario, each agent in the sequence trusts the information transmitted to it by the previous one, and it transmits this information about trust to the next one, until x who transmits it to r. r's confidence in x then allows him to conclude on the information transmitted by the first agent.

Type B scenarios Receiver r knows all the agents in the sequence and trusts the information they transmit to each other. Agent x transmits to r only the sequence of transmissions, starting with the first agent to transmit ϕ. Agent r can then conclude, using the trust he has in each agent, on the information transmitted by the first agent.

Type C scenarios This scenario is similar to type B scenarios, but the information concerning the agents in the sequence is transmitted to r by a referent, rf, whom r trusts.

In the following sections, we first present the formal system for expressing information and the rules for reasoning about this information, as well as a general definition of the different types of trust that an agent can have in another agent. It is within this formal context that the three types of scenarios and their properties are then defined, and the final sections give a comparison with other work in the same field before presenting the conclusion.

2 Formal system

The different scenarios are formalized in the L language of modal logic defined below.

Langage. Definition of langage L.

$$\phi ::= p \mid \neg\phi \mid \phi \vee \phi \mid \phi \wedge \phi \mid Bel_i\phi \mid Inf_{i,j}\phi$$

Where p is an atomic formula, ϕ is a formula of L, and i and j denote agents.

The intuitive meaning of $Bel_i\phi$ is: agent i believes the information represented by formula ϕ.

The intuitive meaning of $Inf_{i,j}\phi$ is: agent i has transmitted to agent j information represented by the formula ϕ.

Axiomatics. The axiomatics of logic expressed in the language L, which will be used in the sequel is defined as follows:

- axiomatics of the Calculus of Propositions (CP).
- for the $Bel_i\phi$ modalities: logic (KD). Let be :
- - (K) $Bel_i(\phi) \wedge Bel_i(\phi \rightarrow \psi) \rightarrow Bel_i(\psi)$
- - (D) $Bel_i(\phi) \rightarrow \neg Bel_i(\neg\phi)$
- for $Inf_{j,i}\phi$: If $\vdash \phi \leftrightarrow \psi$ then $\vdash Inf_{j,i}\phi \leftrightarrow Inf_{j,i}\psi$
- for relationships between $Inf_{j,i}\phi$ and $Bel_i\phi^1$: (IB1) $Inf_{j,i}\phi \rightarrow Bel_i(Inf_{j,i}phi)$

Intuitively (IB1) means that the information transmitted is actually perceived by the recipient.

The meaning of these operators has not been defined by formal semantics. We felt it more intuitive to define the meaning of these operators through the reasoning they enable, as seen in the scenarios presented in the introduction.

3 Trust Definition

There are many definitions of trust (see [1, 2, 3, 4]), here we rely on the definition that is introduced in [5, 6].

The trust of an agent i in another agent j concerning certain properties expresses the fact that agent i believes that agent j satisfies these properties. Formally, trust is represented in a general way as

---

[1]If we accept (IB2) $\neg Inf_{j,i}\phi \rightarrow Bel_i(\neg Inf_{j,i}\phi)$, we can deduce that i believes $\neg Inf_{j,i}\phi$ for all ϕ formulas in the language, except for ψ formulas for which we have $Inf_{j,i}\psi$. So the set of formulas $Bel_i(\neg Inf_{j,i}\phi)$ is not defined because it is not explicitly restricted.

Rejecting IB2 is not contradictory with accepting, in a given situation, this, or that, particular hypothesis of the form: $\neg Inf_{j,i}p \rightarrow Bel_i(\neg Inf_{j,i}p)$, where p is a given formula.

$Bel_i(prop_j)$.[2].

In the context of information communication, the properties we consider are validity, sincerity and competence, and the dual properties of completeness, cooperativeness and vigilance. These properties are formally represented as shown below, and can be illustrated using an example where i is a person in an airport who wants to know whether a certain flight AF001 has arrived, and j is an airport agent who provides information on flight arrivals.

Validity. j validly informs i about ϕ if (if j transmits the information ϕ to i, then ϕ is true). Formally we have :

$Inf_{j,i} \to \phi$

Example. If j announces that flight AF001 has arrived, then it is true that it has arrived.

Sincerity. j is sincere when it informs i about ϕ if (if j transmits the information ϕ to i, then j believes ϕ). Formally we have :

$Inf_{j,i}\phi \to Bel_j(\phi)$

Example. If j announces that AF001 has arrived, then j believes it has arrived.

Competence. j is competent about ϕ if (if j believes ϕ, then ϕ is true). Formally we have :

$Bel_j(\phi) \to \phi$

Example. If j believes that AF001 has arrived, then it is true that it has arrived.

Completeness. j informs i completely about ϕ if (if ϕ is true, then j transmits the information ϕ to i). Formally we have :

$\phi \to Inf_{j,i}\phi$

This can be expressed equivalently as :

$\neg Inf_{j,i}\phi \to \neg\phi$

Example. If flight AF001 has arrived, then j announces that it has arrived. This is equivalent to: if j does not announce that AF001 has arrived, then AF001 has not arrived.

Cooperativity. j is cooperative about ϕ if (if j believes that ϕ

---

[2]It is important not to confuse assumptions that express, in a particular situation, the trust of one agent in another agent, with axiom schemes that would be true in all situations for all agents

is true, then j transmits the information ϕ to i). Formally we have :
$Bel_j(\phi) \rightarrow Inf_{j,i}\phi$

Example. If j believes that AF001 has arrived, then j announces that it has arrived.

Vigilance. j is vigilant about ϕ if (if ϕ is true, then j believes ϕ). Formally we have :
$\phi \rightarrow Bel_j(\phi)$

Example. If AF001 has arrived, then j believes it has arrived.

Note that these properties are not independent. The conjunction of sincerity and competence implies validity, and the conjunction of vigilance and cooperativeness implies completeness.

Of course, an agent may satisfy one of the properties of these conjunctions, but not both. For example, j may be sincere about AF001's arrival, but not competent about it, or he may be cooperative about AF001's arrival, but not vigilant about it.

4 Analysis of scenarios

4.1 Scenarios A

In the following, we'll denote r as the terminal agent receiving the information, and the other agents will be denoted by integers. The agent who transmitted the information to r is agent 1, who in turn received the information from agent 2, who received it from agent 3, and so on, in the general case, up to the first agent who transmitted the information and who is denoted by n. So the information transmitted by 1 expresses everything that the preceding agents have transmitted to each other. In what follows, we'll restrict ourselves to 3 agents.

The example in the introduction can be represented schematically as follows:

$Agent_3 \longrightarrow \phi_3 \longrightarrow Agent_2 \longrightarrow \phi_2 \longrightarrow Agent_1 \longrightarrow \phi_1 \longrightarrow Agent_r$

Where ϕ_3 denotes the proposition: "Vaccine V cures cancer K".

ϕ_2 means that agent 3 has told agent 2 that "Vaccine V cures cancer K", i.e. formally: $\phi_2 = Inf_{3,2}(\phi_3)$

ϕ_1 means that agent 2 has told agent 1 ϕ_2, i.e. formally: $\phi_1 =$

$Inf_{2,1}(\phi_2)$, and according to the definition of
ϕ_2: $\phi_1 = Inf_{2,1}(Inf_{3,2})(\phi_3))$.

And the fact that agent 1 has transmitted to agent r ϕ_1 is represented by: $Inf_{1,r}(\phi_1)$, i.e. $Inf_{1,r}(\phi_1) = Inf_{1,r}(Inf_{2,1}(Inf_{3,2}(\phi_3)))$. This means intuitively that 1 told r that 2 told 1 that 3 told 2 that "vaccine V cures cancer K".

Scenario A 1.

In this first scenario each agent transmits to the next only what the previous agent has transmitted. This is formally represented by :
$$\phi_i \overset{\text{def}}{=} Inf_{i+1,i}(\phi_{i+1}).$$

We therefore have: $\phi_1 \overset{\text{def}}{=} Inf_{2,1}(\phi_2)$ and $\phi_2 \overset{\text{def}}{=} Inf_{3,2}(\phi_3)$.

The scenario is expressed by the hypothesis: 1 has transmitted the information ϕ_1 to r, i.e.: (h) $Inf_{1,r}(\phi_1)$.

We deduce:

(1) $Bel_r(Inf_{1,r}(\phi_1))$ from (h) and IB1

(2) $Bel_r(Inf_{1,r}(Inf_{2,1}(Inf_{3,2}(\phi_3)))))$ from (1) and the definitions of ϕ_1 and ϕ_2.

Conclusion (2) means that r believes that 1 told him that 2 told him that 3 told him ϕ_3. Note that in this scenario r knows the succession of information transmissions, but cannot conclude whether ϕ_3 is true or false.

Scenario A 2.

In this scenario, each agent i transmits to the next the information ϕ_{i+1} that the previous $i + 1$ transmitted to him, and the fact that i has confidence in the validity of $i + 1$ for ϕ_{i+1}.

So we have :
$$\phi_i \overset{\text{def}}{=} Bel_i(Inf_{i+1,i}(\phi_{i+1}) \wedge (Inf_{i+1,i}(\phi_{i+1}) \rightarrow \phi_{i+1}))$$

What i believes about agent $i + 1$ is noted :
$$Inf_{i+1,i}(\phi_{i+1}) \wedge (Inf_{i+1,i}(\phi_{i+1}) \rightarrow \phi_{i+1}).$$

From which: $\phi_i \overset{\text{def}}{=} Bel_i(\psi_{i+1})$

In particular, let :
$$\phi_1 \overset{\text{def}}{=} Bel_1(Inf_{2,1}(\phi_2) \wedge (Inf_{2,1}(\phi_2) \rightarrow \phi_2))$$
$$\psi_2 \overset{\text{def}}{=} Inf_{2,1}(\phi_2) \wedge (Inf_{2,1}(\phi_2) \rightarrow \phi_2)$$
$$\phi_2 \overset{\text{def}}{=} Bel_2(Inf_{3,2}(\phi_3) \wedge (Inf_{3,2}(\phi_3) \rightarrow \phi_3))$$

$\psi_3 \overset{\text{def}}{=} Inf_{3,2}(\phi_3) \wedge (Inf_{3,2}(\phi_3) \rightarrow \phi_3)$

Assumptions. 1 has transmitted to r the information ϕ_1 and r has confidence in the validity of 1 for ϕ_1, i.e. :

(h1) $Inf_{1,r}(\phi_1)$

(h2) $Bel_r(Inf_{1,r}(\phi_1) \rightarrow \phi_1)$

We deduce:

(1) $Bel_r(Inf_{1,r}(\phi_1))$ from (h1) and IB1

(2) $Bel_r(\phi_1)$ from (1) (h2) and (K)

(3) $Bel_r(Bel_1(\psi_2))$ from (2) and the definition of ϕ_1.

(4)[3] $\vdash \psi_2 \rightarrow \phi_2$ from (CP)

(5) $Bel_r(Bel_1(\phi_2))$ from (3) (4) and (KD)

(6) $Bel_r(Bel_1(Bel_2(\psi_3)))$ from (5) and the definition of ϕ_2.

(7) $\vdash \psi_3 \rightarrow \phi_3$ from (CP)

(8) $Bel_r(Bel_1(Bel_2(\phi_3))))$ from (6) (7) and (KD)

Conclusion (8) means that r believes that 1 believes that 2 believes ϕ_3. In this scenario r is informed about what each agent in the sequence believes, but cannot conclude whether ϕ_3 is true or false.

4.1.1 Explicit scenario A.

In this scenario, each agent i transmits to the next the information ϕ_{i+1} that the previous $i+1$ transmitted to him, the fact that i has confidence in the validity of $i+1$ for ϕ_{i+1}, and in the fact that $i+1$ is competent about ψ_{i+2}. This is formally represented by:

$\phi_i \overset{\text{def}}{=} Bel_i(Inf_{i+1,i}(\phi_{i+1}) \wedge (Inf_{i+1,i}(\phi_{i+1}) \rightarrow \phi_{i+1}) \wedge$
$(Bel_{i+1}(\psi_{i+2}) \rightarrow \psi_{i+2}))$

What i believes about agent $i+1$ is noted :
$Inf_{i+1,i}(\psi_{i+1})(Inf_{i+1,i}(\psi_{i+1}) \rightarrow \psi_{i+1})(Bel_{i+1}(\psi_{i+2}) \rightarrow \psi_{i+2})$

Hence: $\phi_i \overset{\text{def}}{=} Bel_i(\psi_{i+1})$.

$\phi_1 \overset{\text{def}}{=} Bel_1(Inf_{2,1}(\phi_2) \wedge (Inf_{2,1}(\phi_2) \rightarrow \phi_2) \wedge (Bel_2(\psi_3) \rightarrow \psi_3))$

$\psi_2 \overset{\text{def}}{=} Inf_{2,1}(\phi_2) \wedge (Inf_{2,1}(\phi_2) \rightarrow \phi_2) \wedge (Bel_2(\psi_3) \rightarrow \psi_3)$

$\phi_1 \overset{\text{def}}{=} Bel_1(\psi_2)$

$\phi_2 \overset{\text{def}}{=} Bel_2(Inf_{3,2}(\phi_3) \wedge (Inf_{3,2}(\phi_3) \rightarrow \phi_3))$

---

[3]The symbol $\vdash \phi$ means that ϕ is a theorem in the logic that has been chosen.

$\psi_3 \overset{\text{def}}{=} Inf_{3,2}(\phi_3) \wedge (Inf_{3,2}(\phi_3) \rightarrow \phi_3)$

$\phi_2 \overset{\text{def}}{=} Bel_2(\psi_3)$

Note: the definition of ψ_3 is different from that of ψ_2 because agent 3 has no predecessor.

Assumptions.

(h1) $Inf_{1,r}(\phi_1)$

(h2) $Bel_r(Inf_{1,r}(\phi_1) \rightarrow \phi_1)$

(h3) $Bel_r(Bel_1(\psi_2) \rightarrow \psi_2)$

In the example presented at the beginning of the article, the meaning of (h1) is : Ines has conveyed to Robert the fact that she believes (John has conveyed to her ϕ_2, and John is valid for ϕ_2 and John is competent about what Kenzo believes).

Assumptions (h2) and (h3) express that r has confidence in the validity of 1 for ϕ_1 and r has confidence in the competence of 1 for ψ_2.

(1) $Bel_r(Inf_{1,r}(\phi_1))$ from (h1) and (IB1)

(2) $Bel_r(Bel_1(\psi_2))$ from (1) (h2) (KD) and the definition of ϕ_1.

(3) $Bel_r(\psi_2)$ from (2) (h3) and (KD)

(4) $\vdash \psi_2 \rightarrow \phi_2$ from (CP)

(5) $Bel_r(\phi_2)$ from (3) (4) and (KD)

(6) $Bel_r(Bel_2(\psi_3))$ from (5) and the definition of ϕ_2.

(7) $\vdash \psi_2 \rightarrow (Bel_2(\psi_3) \rightarrow \psi_3)$ from (CP)

(8) $Bel_r(Bel_2(\psi_3) \rightarrow \psi_3)$ from (3) (7) and (KD)

(9) $Bel_r(\psi_3)$ from (6) (8) and (KD)

(10) $\vdash \psi_3 \rightarrow \phi_3$ from (CP)

(11) $Bel_r(\phi_3)$ from (9) (10) and (KD)

The information ϕ_1 transmitted by 1 expresses the entire succession of information that the agents have transmitted to each other from the first. The reasoning of r consists in breaking down the history of this succession step by step. Thus (3) expresses what r believes about the beliefs transmitted by 2, and (9) expresses what r believes about the beliefs transmitted by 3. Finally, (11) shows that r believes that ϕ_3 is true.

4.1.2 Scenario A implicit

In this scenario, the receiver receives no information and believes that if the information he is interested in were true, agent 1 would have passed it on to him. This allows him to conclude, under certain assumptions, that the information is false.

More precisely, on the basis of certain assumptions, and from what 1 has not transmitted, for each agent $i+1$, r can deduce:

- $i+1$ has not transmitted to the next i the information ϕ_{i+1}, and
- $i+1$ is complete for what it believes (i.e. ϕ_{i+1}), and
- $i+1$ is vigilant for what $i+2$ believes (i.e. ϕ_{i+2}).

Notation: $\phi_i \stackrel{\text{def}}{=} Bel_i(\psi_{i+1})$.

$\neg\psi_{i+1} \stackrel{\text{def}}{=} \neg Inf_{i+1,i}(\phi_{i+1}) \wedge (\neg Inf_{i+1,i}(\phi_{i+1}) \to \neg\phi_{i+1}) \wedge (\neg Bel_{i+1}(\psi_{i+2}) \to \neg\psi_{i+2})$

In particular:

$\phi_1 \stackrel{\text{def}}{=} Bel_1(\psi_2)$

$\neg\psi_2 \stackrel{\text{def}}{=} \neg Inf_{2,1}(\phi_2) \wedge (\neg Inf_{2,1}(\phi_2) \to \neg\phi_2) \wedge (\neg Bel_2(\psi_3) \to \neg\psi_3)$

$\phi_2 \stackrel{\text{def}}{=} Bel_2(\psi_3)$

$\neg\psi_3 \stackrel{\text{def}}{=} \neg Inf_{3,2}(\phi_3) \wedge (\neg Inf_{3,2}(\phi_3) \to \neg\phi_3)$

Note: the definition of $\neg\psi_3$ is different from that of $\neg\psi_2$ because agent 3 has no predecessor.

Assumptions.

(h1) $\neg Inf_{1,r}(\phi_1)$

(k1) $\neg Inf_{1,r}(\phi_1) \to Bel_r(\neg Inf_{1,r}(\phi_1)$

(h2) $Bel_r(\neg Inf_{1,r}(\phi_1) \to \neg\phi_1)$

(h3) $Bel_r(\neg Bel_1(\psi_2) \to \neg\psi_2)$

In the example presented at the beginning, the meaning of (h1) is: Ines didn't pass on to Robert the fact that she believes that (John didn't pass on ϕ_2 and John is complete for ϕ_2 and John is vigilant about what Kenzo believes).

Assumption (k1) can be interpreted as a dual property of (IB1) for the special case where 1 does not inform r about ϕ_1. Assumption (h2) expresses r's confidence in 1's cooperativeness for ϕ_1 and assumption (h3) expresses r's confidence in 1's vigilance for ψ_2.

We have :

(1) $Bel_r(\neg Inf_{1,r}(\phi_1))$ from (h1) (k1) and (CP)

(2) $Bel_r(\neg Bel_1(\psi_2)))$ from (1) (h2) (KD) and the definition of ϕ_1.

(3) $Bel_r(\neg\psi_2)$ from (2) (h3) and (KD)

(4) $\vdash \neg\psi_2 \rightarrow \neg\phi_2$ from (CP)

(5) $Bel_r(\neg\phi_2)$ from (3) (4) and (KD)

(6) $Bel_r(\neg Bel_2(\psi_3)))$ from (5) and the definition of ϕ_2.

(7) $\vdash \neg\psi_2 \rightarrow (\neg Bel_2(\psi_3) \rightarrow \neg\psi_3)$ from (CP)

(8) $Bel_r(\neg Bel_2(\psi_3) \rightarrow \neg\psi_3)$ from (3) (7) and (KD)

(9) $Bel_r(\neg\psi_3)$ from (6) (8) and (KD)

(10) $\vdash \neg\psi_3 \rightarrow \neg\phi_3$ from (CP)

(11) $Bel_r(\neg\phi_3)$ from (9) (10) and (KD)

Comment. r's reasoning is similar to the reasoning presented in explicit scenario A, with the difference that ϕ_1 expresses what agent 1 about what has not been transmitted. So we see that step (3) in the demonstration expresses what r believes about 2, and this belief says: that 1 didn't receive the information ϕ_2 from 2, that 1 is complete with respect to 2 about ϕ_2, and that 2 is vigilant about ψ_3.

4.2 Scenarios B

In these scenarios, we assume that r knows each of the agents in the sequence and trusts their validity for the information they pass on to the next.

4.2.1 Explicit scenario B

Notation.
$$\phi_i \overset{\text{def}}{=} Inf_{i+1,i}(\phi_{i+1})$$

The assumptions about r's confidence in the validity of each agent in the sequence are of the form :

(h_i) $Bel_r(Inf_{i+1,i}(\phi_{i+1}) \rightarrow \phi_{i+1})$

In particular:
$$\phi_1 \overset{\text{def}}{=} Inf_{2,1}(\phi_2)$$
$$\phi_2 \overset{\text{def}}{=} Inf_{3,2}(\phi_3)$$

Assumptions.

(h) $Inf_{1,r}(\phi_1)$

(h1) $Bel_r(Inf_{1,r}(\phi_1) \rightarrow \phi_1)$

(h2) $Bel_r(Inf_{2,1}(\phi_2) \rightarrow \phi_2)$

(h3) $Bel_r(Inf_{3,2}(\phi_3) \rightarrow \phi_3)$

In the example presented at the beginning, the meaning of (h1), (h2) and (h3) is :

Robert believes in Ines' validity about ϕ_1.

Robert believes in John's validity about ϕ_2.

Robert believes in Kenzo's validity about ϕ_3.

We have:

(1) $Bel_r(Inf_{1,r}(\phi_1))$ from (h) and (IB1)

(2) $Bel_r(\phi_1)$ from (1) (h1) and (KD)

(3) $Bel_r(Inf_{2,1}(\phi_2))$ from (2) and the definition of ϕ_1.

(4) $Bel_r(\phi_2)$ from (3) (h2) and (KD)

(5) $Bel_r(Inf_{3,2}(\phi_3))$ from (4) and the definition of ϕ_2.

(6) $Bel_r(\phi_3)$ from (5) (h3) and (KD)

4.2.2 Implicit scenario B.

In this scenario, r trusts the completeness of the agents.

We have the definitions:

$$\phi_i \stackrel{\text{def}}{=} Inf_{i+1,i}(\phi_{i+1})$$

Assumptions about r's confidence in the completeness of each agent in the sequence are of the form:

(h_i) $Bel_r(\neg Inf_{i+1,i}(\phi_{i+1}) \rightarrow \neg\phi_{i+1})$

In particular, let :

$$\phi_1 \stackrel{\text{def}}{=} Inf_{2,1}(\phi_2)$$

$$\phi_2 \stackrel{\text{def}}{=} Inf_{3,2}(\phi_3)$$

Assumptions.

(h) $\neg Inf_{1,r}(\phi_1)$

(k) $\neg Inf_{1,r}(\phi_1) \rightarrow Bel_r(\neg Inf_{1,r}(\phi_1)$

(h1) $Bel_r(\neg Inf_{1,r}(\phi_1) \rightarrow \neg\phi_1)$

(h2) $Bel_r(\neg Inf_{2,1}(\phi_2) \rightarrow \neg\phi_2)$

(h3) $Bel_r(\neg Inf_{3,2}(\phi_3) \rightarrow \neg\phi_3)$

We have:

(1) $Bel_r(\neg Inf_{1,r}(\phi_1))$ from (h) (k) and (CP)

(2) $Bel_r(\neg\phi_1)$ from (1) (h1) and (KD)

(3) $Bel_r(\neg Inf_{2,1}(\phi_2))$ from (2) and the definition of ϕ_1.

(4) $Bel_r(\neg\phi_2)$ from (3) (h2) and (KD)

(5) $Bel_r(\neg Inf_{3,2}(\phi_3))$ from (4) and the definition of ϕ_2.

(6) $Bel_r(\neg\phi_3)$ from (5) (h3) and (KD)

4.3 Scenarios C.

It is assumed that r does not know the agents in the sequence, but r knows an agent rf who acts as a referent to inform him about the agents in the sequence, and r trusts the information that the referent rf gives him.

4.3.1 Scenario C explicit

We have the definitions:

$k_0 \stackrel{\text{def}}{=} Inf_{1,r}(\phi_1) \to \phi_1$

For $i \in [1, n]$: $k_{i+1} \stackrel{\text{def}}{=} Inf_{i+1,i}(\phi_{i+1}) \to \phi_{i+1}$.

$k \stackrel{\text{def}}{=} \bigwedge_{i\in[0,n]} k_i$

Intuitively k expresses that all agents in the sequence are valid for the information they transmit to the next.

In particular, let :

$k_0 \stackrel{\text{def}}{=} Inf_{1,r}(\phi_1) \to \phi_1$

$k_1 \stackrel{\text{def}}{=} Inf_{2,1}(\phi_2) \to \phi_2$

$k_2 \stackrel{\text{def}}{=} Inf_{3,2}(\phi_3) \to \phi_3$

Assumptions.

(h') $Inf_{rf,r}(Bel_{rf}(k))$

(h'1) $Bel_r(Inf_{rf,r}(Bel_{rf}(k)) \to Bel_{rf}(k))$

(h'2) $Bel_r(Bel_{rf}(k) \to k)$

Assumption (h'1) expresses that r has confidence in the validity of referent rf about what the referent believes about the validity of the agents in the sequence about what they transmit to the next. Hypothesis (h'2) expresses that r has confidence in rf's competence about what he believes about the validity of the agents in the sequence.

If we call Sapiens the referent rf, with the example at the beginning of the article the meaning of (h'1) is: Robert trusts Sapiens' validity about what Sapiens believes about the validity of Ines, John and Kenzo.

We have:

(1) $Bel_r(Inf_{rf,r}(Bel_{rf}(k)))$ from (h') and (IB1)

(2) $Bel_r(Bel_{rf}(k))$ from (1) (h'1) and (KD)

(3) $Bel_r(k)$ from (2) (h'2) and (KD)

(h1) $Bel_r(Inf_{1,r}(\phi_1) \to \phi_1)$ from (3) (KD) and the definition of k_0.

(h2) $Bel_r(Inf_{2,1}(\phi_2) \to \phi_2)$ from (3) (KD) and the definition of k_1.

(h3) $Bel_r(Inf_{3,2}(\phi_3) \to \phi_3)$ from (3) (KD) and the definition of k_2.

Conclusions (h1), (h2) and (h3) show that referent rf has passed on to r the assumptions that r accepts in explicit scenario B to deduce $Bel_r(\phi_3)$.

4.3.2 Scenario C implicit.

We have the definitions:

$k_0 \stackrel{\text{def}}{=} \neg Inf_{1,r}(\phi_1) \to \neg\phi_1$

For $i \in [1,n]$: $k_{i+1} \stackrel{\text{def}}{=} \neg Inf_{i+1,i}(\phi_{i+1}) \to \neg\phi_{i+1}$.

$k \stackrel{\text{def}}{=} \bigwedge_{i \in [0,n]} k_i$

In particular, let :

$k_0 \stackrel{\text{def}}{=} \neg Inf_{1,r}(\phi_1) \to \neg\phi_1$

$k_1 \stackrel{\text{def}}{=} \neg Inf_{2,1}(\phi_2) \to \neg\phi_2$

$k_2 \stackrel{\text{def}}{=} \neg Inf_{3,2}(\phi_3) \to \neg\phi_3$

Assumptions.

(h') $Inf_{rf,r}(Bel_{rf}(k))$

(h'1) $Bel_r(Inf_{rf,r}(Bel_{rf}(k)) \to Bel_{rf}(k))$

(h'2) $Bel_r(Bel_{rf}(k) \to k)$

We have:

(1) $Bel_r(Inf_{rf,r}(Bel_{rf}(k)))$ from (h') and (IB1)

(2) $Bel_r(Bel_{rf}(k))$ from (1) (h'1) and (KD)

(3) $Bel_r(k)$ from (2) (h'2) and (KD)

(h1) $Bel_r(\neg Inf_{1,r}(\phi_1) \rightarrow \neg\phi_1)$ from (3) (KD) and the definition of k_0.

(h2) $Bel_r(\neg Inf_{2,1}(\phi_2) \rightarrow \neg\phi_2)$ from (3) (KD) and the definition of k_1.

(h3) $Bel_r(\neg Inf_{3,2}(\phi_3) \rightarrow \neg\phi_3)$ from (3) (KD) and the definition of k_2.

The conclusions (h1), (h2) and (h3) are the assumptions used by r in the implicit B scenario to deduce $Bel_r(\neg\phi_3)$.

5   Graded Trust

In the foregoing, we have considered only two types of situation: agent i trusts agent j, or i does not trust j.

In practice, however, there are many situations in which an agent has more or less trust in another agent. It is these situations that we will analyze next.

Since trust is essentially based on the notion of belief, we will restrict ourselves to the study of graded beliefs.

We consider degrees of belief represented by qualitative levels that are in total linear order. From an intuitive point of view, the highest level can be interpreted as the fact, for example, that agent i is certain that ϕ is true. A lower level can be interpreted as i being almost certain that ϕ is true, and so on. A very low level can be interpreted as i's belief being very uncertain.

Finally, the lowest level can be interpreted as i having no opinion that ϕ is true, i.e. being in the dark. Note that it would be wrong to deduce that i has a certain level of belief that ϕ is false. Indeed, these levels of belief represent an agent's **subjective**[4] attitude to the

---

[4]We would like to thank Cristiano Castelfranchi for drawing our attention to the subjective nature of these beliefs during our presentation of [15].

fact that ϕ is true, but they are not based on counting a number of situations where ϕ is true, as is the case with probabilities.

Formalisation.

The language L is completed by modalities of the form: $Bel_i^x\phi$, whose intuitive meaning is: the strength of agent i's belief that ϕ is true is x.

Axiomatics.

In addition to the axioms introduced earlier, we have:

(K') $Bel_i^x\phi \wedge Bel_i^y(\phi \to \psi) \to Bel_i^z\psi$ and z = min{x,y}

We'll show examples of reasoning in this logic by considering, for simplicity, only two agents: agent j, who transmits information, and agent i, who receives information.

Assumptions.

(h1) $Inf_{j,i}\phi$

(h2) $Inf_{j,i}\phi \to Bel_i^g(Inf_{j,i}\phi)$

(h3) $Bel_i^h(Inf_{j,i}\phi \to Bel_j^k\phi)$

(h4) $Bel_i^{h'}(Bel_j^k\phi \to \phi)$

(1) $Bel_i^g(Inf_{j,i}\phi)$ from (h1) (h2) and (CP)

(2) $Bel_i^l(Bel_j^k)$ and l = min{g,h} from (1) (h3) and (K')

(3) $Bel_i^{l'}(\phi)$ and l' = min {l,h'} from (2) (h4) and (K')

Assumption (h3) expresses i's graded confidence in j's sincerity, and (h4) expresses i's confidence in j's competence.

Example. It is assumed that j is a doctor, i is a patient who consults j, and the meaning of ϕ is: "i has cancer".

The strengths of the beliefs are noted: 1, 2, 3, 4 and 5, where the maximum value is 5 and the minimum value 1 (these values are ordinal only).

In this scenario, where information is transmitted explicitly (h1) means that j has told i that he has cancer. It is assumed that the transmission works normally, so in (h2) we have g = 5. Assumption (h3) expresses i's confidence in j's sincerity, and it is assumed that i has high confidence in this sincerity, hence h = 4 and k = 4. Assumption (h4) expresses i's confidence in j's competence. It is assumed that this confidence is strong and we have h' = 4.

314

Under these assumptions, l' = 4.

In the case below, assumption (h3) expresses i's graduated confidence in j's cooperativeness, and (h4) expresses i's confidence in j's vigilance.

Assumptions.

(h1) $\neg Inf_{j,i}\phi$

(h2) $\neg Inf_{j,i}\phi \rightarrow Bel_i^g(\neg Inf_{j,i}\phi)$

(h3) $Bel_i^h(\neg Inf_{j,i}\phi \rightarrow \neg Bel_j^k\phi)$

(h4) $Bel_i^{h'}(\neg Bel_j^k\phi \rightarrow \neg\phi)$

.

(1) $Bel_i^g(\neg Inf_{j,i}\phi)$ from (h1), (h2) and (CP)

(2) $Bel_i^l(\neg Bel_j^k)$ and l = min{g,h} from (1), (h3) and (K')

(3) $Bel_i^{l'}(\neg\phi)$ and l' = min {l,h'} from (2), (h4) and (K')

Example. In this scenario, the information is not explicitly transmitted. Patient j has no doubt that the information has not been transmitted, so g = 5. Hypothesis (h3) expresses i's confidence in j's cooperativeness. i's confidence is assumed to be low because i thinks it's possible that j won't tell i that he has cancer so as not to traumatize him. We therefore have h = 3. Hypothesis (h4) expresses j's vigilance, and i has good confidence in j's vigilance. Therefore h' = 4.

Under these assumptions, l' = 3.

6 Related Works

Work on trust in the cognitive sciences [1, 2] and in philosophy [7, 4, 3] shows that there are many definitions of trust. In particular, in [2], Castelfranchi and Falcone give a comprehensive presentation of the important issues surrounding the notion of trust and their possible solutions.

In more detail, [8] shows that trust can be based on both a mental and a social attitude, and that it is founded on both morality and reputation. In [9, 10] trust involves the different categories of trustee.

The formalization is expressed in predicate logic in the first case, and in fuzzy sets in the second.

The work presented in [11, 12] is more directly related to what we have proposed in this paper, as it concerns different notions of trust transitivity. From a formal point of view, these notions are expressed in modal logics. In [1], on the other hand, the level of trust is defined by subjective probabilities, i.e. probabilities based solely on the beliefs of the trusting agent. Note that, in this approach, if the probability of ϕ is low, the probability of $\neg\phi$ is necessarily high, which means that the agent is never in the dark about whether ϕ is true or false. Furthermore, in [13] it is shown that confidence does not simply boil down to subjective probability.

We will now look at works in which the definitions adopted to formalize transitivity have certain points in common with what we have presented.

In [14, 5] trust and transitivity have the same definition as presented above, but only type A scenarios are formalized. On the other hand, they are formalized for any number of agents, and Theorems 4, 5 and 6 generalize what was presented for 3 agents. For graded beliefs, the definitions we have given in [15, 16] correspond to "objective" levels, unlike those given above.

In [17] Huang and Fox propose a notion of transitivity based on a different definition of trust. Two kinds of trust are presented. The first means that the agent has confidence in the trustee's competence. The second expresses confidence in the fact that the present situation has been caused by the trustee. But these definitions are problematic because they assume that trust in the trustee applies to any proposition. This is unlikely to be the case in real-life situations.

7 Conclusion

Modal logic provides precise and detailed definitions of trust. We can choose other definitions than those we have presented, but if we accept these definitions we can rigorously deduce the consequences of hypotheses that describe complex situations.

More specifically, we have considered situations where an agent receives information transmitted by a sequence of agents. Several scenarios have been considered, depending on whether the last receiver has this or that type of trust in the last agent who transmitted the information, or when it has certain types of trust in each of the agents who intervened in the sequence, or in a trusted referent who informs it about the agents in the sequence.

In the analysis we have presented, an agent's trust in another agent concerns a particular piece of information represented by ϕ. More often than not, trust concerns a set of information related to a theme. For example, trusting a doctor's competence for all cancer-related information. This extension could be made using the results presented in [18]. Another extension would be to make explicit in the definitions the fact that trust only applies in a particular context.

References

[1] C. Castelfranchi and R. Falcone. Social trust: a cognitive approach. In C. Castelfranchi and Y-H. Tan, editors, *Trust and Deception in Virtual Societies*. Kluwer Academic Publisher, 2001.

[2] C. Castelfranchi and R. Falcone. *Trust Theory: A Socio-Cognitive and Computational Model*. Wiley, 2010.

[3] A.J.I. Jones. On the concept of trust. *Decision Support Systems*, 33, 2002.

[4] A.J.I. Jones and B.S. Firozabadi. On the characterisation of a trusting agent. Aspects of a formal approach. In C. Castelfranchi and Y-H. Tan, editors, *Trust and Deception in Virtual Societies*. Kluwer Academic Publisher, 2001.

[5] R. Demolombe. To trust information sources: a proposal for a modal logical framework. In C. Castelfranchi and Y-H. Tan, editors, *Trust and Deception in Virtual Societies*. Kluwer Academic Publisher, 2001.

[6] R. Demolombe. Reasoning about trust: a formal logical framework. In C. Jensen, S. Poslad, and T. Dimitrakos, editors, *Trust management: Second International Conference iTrust (LNCS 2995)*. Springer Verlag, 2004.

[7] M. Bacharach and D. Gambetta. Trust as type detection. In C. Castelfranchi and Y-H. Tan, editors, *Trust and Deception in Virtual Societies*. Kluwer Academic Publisher, 2001.

[8] C. Castelfranchi and R. Falcone. Social Trust: cognitive anatomy, social importance, quantification and dynamics. In *Workshop on "Deception, Fraud and Trust in Agent Societes", at Autonomous Agents* . 1998.

[9] R. Falcone, A. Sapienza, and C. Castelfranchi. The relevance of categories for trusting information sources. *ACM Trans. Internet Techn.*, 2015.

[10] C. Castelfranchi, R. Falcone, and G. Pezzulo. Trust in information sources as a source for trust: a fuzzy approach. In *The Second International Joint Conference on Autonomous Agents & Multiagent Systems, AAMAS*, 2003.

[11] R. Falcone and C. Castelfranchi. Trust and transitivity: How trust-transfer works. Volume 156 of *Advances in Intelligent and Soft Computing*, 2012.

[12] D. Adamatti, C. Castelfranchi, and R. Falcone. Structural transitivity of trust in academic social networks using agent-based simulation. In *(EUMAS 2013) 2013*.

[13] C. Castelfranchi and R. Falcone. Trust is much more than subjective probability: Mental components and sources of trust. In *Hawaii International Conference on System Sciences (HICSS-33)*, 2000.

[14] R. Demolombe. Transitivity and propagation of trust in information sources. An Analysis in Modal Logic. In J. Leite and P. Torroni and T. Agotnes and L. van der Torre, editor, *Computational Logic in Multi-Agent Systems, LNAI 6814*. Springer Verlag, 2011.

[15] R. Demolombe. Graded Trust. In R. Falcone and S. Barber and J. Sabater-Mir and M.Singh, editor, *Proceedings of the Trust in Agent Societies Workshop at AAMAS 2009*, 2009.

[16] R. Demolombe and C-J. Liau. A logic of graded trust and belief fusion. In C. Castelfranci and R. Falcone, editors, *Proc. of 4th Workshop on Deception, Fraud and Trust*, 2001.

[17] Jingwei Huang and Mark S. Fox. An ontology of trust: formal semantics and transitivity. In *International Conference on Evolutionary Computation*, 2006.

[18] R. Demolombe and L. Fariñas del Cerro. Information about a given entity: from semantics towards automated deduction. *Journal of Logic and Computation*, 20(6), 2010.

TRUST AND COGNITION IN HUMAN AND HYBRID SOCIETIES

FILIPPO CANTUCCI

Institute of Cognitive Sciences and Technologies, National Research Council of Italy (ISTC-CNR), 00196 Rome, Italy

filippo.cantucci@istc.cnr.it

ALESSANDRO SAPIENZA

Institute of Cognitive Sciences and Technologies, National Research Council of Italy (ISTC-CNR), 00196 Rome, Italy

alessandro.sapienza@istc.cnr.it

Abstract

Trust is a complex mental state that acts as the glue for any form of collaboration, both among humans and between humans and machines. In human societies, trust is essential for social cohesion, cooperation, and the smooth functioning of communities. Similarly, as artificial systems have evolved beyond their initial complexity, this evolution has profoundly changed human expectations during interactions with these systems. Consequently, this has necessitated the evolution of computational models based on trust, both at the multi-agent system (MAS) level and the human-agent interaction (HAI) level. A socio-cognitive approach to trust modeling has become increasingly necessary in the definition and design of these computational

This work has been partially supported by the project FAIR - Future Artificial Intelligence Research (MIUR-PNRR) -, by the project TrustPACTX - Design of the Hybrid Society Humans-Autonomous Systems: Architecture, Trustworthiness, Trust, EthiCs, and EXplainability (the case of Patient Care) (MIUR-PRIN) -, and by the ICRAS project - Behavioral Interventions for Resilience to Environmental and Health Risks (MIUR-PNRR).

models. In this work, we analyzed some of our main contributions to modeling trust in MAS and HAI, particularly in Human-Robot Interaction (HRI), conducted under the guidance of Rino Falcone, whose contributions have been fundamental in understanding trust and cognition in the fields of multi-agent systems and human-artificial agent collaboration.

1 Introduction

Trust has always been a key concept in both social and non-social interactions, whether among humans or artificial systems, as evidenced by the extensive body of literature on this concept [1, 2, 3]. In human societies, trust is a foundational element that supports cooperation, social cohesion, and the functioning of institutions. It underpins interactions between individuals and groups, helping to reduce uncertainty and fostering collaboration, which is essential for the stability and progress of any society. In fact, this applies not only to human societies but also to interactional dynamics with and between artificial agents.

In recent years, the application of the concept of trust has exponentially increased, also thanks to the relatively recent emergence of trustworthy AI. The integration of artificial systems, such as AI-based agents, virtual assistants, and robots, into daily life raises profound questions about the nature and significance of trust [4, 5]. As these systems become integral to human environments, establishing trust-based interactions significantly influences their acceptance, utilization, and overall effectiveness. Artificial systems have evolved, leading to changes in human expectations during interactions with these systems. People now expect intelligent systems to be proactive, similar to human collaboration, where partners can autonomously take action and adapt to varying situations, especially within social contexts. Consequently, adaptive autonomy and initiative have become essential, even though they introduce challenges related to failures, errors, conflicts, and vulnerability. This ongoing evolution of robots and AI adds new perspectives and challenges to the understanding of trust within Multi-Agent Systems (MAS) [6] and Human-Agent Interaction

(HAI) [7].

Introducing trust in these contexts is crucial for managing interaction and cooperation. However, this raises several issues: how should trust be managed and modeled? When and how is it appropriate to trust? What are the reasons for trust? In general, the need for sophisticated trust modeling becomes increasingly pressing.

In this context, the socio-cognitive model of trust, formalized by Cristiano and Rino, has been a cornerstone for the entire scientific community working on trust in MAS and HAI. Throughout his prolific research career, Rino has played a pivotal role in advancing our understanding of trust in intelligent systems. We have had the pleasure and honor of collaborating with him for many years—Alessandro for over 11 years and Filippo for about 8 years. Together, we have explored various aspects of trust, leading to significant advancements in the field. Our theoretical reference for any investigation, whether in MAS, HAI, or specifically in Human-Robot Interaction (HRI), has always been this socio-cognitive model of trust. This model represents a powerful tool for define the concept of trust in collaborative contexts, encompassing and formalizing many of the cognitive constructs involved when two or more agents interact with each other. In this work, we analyze some of our main contributions, developed in collaboration with Rino, that have had a notable impact on the field.

2 The role of trust in hydrogeological and health risk management

An important line of research we have pursued with Rino, which has received significant scientific and social attention, concerns the complex relationship of trust between citizens and public institutions in the contexts of natural and health-related risks.

Indeed, these are two closely related issues which have similar characteristics. First and foremost, they require cooperative solutions, since the contributions of individual citizens and involved institutional entities are crucial for ensuring community resilience. However, many of these risks arise in environments where there are incentives for unilateral deviation from cooperative norms, exploiting the cooperative efforts of others for personal gain. This phenomenon is known in the literature as free riding. Hence, there is a need to address human factors such as mutual trust, particularly towards institutions, as these are key elements influencing individual decisions and consequently the success of implemented measures.

As far as it concerns the hydrogeological risk, we have worked for several years on natural hazards within the PON CLARA project [1]. The topic of natural disasters (particularly flooding events) has always been of considerable importance due to the high risks involved [8], both in terms of potential economic and social damages, as well as the danger to human lives [9, 10]. The increasing impact of climate change further underscores the relevance of this topic [11, 12]. Within this context, our work [13, 14, 15, 16] aimed to contribute to evaluating the role human factors play in response to critical hydrogeological phenomena.

Literature indicates that citizen decisions and behaviors can reduce flood damage by up to 80% [17]. Therefore, it is essential to focus on this component. Effective communication in these contexts is crucial to ensure accurate risk perception among citizens. A necessary prerequisite for this is a sufficient level of trust in institutions.

In our research, we developed an agent-based platform to analyze social behaviors in these critical situations. This platform was subsequently released as one of the primary deliverables of the above-mentioned CLARA project. In our experiments, we simulate a population facing the risk of a potentially catastrophic event. In this scenario, we investigate how citizens (modeled as cognitive agents) utilize various available information sources, including authoritative guidance, to accurately assess the risks they face and effectively respond to critical events.

Specifically, our innovative contributions to the literature include:

1. A comprehensive analysis of the behavioral profiles of both citizens and authorities. Specifically, regarding authorities, we explored punitive, incentive-based, and mixed strategies.

---

[1]http://www.smartcities-clara.eu/

2. The potential to use social structure as a "social absorber of risk", enabling support for higher risk levels. The significance and effectiveness of this phenomenon is closely tied to the presence of free riders; if their prevalence within the network is too high, this positive effect diminishes.

Similarly, we addressed this issue within the healthcare context, particularly during the COVID-19 pandemic. The results obtained during this period are of particularly significant relevance due to both the critical nature of the issues addressed and their scientific and social value. Specifically, by conducting two large-scale national experimental surveys, we investigated the role of trust in its various forms throughout different stages of the pandemic.[18].

The originality and strength of our contributions lie in the socio-cognitive perspective of the analysis, which is grounded in the Castelfranchi-Falcone socio-cognitive model of trust[19]. This approach has enabled us not only to quantify trust but also to explore the underlying cognitive and social factors that determine and influence it.

By implementing an opinion dynamics model[20], we were able to explain how the virtuous use of information positively influenced citizens' decisions, and we attempted to quantify its impact by comparing it with pre-pandemic data. Furthermore, utilizing agent-based simulation, we investigated potential variables that could affect the acceptance of imposed restrictions, such as trust in institutions, risk perception, public communication, and social dynamics[21].

Additionally, we explored the interaction between institutional trust, trust in COVID-19 vaccines, information habits, personal motivations, and underlying beliefs about the pandemic in determining vaccine willingness[22]. The SEM analysis, guided by our theoretical approach, revealed key predictors of vaccination intent: trust in vaccine manufacturers (which is itself supported by trust in regulators), collectivist goals, self-perceived knowledge, reliance on traditional media for information, and trust in institutional and scientific sources.

At the conclusion of this significant research cycle, given our central role in the topic, we conducted a literature review aimed at providing an overview of the role of trust in determining COVID-19 vaccine acceptance[23]. The objective was to shed light on the various forms of trust that have been considered, in the different studies, and how these influenced citizens' acceptance of the vaccine. The analysis revealed that trust played a crucial role, particularly with respect to the COVID-19 vaccine, governments, manufacturers, healthcare systems, and science. The scientific and social impact of this line of research has been particularly notable. Internationally, the article was even cited in guidelines released by the WHO[2], while nationally, it was referenced by INAIL[3] and featured in several newspapers.

3 Trust and knowledge generalization within virtual social networks

In this research topic [24, 25], we have focused on modeling trust within virtual social networks composed of human or artificial cognitive agents. As digital infrastructures continue to spread, social interaction is undergoing a paradigm shift [26, 27]. We are experiencing a major change, which began decades ago with the introduction of the Web, and which will increasingly transform interactional and communicative modalities as well as their actors (human or artificial). Thanks to the Internet, more and more people are connecting with each other, creating virtual societies that "facilitate global and simultaneous interaction, create a common context for collaboration, combine different tools for communication and enhance knowledge and knowing processes" [28]. Remarkably, in many environments, a significant portion of communications occurs between individuals (human or artificial agents) who know little or nothing about each other [29, 30, 31]. Indeed, this is the case of e-markets, forums, cloud systems, and multi-agent systems in general. Identifying methodologies capable of evaluating our interlocutors [32] and selecting appropriate partners, even in this context, becomes crucial.

If in the classic social paradigm we rely mainly on direct expe-

---

[2]https://apps.who.int/iris/bitstream/handle/10665/362317/
WHO-EURO-2022-6045-45810-65956-eng.pdf?sequence=1&isAllowed=y)
[3]https://www.facebook.com/share/p/BQLBZvYog7JBnsM6/

rience, there can be situations where this dimension might not be sufficient. Indeed, the more populated the network, the more complicated it becomes to know its members and their performance well enough. In this context, characterized by a high number of agents and high dynamism, practical alternatives are needed to further generalize the knowledge we have.

At the same time, a high level of interconnection is not necessarily counterproductive, as it increases the possibility of retrieving indirect information, such as trust evaluations produced by other agents in the network. In this regard, several trust models have emphasized the importance of combining inferential processes and recommendations [33, 34, 35], i.e., category recommendations. This represents a powerful tool that in many situations can be even more effective than simple individual recommendations. For a more precise definition of category, we refer to [36, 37].In short, a category is defined by a set of specific characteristics and distinguishing traits, which can be determined through a set of visible and non-deceptive signs. The specific categories to consider change with the context and the task of interest. Regarding virtual societies, practical examples of categories can be related to the type of service offered, experience, but also role, age, gender, economic status, education level, and so on. An agent belongs to a given category if it possesses the characteristics that define it. Recommending a category implies suggesting to a potential trustor to interact not with a specific agent, but with one of the members of the identified category.

While the state of the art just showed that their overall performance is better when category recommendations are introduced, not delving into the relationship between individual and category recommendations within their model, within our studies, we focused on the limits and the situations in which this approach is actually advantageous, compared to classical recommendation.

Starting from the assumption that direct experience is an irreplaceable resource, we have investigated the use of trust in the absence of direct experience. In this case, the trustor, i.e., the agent that needs to find a partner to rely on for a specific task, is forced to make use of external information. This is a fairly common situation, which occurs, for example, whenever the trustor joins a new network, or even when they have already interacted with agents in the network but for tasks completely different from those currently needed, making it impossible to generalize previous knowledge. Under these conditions, we identified when it is more convenient to use individual recommendations rather than category recommendations. Specifically, we focused on two specific parameters:

1. the network turnover, i.e., the frequency with which the agents present leave the network and are replaced by new agents

2. the reliability of the recommenders, which models how faithfully the recommenders report the information they have estimated.

For the purpose of comparing the two approaches, the metric category value CV was defined, i.e., the percentage of times when category recommendations allow interaction with agents that provide better performance than individual recommendations. We therefore exploited an agent-based simulation to investigate these concepts within an heterogeneous population of agents, obtain the following results:

1. The CV increases with the increase in network turnover, so indeed the dynamism of the network positively influences the value of the categories. The results clearly show that the importance of categories here increases further. In particular, the higher the turnover in the network, the more convenient it becomes to rely on categories. This effect, while not surprising, represents an additional element in support of the usefulness of category recommendations within virtual societies, precisely because these are often characterized by high turnover.

2. The CV notably increases when the reliability of the recommenders is unknown. This finding introduces a novel insight into the literature, as it demonstrates that category recommendations offer a more robust approach in the presence of potentially unreliable recommenders. This effect is further amplified

when the reliability of the recommenders is unknown to the trustor.

Thus, consistent with the literature, our results revealed that, under specific conditions, category recommendations can indeed perform on par with or even exceed individual recommendations. While direct experience is undeniably important, it is not always available for evaluating potential partners. In this context, having mechanisms for generalizing knowledge provides a significant advantage.

4 Trust in Human-Robot Interaction: a socio-cognitive perspective

A major challenge in the field of HRI is designing autonomous robots capable of an effective and trustworthy collaboration with humans [38]. Robots will be our collaborators in multiple environments, such as hospitals, offices, and tourist settings. In these social contexts, is required that they leverage on their autonomy, anthropomorphic features (e.g. verbal and non-verbal communication skills, personality traits, emotions) to interact effectively with a wide range of human users, who may not be able or willing to adapt their interaction level to suit the machine's requirements. Effective collaboration and acceptability require finding an appropriate balance between human's trust and control, where humans can confidently delegate tasks, reducing their control over the robots, while robots can autonomously adopt those tasks in a way that builds and maintains human's trust. In this sense, the concept of robot's trustworthiness extends beyond safety and predictability, to include behavior adaptability and transparency.

A first challenging line of research that we have pursued over the years focused on the design of new cognitive architectures [39, 40] to enhance a real sense of *dependability* and *trust* in intelligent robots. In particular, our main goal has been, and still is, to design cognitive robots [41] capable to show a behavior:

- *Useful*: humans exploit interaction with robots in order to achieve their own goals, both explicits and implicits;

- *Effective*: interaction realize those results, in times and ways that are satisfying for the specific human subject;

- *Acceptable*: humans recognize and agree with the role played by robots in the interaction, and they accept the robot's prerogatives in assuming that role;

- *Trustworthy*: humans *trust* the robot, in order to start the interaction and recognize its potential benefits. This entails the human perceiving the robot as possessing both *competence* and *willingness* to pursue the task that will yield the aforementioned benefits during the interaction.

Since our initial work [42], we proposed a computational cognitive model that progressively integrated the theoretical principles outlined in the theory of trust formalized by Cristiano and Rino [19], which summarizes their decades-long research on MAS and AI. This theoretical framework has provided a solid theoretical foundation for exploring various dimensions of trust-based interactions (trust, autonomy, task delegation and adoption and so on). In particular, we focused on the development of computational cognitive models capable of intelligently managing collaboration based on task delegation and adoption. As demonstrated in the theory of trust, there is a strong relationship between trust and delegation: trust is the mental counterpart of task delegation. Therefore, an effective way to promote trust-based interaction between humans and robots is to equip robots with the cognitive capabilities to intelligently adopt a task delegated by a human. But what specific cognitive capabilities are required? First of all, we considered robots as intentional systems, capable of representing information about the world and interacting agents in terms of mental states, such as beliefs, goals, plans, intentions, and so on. This approach has allowed us to manage the robot's internal knowledge representation at a level of abstraction similar to that typically used by humans. In an early version [39], we designed a knowledge-driven, plan-based, cognitive architecture that allowed a BDI robot to dynamically modulate its own level of social autonomy every time a human user delegated to it a task to achieve. The task adoption process leveraged on the robot's capability to build a complex representation of the world, to create a profile of the human user and to classify her. In particular, the profile of the user has been built on the basis of the robot's capability to perceive and infer specifics human features and mental states (Theory of Mind). On the basis of these representations the robot was able to adjust its own level of collaborative autonomy by adopting the task at the different levels of help. Furthermore, the architecture gave to the robot the ability to self-evaluate (guided by specifics feedback provided by the user) the trustworthiness of the skills it relies on for profiling the human user. Figure 1 represents the first version of the computational cognitive model.

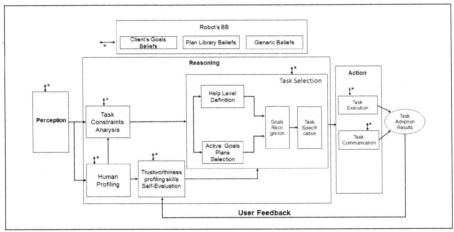

Figure 1: An early version of the cognitive model

This cognitive architecture was tested in a simulated environment, with the goal of providing a proof of concept that demonstrated how it could support adaptive and trustworthy behavior in HRI scenarios [43, 44]. For its implementation and simulation, we exploited JaCaMo, a software for multi-agent programming that integrates different levels of multi-agent oriented programming (MAOP). Although the results were obtained in a simulation, they motivated us to investigate different versions of the model with the aim of testing it in real-world contexts, using a physical robot. A new version of the model was presented in [40].

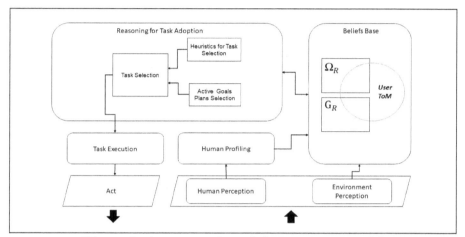

Figure 2: A new version of the cognitive model

Figure 2 shows the new model. Similarly to the version proposed in [39], we designed a model that enabled the robot to i) build a profile of the interacting user and classify her, ii) model the mental states of the interacting human user, iii) model the beliefs, goals, and plans of other agents involved in the interaction, and iv) dynamically adjust its own level of collaborative autonomy by restricting or expanding a delegated task based on an internal negotiation process that considers various contextual factors, such as needs or goals that may not necessarily be declared by the interacting human user, as well as the constraints or needs of other agents involved in the interaction. In addition to this, we extended the model by introducing a new cognitive capability: generating an explanation of the robot's adoption strategy and justifying any possible changes in relation to the explicit agreements. We provide the results of an HRI pilot study [45, 40] in which we recruited real participants that have interacted with the humanoid robot Nao, widely used in HRI scenarios. The robot played the role of a museum assistant with the main goal to provide the user the most suitable museum exhibition to visit.

Alongside our theoretical line of research, we recently conducted an experiment aimed at investigating the concept of trust in real-world HRI scenarios [46, 47]. Specifically, our objective was to study the interplay between robot trustworthiness, user acceptance, and the cognitive capabilities underlying previously defined cognitive models. In particular, we continued to explore the museum settings scenario, as this represents a social context in which robots are often employed as intelligent assistance systems. We exploited the Nao robot due to its advanced verbal and non-verbal communication skills, including speech, body movements, and gestures. The experiment was designed and conducted in a real museum, named Palazzo delle Esposizioni, in Rome. A total of 84 museum visitors participated in the study and provided informed consent. We conducted the experiment in an exhibition setting where participants were naturally present for a similar purpose, creating a more authentic environment than a laboratory or online survey could provide. In the study, the humanoid robot Nao played as a museum assistant, engaging with actual museum visitors. Visitors assigned the robot the task of guiding them through a virtual tour that highlighted artworks from an explicitly preferred artistic period. The robot's task adoption involved suggesting the most suitable virtual tour based on visitors' needs and subsequently guiding them through the selected tour. The task adoption could also lead to outcomes not explicitly expected by the visitors, as it considered various contextual elements and mental states implicitly attributed to the visitors themselves, based on their profile, perceived by the robot at the beginning of the interaction.

The results of this experiment have been analyzed in two separate works. In the first study [46] we focused on analyzing the impact of intelligent assistance on visitor satisfaction and the level of surprise as indicators of the robot's ability to anticipate visitors' needs, even when those needs were not explicitly stated. Specifically, we investigated visitor satisfaction with (1) the suggested tour (adoption results) and (2) the robot's behavior (adoption strategy). The study highlights that user satisfaction increased when the robot adapted its behavior to unexpressed user needs, incorporating implicit mental states. This supported the importance of autonomous robots as cognitive agents capable of adjusting their actions based on a representation of the user's mind. However, while a critical form of help led to higher satisfaction with the final outcome, users showed some reluctance, preferring a more literal form of help, due to potential risks of misunderstandings or conflicts when the robot displayed too much autonomy. The results affirmed the nuanced impact of adjustable social autonomy, where the robot's role transitions from a task-executing system to a collaborative agent, though this shift could sometimes create uncertainty in the user's perception of the robot's initiative.

In a second work [47] we explored two further dimensions within the field of human-robot interaction (HRI), by exploiting the same experiment described above: the robot's trustworthiness and the role of explainability. In particular, we examined the interplay between the robot's trustworthiness, based on competence and willingness, and its ability to explain the reasons behind the strategy adopted for the human-delegated task. Findings demonstrated that an adaptive robot that deviates from the human's explicitly stated expectations, is perceived as trustworthy if evaluated competent in the collaboration, regardless of the explanation provided. Additionally, a robot evaluated willing to interact with the human for achieving the delegated task is perceived as trustworthy when it is able to explain the mechanism beyond the task adoption process. Furthermore, we found that a human who accepts and comprehends the logic underlying the robot's explanation tends to evaluate the provided explanation positively. Taken together, these results contributed to assess the relationship between trustworthiness, competence, willingness, and explainability in adaptive robots.

5 Conclusions

In conclusion, our exploration of trust in human and artificial societies, conducted together with Rino and strongly based on the socio-cognitive theory of trust, has allowed us to produce particularly significant contributions to the state of the art. His work has significantly contributed to shaping our understanding of how trust operates in increasingly complex, adaptive artificial systems. Together, we have demonstrated how trust can be modeled and applied to various contexts, from AI-based agents to Human-Robot Interaction, always keeping in mind the intricate balance between autonomy and cooperation. As we look back on our years of collaboration with Rino we are filled with both pride and gratitude. His insights, mentorship, and friendship have been instrumental in our collective journey, and we are honored to have contributed to the ongoing development of trust in intelligent systems alongside him. Trust will remain a cornerstone of human-technology relationships, and we are excited to see how the next generation of researchers will build upon these contributions. To Rino, we offer our heartfelt thanks, not only for the scientific achievements we have shared but also for the spirit of collaboration that has defined our partnership.

References

[1] N. Luhmann, *Trust and power*. John Wiley & Sons, 2018.

[2] D. Gambetta, "Trust: Making and breaking cooperative relations," 1988.

[3] A. Baier, "Trust and antitrust," in *Feminist Social Thought*, pp. 604–629, Routledge, 2014.

[4] M. Brundage, S. Avin, J. Wang, H. Belfield, G. Krueger, G. Hadfield, H. Khlaaf, J. Yang, H. Toner, R. Fong, *et al.*, "Toward trustworthy ai development: mechanisms for supporting verifiable claims," *arXiv preprint arXiv:2004.07213*, 2020.

[5] L. Floridi, "Establishing the rules for building trustworthy ai," *Ethics, Governance, and Policies in Artificial Intelligence*, pp. 41–45, 2021.

[6] S. D. Ramchurn, D. Huynh, and N. R. Jennings, "Trust in multi-agent systems," *The knowledge engineering review*, vol. 19, no. 1, pp. 1–25, 2004.

[7] E. J. De Visser, R. Pak, and T. H. Shaw, "From 'automation'to 'autonomy': the importance of trust repair in human–machine interaction," *Ergonomics*, vol. 61, no. 10, pp. 1409–1427, 2018.

[8] S. N. Jonkman, "Global perspectives on loss of human life caused by floods," *Natural hazards*, vol. 34, no. 2, pp. 151–175, 2005.

[9] J. Cunado and S. Ferreira, "The macroeconomic impacts of natural disasters: New evidence from floods," 2011.

[10] D. Guha-Sapir, P. Hoyois, P. Wallemacq, and R. Below, "Annual disaster statistical review 2016," *The numbers and trends*, pp. 1–91, 2017.

[11] P. C. D. Milly, R. T. Wetherald, K. Dunne, and T. L. Delworth, "Increasing risk of great floods in a changing climate," *Nature*, vol. 415, no. 6871, pp. 514–517, 2002.

[12] C. Wasko, R. Nathan, L. Stein, and D. O'Shea, "Evidence of shorter more extreme rainfalls and increased flood variability under climate change," *Journal of Hydrology*, vol. 603, p. 126994, 2021.

[13] A. Sapienza and R. Falcone, "Flood risk and preventive choices: A framework for studying human behaviors," *Behavioral Sciences*, vol. 14, no. 1, p. 74, 2024.

[14] R. Falcone and A. Sapienza, "Using sources trustworthiness in weather scenarios: The special role of the authority," in *AI* IA 2017 Advances in Artificial Intelligence: XVIth International Conference of the Italian Association for Artificial Intelligence, Bari, Italy, November 14-17, 2017, Proceedings 16*, pp. 3–16, Springer, 2017.

[15] R. Falcone and A. Sapienza, "Interactions among information sources in weather scenarios: The role of the subjective impulsivity," in *Advances in Practical Applications of Cyber-Physical Multi-Agent Systems: The PAAMS Collection: 15th International Conference, PAAMS 2017, Porto, Portugal, June 21-23, 2017, Proceedings 15*, pp. 56–69, Springer, 2017.

[16] R. Falcone, A. Sapienza, and C. Castelfranchi, "Trusting different information sources in a weather scenario: A platform for computational simulation.," in *ICAART (1)*, pp. 165–172, 2016.

[17] T. Grothmann and F. Reusswig, "People at risk of flooding: Why some residents take precautionary action while others do not," *Natural hazards*, vol. 38, pp. 101–120, 2006.

[18] R. Falcone, E. Colì, S. Felletti, A. Sapienza, C. Castelfranchi, and F. Paglieri, "All we need is trust: How the covid-19 outbreak reconfigured trust in italian public institutions," *Frontiers in psychology*, vol. 11, p. 561747, 2020.

[19] C. Castelfranchi and R. Falcone, *Trust theory: A socio-cognitive and computational model*. John Wiley & Sons, 2010.

[20] R. Falcone and A. Sapienza, "How covid-19 changed the information needs of italian citizens," *International Journal of Environmental Research and Public Health*, vol. 17, no. 19, p. 6988, 2020.

[21] R. Falcone and A. Sapienza, "An agent-based model to assess citizens' acceptance of covid-19 restrictions," *Journal of Simulation*, vol. 17, no. 1, pp. 105–119, 2023.

[22] R. Falcone, A. Ansani, E. Colì, M. Marini, A. Sapienza, C. Castelfranchi, and F. Paglieri, "Trusting covid-19 vaccines as individual and social goal," *Scientific reports*, vol. 12, no. 1, p. 9470, 2022.

[23] A. Sapienza and R. Falcone, "The role of trust in covid-19 vaccine acceptance: considerations from a systematic review," *International journal of environmental research and public health*, vol. 20, no. 1, p. 665, 2022.

[24] A. Sapienza and R. Falcone, "Evaluating agents' trustworthiness within virtual societies in case of no direct experience," *Cognitive Systems Research*, vol. 64, pp. 164–173, 2020.

[25] R. Falcone, A. Sapienza, and C. Castelfranchi, "Recommendation of categories in an agents world: The role of (not) local communicative environments," in *2015 13th Annual Conference on Privacy, Security and Trust (PST)*, pp. 7–13, IEEE, 2015.

[26] Z. Liang and W. Shi, "Analysis of ratings on trust inference in open environments," *Performance Evaluation*, vol. 65, no. 2, pp. 99–128, 2008.

[27] A. Martin, "Digital literacy and the "digital society"," *Digital literacies: Concepts, policies and practices*, vol. 30, no. 151, pp. 1029–1055, 2008.

[28] J. Mueller, K. Hutter, J. Fueller, and K. Matzler, "Virtual worlds as knowledge management platform–a practice-perspective," *Information Systems Journal*, vol. 21, no. 6, pp. 479–501, 2011.

[29] C. Haythornthwaite, "Social networks and internet connectivity effects," *Information, Community & Society*, vol. 8, no. 2, pp. 125–147, 2005.

[30] X. Liu, A. Datta, K. Rzadca, and E.-P. Lim, "Stereotrust: a group based personalized trust model," in *Proceedings of the 18th ACM conference on Information and knowledge management*, pp. 7–16, 2009.

[31] P. Resnick and R. Zeckhauser, "Trust among strangers in internet transactions: Empirical analysis of ebay's reputation system," in *The Economics of the Internet and E-commerce*, pp. 127–157, Emerald Group Publishing Limited, 2002.

[32] S. Abar, G. K. Theodoropoulos, P. Lemarinier, and G. M. O'Hare, "Agent based modelling and simulation tools: A review of the state-of-

art software," *Computer Science Review*, vol. 24, pp. 13–33, 2017.

[33] R. Conte and M. Paolucci, *Reputation in artificial societies: Social beliefs for social order*, vol. 6. Springer Science & Business Media, 2002.

[34] P. De Meo, L. Fotia, F. Messina, D. Rosaci, and G. M. Sarné, "Providing recommendations in social networks by integrating local and global reputation," *Information Systems*, vol. 78, pp. 58–67, 2018.

[35] P. Yolum and M. P. Singh, "Emergent properties of referral systems," in *Proceedings of the second international joint conference on autonomous agents and multiagent systems*, pp. 592–599, 2003.

[36] R. Falcone, A. Sapienza, and C. Castelfranchi, "The relevance of categories for trusting information sources," *ACM Transactions on Internet Technology (TOIT)*, vol. 15, no. 4, pp. 1–21, 2015.

[37] A. Sapienza and R. Falcone, "A bayesian computational model for trust on information sources.," in *WOA*, pp. 50–55, 2016.

[38] R. Falcone, A. Sapienza, F. Cantucci, and C. Castelfranchi, "To be trustworthy and to trust: The new frontier of intelligent systems," *Handbook of Human-Machine Systems*, pp. 213–223, 2023.

[39] F. Cantucci and R. Falcone, "Towards trustworthiness and transparency in social human-robot interaction," in *2020 IEEE International Conference on Human-Machine Systems (ICHMS)*, pp. 1–6, IEEE, 2020.

[40] F. Cantucci and R. Falcone, "Collaborative autonomy: Human–robot interaction to the test of intelligent help," *Electronics*, vol. 11, no. 19, p. 3065, 2022.

[41] F. Cantucci and R. Falcone, "A cognitive approach to model intelligent collaboration in human-robot interaction.," in *WOA*, pp. 138–150, 2023.

[42] F. Cantucci and R. Falcone, "A computational model for cognitive human-robot interaction: An approach based on theory of delegation.," in *WOA*, pp. 127–133, 2019.

[43] F. Cantucci, R. Falcone, and C. Castelfranchi, "Robot's self-trust as precondition for being a good collaborator.," in *TRUST@ AAMAS*, 2021.

[44] F. Cantucci, R. Falcone, and C. Castelfranchi, "Human-robot interaction through adjustable social autonomy," *Intelligenza Artificiale*, vol. 16, no. 1, pp. 69–79, 2022.

[45] F. Cantucci and R. Falcone, "Autonomous critical help by a robotic assistant in the field of cultural heritage: A new challenge for evolving human-robot interaction," *Multimodal Technologies and Interaction*,

vol. 6, no. 8, p. 69, 2022.

[46] F. Cantucci, R. Falcone, and M. Marini, "Redefining user expectations: The impact of adjustable social autonomy in human–robot interaction," *Electronics*, vol. 13, no. 1, p. 127, 2023.

[47] F. Cantucci, M. Marini, and R. Falcone, "Trustworthiness assessment of an adaptive and explainable humanoid robot in a real environment," *The International Journal of Social Robotics*, 2024.

Trust the Process: Exploring Trust as a Foundational Element in Open Dialogue

FRANCESCA CAMILLI

Institute of Cognitive Science and Technologies, CNR, Rome, Italy

`francesca.camilli@istc.cnr.it`

TAREK EL SEHITY

Institute of Cognitive Science and Technologies, CNR, Rome, Italy
Sigmund Freud PrivateUniversity Vienna, Institute of Psychology

`Tarek.el-Sehity@istc.cnr.it`

RAFFAELLA POCOBELLO

Institute of Cognitive Science and Technologies, CNR, Rome, Italy

`raffaella.pocobello@istc.cnr.it`

Abstract

The focus of this chapter is to approach trust as a core ingredient in OD, emphasizing its importance in developing therapeutic relationships and promoting client recovery. Trust, being an implicit concept in most OD literature, can be considered an indispensable component in the regeneration of social and self-trust for those experiencing a mental health crisis. The OD approach, which originated in Finland, is underpinned by principles of immediate support, inclusiveness, and dialogism, all combining to foster an atmosphere that is warm and client-centered. The present contribution assumes that, within OD, trust represents relational capital, where clients' transformative changes in treatment could be enabled through sustained and transparent interaction. The creation of a safe space for mutual vulnerability and power-sharing give clients an opportunity to recover their agency and their self and social trust. In our contribution, inspired by the insights from the Theory of Trust developed by Rino Falcone and Cristiano Castelfranchi, we underlined some trust elements in OD, like the "holding space" approach and the transitivity of trust among the practitioners, clients and their networks. The hypotheses formulated in this chapter highlight promising areas for future theoretical development and empirical research. Exploring how trust is established, maintained, and expanded within OD—and its impact on long-term client outcomes—could significantly enhance our understanding of the dialogical process and its efficacy.

1 Introduction

In mental health services, trust has long been considered essential for successful therapeutic relationships between clients and their professionals (e.g.,[1, 2]). However, emerging literature suggests that trust is not merely an ingredient for therapeutic relationships but is at the very heart of the client's experience of mental health crises. Particularly, the erosion of self-trust - diminished confidence in one's perceptions, feelings, and decisions - often characterizes and intensifies mental health issues, presenting a foundational challenge for both clients and practitioners ([3]). A recent meta-analysis indicates that individuals with psychosis or at risk of it exhibit less baseline trust than healthy controls. They struggle with adjusting trust behaviour according to context, unlike healthy controls. These impairments may inhibit social functioning due to the crucial role of trust in social interactions ([4]). Thus, a substantial part of mental health intervention should aim to restore trust, rebuilding the individual's ability to rely on their inner judgments and significant others as an essential component of their recovery journey.

The key transformative components recognized of therapeutic relationships in mental health services are power, safety, and identity, which can influence the psychological change of clients, suggesting that mental health professionals should focus on care rather than control in their relationships with clients to improve psychological distress and promote recovery ([5, 6]). For example, patients experiencing psychosis not only express a desire for greater participation and respectful, humane treatment but also demonstrate the ability to build trust over time through consistent and respectful collaboration ([7, 8]). Such frameworks shift the focus from merely treating symptoms to restoring the client's agency and autonomy within a socially supportive environment. This chapter argues that in approaches that emphasize collaborative therapeutic environments, such as Open Dialogue (OD), the restoration of trust is actively fostered through its core principles. In our view, meetings involving clients, their network, and mental health professionals strive to build trust and establish mutual understandings, thus activating the psychological resources of both the individual and the network ([9]). In addition, through dialogic practices that prioritize openness, transparency, and the active role of the client, OD facilitates the regeneration of both social trust and self-trust, encouraging clients to regain ownership of their mental health journey. However, in the field of OD, the study of trust has remained largely implicit in the literature.

The first part of the chapter will review how trust is addressed, often indirectly, in the existing literature on OD. In the second part, we will present our working hypothesis: trust functions as a central therapeutic mechanism in OD, not only addressing acute symptoms but also facilitating lasting, empowering transformations in individuals facing severe mental health crises. These transformations are achieved through the re-establishment of social trust, which in turn fosters the restoration of self-trust. To substantiate this hypothesis, we will engage with trust theory, particularly drawing on the contributions of our dear colleagues Rino and Cristiano.

2 Open Dialogue

Open Dialogue offers a human rights-based framework for mental health care that emphasizes dialogical communication and shared responsibility ([10]). Developed in northern Finland in the 1980s, OD was described in 1995 through seven guiding principles aimed at fostering an inclusive and collaborative therapeutic environment for individuals experiencing psychological distress. Central to OD is understanding that mental health crises do not exist in isolation but are interconnected with the social system of which the individual is a part, a notion reflected in its systemic and social network-based approach ([11]). Furthermore, OD is grounded in the humanistic approach to psychosis known as Need-Adapted Treatment (NAT) ([12]). The NAT fundamental concept is the adaptation of therapeutic activities according to the specific characteristics of each case and the integration of various forms of treatment as opposed to their exclusive selection.

In Finland, OD has been shown to be an effective intervention for the recovery of patients with psychotic experiences, as evidenced by a reduction in symptoms and an increased probability of returning to work or studies ([13]). Furthermore, the approach has been shown to reduce the need for psychiatric services and antipsychotic medications, including long-term ([13]; [14]). These findings have provided empirical support for the notion that individuals experiencing psychotic symptoms can recover with the assistance of home-based interventions. In addition, these favorable results have facilitated the global dissemination of OD services in more than 20 countries ([15]).

At the heart of OD practice are "network meetings," a therapeutic space where individuals in crisis, their families, and relevant social network members—such as friends, colleagues, or even professionals from employment services—come together to create a shared dialogue around the client's experience. This "social network perspective," formalized as the second principle of OD, shifts the focus from individual treatment to a relational and community-based approach that seeks to harness the support and understanding of the person's immediate social network ([16, 17]). The goal is to promote a sense of shared accountability and trust within the network, positioning mental health treatment as a collective endeavor.

The first principle of OD, "immediate help," requires that the mental health service organize a network meeting within 24 hours after being contacted, underscoring the approach's commitment to rapid and community-oriented crisis intervention. This principle is intended to mitigate the need for hospitalization by providing prompt community-based support ([16]). The principle of 'flexibility and mobility' further supports this by allowing network meetings to be held at locations that best serve the patient's comfort and accessibility, often at the family home, thus strengthening trust and collaboration in familiar and supportive settings ([17]).

The OD model also emphasizes "responsibility" (principle four), which ensures continuity and accountability by assigning the initial responder to oversee the first network meeting and subsequent care [16]. The principle of 'psychological continuity' (principle five) requires that the same team of professionals remains involved throughout the treatment of the client, establishing a stable foundation of trust and familiarity critical to the therapeutic relationship ([16, 17]).

The final principles of OD, "tolerance of uncertainty" and "dialogism," encapsulate its open-ended, nondirective philosophy. "Tolerance of uncertainty" encourages professionals to adopt a "not-knowing" stance, remaining open to unfolding situations without preconceived solutions or premature treatment decisions, such as the use of neuroleptics, which must be deliberated over multiple meetings. This acceptance of uncertainty creates a space where trust can develop organically, as clients and their networks sense that the therapeutic journey is patient-led and free of coercion ([17]).

Finally, "dialogism" anchors OD in the belief that authentic, multi-voiced dialogue—where all participants, especially the client, are encouraged to speak and be heard—can foster new insights and collective understanding. This dialogical process serves not just as a tool but as the fundamental mechanism of change to facilitate shifts in perception, understanding and trust within the network. Rather than prioritizing symptom alleviation, OD centers on the dialogue that can transform the relational and communicative landscape around the client, allowing trust and self-trust to emerge as natural byproducts of a respectful, inclusive, and participatory therapeutic setting ([16]).

"It can be argued that the ODA has been helpful, at least in moving the commencement of treatment in a less chronic direction. It may have even increased social capital in the entire psychiatric catchment area, and promote mutual trust between the general population and the psychiatric services." ([18] p.179).

The remainder of this chapter will delve into how these principles specifically address trust, showing how OD not only treats mental health crises but actively supports the re-establishment of self-trust, laying a foundation for meaningful and sustained recovery.

3   Trust in the process

The role of trust within OD extends beyond a mere therapeutic tool, embedding itself deeply in the dialogical process at the core of its philosophy. This "trust in the process" suggests a commitment by both practitioners and participants to a collaborative, unfolding journey rather than a directive, goal-oriented outcome. Defined by [17], "trust in the process" holds a dual meaning. First, it reflects an understanding that the goal of each dialogical meeting is the evolution of the process itself—a continual, shared endeavour to address the complexities of the client's experience. Second, it encompasses the practitioners' confidence, shaped by experience, that positive outcomes can emerge through a collective commitment to OD's principles and patient-centred dialogue. In Western Lapland, where OD was developed and extensively practised, this embedded trust is a natural result of the long-term application and consistent, positive outcomes. However, fostering such trust may present unique challenges in regions where OD is less established, and practitioners still struggle to implement the approach's principles.

Central to this process-oriented trust is the transformative shift from monological to dialogical communication. Rather than attempting to "fix" problems, practitioners focus on facilitating a dialogical space where all voices are welcomed and explored. This collaborative openness allows meaning to emerge naturally from within the group, an approach that requires practitioners to release control and avoid imposing solutions prematurely ([19]). As Finnish family therapist

Mia Kurtti notes, trust and hope often emerge from this patient-centered space, as participants sense that "everything is possible" if time and space are granted for authentic expression, free from restrictive categorization ([20]).

Deep listening is essential for creating an atmosphere where participants feel safe to express themselves genuinely. By "holding space," practitioners enable clients and their networks to explore their experiences without fear of judgment. This concept of "therapeutic presence," rooted in Carl Rogers' humanistic psychology, encourages practitioners to engage fully with clients, setting aside self-centered goals to be wholly present in the moment ([21]). Such presence invites participants into a dialogical process in which their contributions are valued, and their experiences are validated, thus cultivating the essential conditions for social and self-trust to take root.

In addition to deep listening, reflecting on conversations between practitioners further reinforces this environment of trust. During these exchanges, practitioners share personal and professional perspectives in the presence of the client and their social network, modelling a non-hierarchical approach to dialogue. Here, insights are not presented as definitive answers but as collaborative reflections, open to discussion and interpretation. This practice, described as a way to model trust and openness, demonstrates to families how differing perspectives can coexist within a shared trust framework ([20]). Reflecting conversations foster not only the therapeutic alliance but also encourage participants to approach varying viewpoints with curiosity rather than defensiveness. Research indicates that greater trust between co-therapists positively impacts the overall atmosphere of the network meeting, enhancing trust in the process for both practitioners and clients alike ([22]).

In sum, trust in OD is cultivated through a process-oriented approach that privileges authentic dialogue over resolution, ensuring that clients and their networks feel heard, respected, and empowered. This trust-centered dynamic within OD allows participants to engage deeply with their own experiences and with each other, establishing a foundation of mutual trust and understanding. In our view, by modelling these dialogical principles, practitioners support clients not only in addressing immediate mental health challenges but in rebuilding the self-trust and social trust necessary for sustainable recovery.

4 Trust Between Practitioners

In our view, trust in OD fundamentally rests on the mutual trust between co-therapists. This foundational trust shapes the quality of interactions within network meetings and sets the tone for how participants engage with vulnerability and openness. Practitioners may hold varied perspectives on what trust with their co-therapist entails, often influenced by personal conceptions of vulnerability and their experiences in dialogical practice ([22, 20]). For some, this trust extends beyond mere reliance on professional competency; it embodies a deeper interpersonal reliance, where the co-therapists presence provides emotional and practical support, reinforcing that their contributions will not go unheard or ignored ([20]). In this way, trust is not only about professional collaboration but about creating a respectful space where each practitioner's words hold significance.

This sense of reliance takes on a nuanced form for others, who view trust as an assurance that their co-therapist will respond gently and with positive intent, even during challenging conversations ([22]). Finnish family therapist Elina Löhönen emphasizes the importance of this respect, suggesting that, even if certain words are not directly helpful, they still deserve to be treated with care and acknowledgement. This approach models a respectful atmosphere for clients and their networks, demonstrating that all contributions—whether harmonious or discordant—have value. This respect fosters a therapeutic environment that reassures clients and their families, encouraging them to share openly and freely, mirroring the vulnerability and openness exhibited by the practitioners ([20]).

Furthermore, when practitioners openly express different or conflicting perspectives during network meetings, they model how to engage constructively with divergent views. This polyphonic approach, especially useful in difficult cases, reinforces that disagreements can enrich the dialogue, adding layers to the understanding of complex issues (Brown et al., 2015). However, practitioners risk experiencing a form of professional vulnerability when their perspectives might be dismissed or met with disrespectful tones by their colleagues. In such instances, practitioners rely on the expectation that their colleagues will reciprocate a "holding space" stance, offering them the same respect they extend to clients and families ([22]).

Vulnerability, therefore, emerges as a complex facet of trust between co-therapists. As [23] note, vulnerability involves "being capable of being physically or emotionally wounded," a risk practitioners undertake when they share openly with colleagues. This vulnerability can affect both the personal and professional identities of practitioners, as they may feel exposed not only as individuals but also as professionals uncertain of answers or solutions. By trusting their co-therapists to support them in this openness, practitioners create an environment where no single person is responsible for "fixing" the client's situation. Instead, they commit to collectively navigating uncertainty, differentiating their role from traditional psychiatric models and reinforcing the collaborative nature of OD ([24]).

Mutual trust among practitioners is also fostered through structured training that includes personal sharing exercises, such as "family of origin" days. During these sessions, practitioners are invited to explore their own family dynamics to be discussed within their training group ([15]). This process encourages the sharing of intimate aspects of their lives, promoting a deeper understanding and trust among group members. The expectation is that these personal disclosures will be handled professionally and contribute to strengthened teamwork despite the inherent risks of revealing vulnerabilities to colleagues of varying ranks ([22]). Such exercises highlight the holistic commitment OD practitioners make to trust-building, which underpins both their professional relationships and their efficacy in modelling trust and openness to clients.

Thus, trust between co-therapists in OD forms the bedrock of the dialogical process, allowing practitioners to bring their full, authentic selves into therapeutic work. This trust supports the fluid, open-ended nature of dialogical interactions, modelling for clients and families the power of vulnerability and the safety of a non-judgmental, supportive environment. Through this foundation, OD practitioners foster an atmosphere that not only addresses immediate mental health crises but also cultivates enduring social and self-trust among all participants.

5 Trust from the perspective of patients and network members

Once trust is firmly established between co-therapists in OD, a central question arises: how is this trust then transferred to patients and their social network? Our hypothesis is that this transfer relies heavily on modelling—where professionals' openness, respect, and willingness to share are observed and mirrored by other participants. Research indicates that when professionals share their thoughts and emotions, patients often perceive the communication as more genuine and meaningful. In settings similar to OD, patients have identified mutuality and transparency as pivotal to building trust, often describing a sense of "relief" when communication is open and reciprocal ([25]). We believe that such mutual communication, combined with shared power and a supportive attitude from practitioners, enhances trust within the network.

In our view, trusting the OD process inherently involves trusting the participants in it, which means seeing patients and their network members as valuable, capable contributors to the therapeutic dialogue. Patients in OD often report feeling truly heard and involved—contrasting sharply with previous experiences in conventional psychiatric settings, where they sensed a lack of trust and autonomy ([26, 27]). We suppose that this engagement fosters a sense of empowerment and personal agency, enabling patients to feel trusted and respected within the therapeutic context. These experiences **might/-could** reinforce trust in the process itself, where patients come to believe in the value and effectiveness of the dialogical approach through the affirmation of their own roles and contributions.

We also believe that the impact of trust within OD extends to cultivating self-trust. As patients experience being trusted, they often report increased self-confidence and a stronger sense of their own resources and capacities. This implicit self-trust aligns with the broader OD philosophy, which seeks not only to support patients in overcoming immediate crises but also to encourage sustained personal growth and resilience through relational empowerment. However, practitioners **might/could** face a potential challenge: placing too much trust in "OD as a method" rather than in the individuals participating in it. Over-reliance on OD's principles, while valuable, risks underestimating the unique insights and strengths of each participant, thereby diluting the genuinely dialogical spirit of OD ([26]).

Another essential aspect of building trust in the context of OD can be the use of non-medicalized, everyday language. This approach, as highlighted in a previous conceptual paper ([28]), allows participants to express their thoughts and experiences without the constraints of diagnostic labels, fostering a polyphonic dialogue where multiple perspectives can coexist. By focusing on personal stories and avoiding psychiatric terminology, the process encourages mutual understanding. This shared language, grounded in the everyday lives of patients and network members, can be described as "a tool for trust" as it validates individual experiences and facilitates open, collaborative meaning-making within the network.

Concluding our literature analysis, understanding trust within OD requires recognizing it as a dynamic, relational force that radiates outward from co-therapists to patients and network members, cultivated through openness, shared power, and mutual respect. This inclusive approach positions trust as both foundational and transformative, enabling each participant to engage authentically and fostering a therapeutic environment where self-trust and collective trust mutually reinforce one another. However, to truly appreciate the therapeutic power of OD, we will attempt to adopt some elements of the theoretical framework of trust advanced by our colleagues Rino Falcone and Cristiano Castelfranchi.

6 Trust theory in the context of Open Dialogue

In this section, we will consider the socio-cognitive model of trust developed and presented in the works by Rino Falcone and Cristiano Castelfranchi (i.e. [29, 30, 31, 32, 33, 34, 35]) for analyzing the role of trust in the context of the OD approach. We will refer to this model as "Trust theory" in the text ([33]).

In this theory, trust is both a cognitive and relational construct, in which an agent expresses confidence in another agent's behaviour, supporting their goals in situations of uncertainty. Trust involves several **key elements**:

- *Goal Adoption Belief*: The trustor believes that the trustee will adopt goals congruent with their own.

- *Intentions and Motivations*: The trustor perceives the trustee as genuinely willing to act in a way that benefits them, motivated by a real desire rather than obligation.

- *Competence and Reliability*: The trustor believes that the trustee has the ability to achieve the desired goal or perform the required task.

In the context of OD, the Goal Adoption Belief can be related to how the client's needs are addressed through the first three core principles: immediate help, inclusion of the client's social network, flexibility, and mobility. Additionally, the principle of tolerating uncertainty creates conditions where the trustor (the client seeking help) can transform into the trustee, trusting that they have the capacity to manage their situation and that it is safe to do so.

The elements of Intentions and Motivations and Competence and Reliability can be seen as closely connected to the principles of responsibility, continuity of care and dialogism as they foster the conditions for shared knowledge and joint action. This collaborative process strengthens the relational trust between all participants.

6.1 Trust Responsiveness and Relational Influence in OD

In an early contribution of [36], the dynamic forces of trust are conceptualized as well as the way they influence the trustworthiness and trust dynamics among individuals within a network. They propose that when one agent (A) trusts another (B), it may lead B to respond by increasing their trust or commitment to A, creating a virtuous cycle where trust reinforces itself. This virtuous nature of trust, wherein A's trust in B can promote B's positive response, is critical in building sustainable trust within social and therapeutic interactions. OD emphasizes non-hierarchical and dialogical communication, where professionals actively engage in a transparent exchange, encouraging clients and their social networks to participate openly. Trust may be conceived as the foundation for fostering a shared space that is perceived as safe and supportive. This reflects the OD principles of "dialogism" and "tolerance of uncertainty," both of which require a willingness to relinquish control and to respect the unique contributions of each individual involved in the process.

6.2 Trust as Relational Capital in OD

Trust is also a form of relational capital, a resource within social networks that supports cooperation, negotiation, and influence. In OD, trust as relational capital is cultivated through the collaborative environment of network meetings, where each participant—therapist, client, and family member—builds trust through respectful and open dialogue. According to Trust theory, relational capital is also contingent upon the accumulation of trustworthiness and reputation within a network. This aligns with the emphasis of OD on sustained relationships throughout the course of network meetings, taking care of the OD principle of 'Psychological Continuity', where trust capital can grow over time. Trust as relational capital in OD not only benefits individual participants but also strengthens the collective therapeutic environment, allowing for deeper engagement and a more resilient support system for the client ([2]).

This element of trust is particularly important for psychotic patients who often lack opportunities to offer social support within their networks (Hamilton et al., 1989). Emphasizing mutuality in treatments like OD helps them both give and receive in relationships, reducing their exclusive identification as patients. In this respect, comprehensive, family-oriented interventions can enhance social capital by fostering mutual trust between the community and the Need-Adapted treatment system. ([18], p. 183)

6.3 Trust Transitivity and Trust-Building in OD

[29] discuss the concept of transitivity of trust, how trust in an individual can be extended to others within a network. This transitivity is crucial in OD, where a client's initial trust in a primary therapist is reflected to the clients and their network, through the OD-principle of 'Tolerance of Uncertainty' which aims at "building up a sense of trust for the joint process. " ([37], p.250).

An essential key for the trust-building process is the process of reflection among OD practitioners: "*You know who you are working with, trust that colleague, trust that whatever comes we handle. Sometimes, the client is locked, with no feelings and very concrete. I question, I wonder why. The client starts to listen to us, to our reflections together, like tasting our views.*" ([20] p.62) creating a cohesive and supportive therapeutic alliance. More generally, trust transitivity in OD is fostered by practitioners who practice openness, respect, and commitment during the therapeutic process. When clients see that their network members consistently work together with practitioners, the trust of their network members can enhance their own engagement with the therapeutic process.

6.4 Trust, Vulnerability, and Reciprocity in OD

Trust theory distinguishes between trust and reciprocity, arguing that genuine trust is not necessarily tied to expectations of mutual benefit. This distinction is crucial in OD, where the therapeutic process prioritizes genuine, non-reciprocal trust over transactional exchanges. In

OD, practitioners engage with clients through goal-adoption—where they adopt the client's well-being as their primary objective without expecting immediate therapeutic success or reciprocation. This aligns with the assertion that trust involves a willingness to rely on others and to accept vulnerability. OD leverages this vulnerability by inviting clients and practitioners alike to engage in open, exploratory dialogue without fixed outcomes, fostering a space where trust can grow organically through shared experiences rather than through expected exchanges ([37]).

7 Conclusion

In this chapter, we attempted to honor Rino's work by exploring the role of trust in Open Dialogue. We began by reflecting on the literature related to OD and then extended that discussion by integrating key insights from Trust Theory. Although our contribution is modest, the process of examining OD through the lens of trust has been both stimulating and eye-opening. It has also made us aware of how preliminary our reflections still are, and we hope to continue deepening them in the years to come, ideally with Rino's guidance.

From our engagement with Trust Theory, we have developed several working hypotheses regarding the role of trust in the OD framework. We hypothesize that trust functions as a form of relational capital—an evolving resource that is critical not only for therapeutic engagement but also as a foundational element of the dialogical process itself. We suggest that trust, nurtured through mutual vulnerability and authentic dialogue, plays a central role in helping clients rebuild both social trust and self-trust. These working hypotheses offer potential pathways for understanding how trust enables more meaningful, transformative therapeutic relationships, which, in turn, foster resilience and recovery for individuals facing severe mental health crises. While these ideas remain hypotheses for now, they highlight promising areas for future theoretical development and empirical research. Exploring how trust is established, maintained, and expanded within Open Dialogue—and its impact on long-term client outcomes—could greatly enhance our understanding of the dialogical process and its efficacy.

Finally, we conclude on a personal note, particularly from Raffaella: without Rino's trust, this entire line of research on OD might never have found its place within our institute. What began as an initial act of trust has since evolved into the coordination of one of the largest international project on this subject. For this, we are deeply grateful, and we recognize how essential trust has been—not just in our research, but in the collaborative spirit that is allowing this work to flourish.

References

[1] C. R. Rogers, "The necessary and sufficient conditions of therapeutic personality change.," vol. 21, no. 2, pp. 95–103.

[2] S. Messer, "The real relationship in psychotherapy: The hidden foundation of change," vol. 22, no. 3, pp. 363–365.

[3] R. Pocobello and T. el Sehity, "The recovered subject: A socio-cognitive snapshot of a new subject in the field of mental health," in *The Goals of Cognition - Essays in honour of Cristiano Castelfranchi* (F. Paglieri, L. Tummolini, R. Falcone, and M. Miceli, eds.), vol. 7 of *Tributes*, pp. 359–380, Colledge Publications.

[4] A. Prasannakumar, V. Kumar, and N. P. Rao, "Trust and psychosis: a systematic review and meta-analysis," vol. 53, no. 11, pp. 5218–5226.

[5] K. Bacha, T. Hanley, and L. A. Winter, "'like a human being, i was an equal, i wasn't just a patient': Service users' perspectives on their experiences of relationships with staff in mental health services," *Psychology and Psychotherapy: Theory, Research and Practice*, vol. 93, no. 2, pp. 367–386, 2020.

[6] R. Pocobello, T. el Sehity, L. Negrogno, C. Minervini, M. Guida, and C. Venerito, "Comparison of a co-produced mental health service to traditional services: A co-produced mixed-methods cross-sectional study," vol. 29, no. 3, pp. 460–475.

[7] A.-K. J. Fett, S. S. Shergill, N. Korver-Nieberg, F. Yakub, P. M. Gromann, and L. Krabbendam, "Learning to trust: trust and attachment in early psychosis," vol. 46, no. 7, pp. 1437–1447.

[8] A. Villena-Jimena, J. M. Morales-Asencio, C. Quemada, and M. M. Hurtado, ""it's that they treated me like an object": A qualitative study on the participation of people diagnosed with psychotic disorders in their health care," vol. 20, no. 5, p. 4614.

[9] J. Seikkula, J. Aaltonen, B. Alakare, K. Haarakangas, J. Keränen, and K. Lehtinen, "Five-year experience of first-episode nonaffective psychosis in open-dialogue approach: Treatment principles, follow-up outcomes, and two case studies," vol. 16, no. 2, pp. 214–228.

[10] S. von Peter, V. Aderhold, L. Cubellis, T. Bergström, P. Stastny, J. Seikkula, and D. Puras, "Open dialogue as a human rights-aligned approach," vol. 10.

[11] J. Seikkula, J. Aaltonen, K. Haarakangas, J. Keränen, and M. Sutela, "Treating psychosis by means of open dialogue," in *The reflecting team in action: Collaborative practice in family therapy.* (S. Friedman, ed.), The Guilford family therapy series, pp. 62–80, Guilford Press.

[12] Y. O. Alanen, K. Lehtinen, V. Räkköläinen, and J. Aaltonen, "Need-adapted treatment of new schizophrenic patients: experiences and results of the turku project," vol. 83, no. 5, pp. 363–372. Publisher: John Wiley & Sons, Ltd.

[13] J. Seikkula, B. Alakare, and J. Aaltonen, "The comprehensive open-dialogue approach in western lapland: II. long-term stability of acute psychosis outcomes in advanced community care," vol. 3, no. 3, pp. 192–204.

[14] T. Bergström, B. Alakare, J. Aaltonen, P. Mäki, P. Köngäs-Saviaro, J. J. Taskila, and J. Seikkula, "The long-term use of psychiatric services within the open dialogue treatment system after first-episode psychosis," vol. 9, no. 4, pp. 310–321.

[15] R. Pocobello, F. Camilli, M. Alvarez-Monjaras, T. Bergström, S. von Peter, M. Hopfenbeck, V. Aderhold, S. Pilling, J. Seikkula, and T. J. el Sehity, "Open dialogue services around the world: a scoping survey exploring organizational characteristics in the implementation of the open dialogue approach in mental health services," vol. 14, pp. 1–16. Publisher: Frontiers.

[16] J. Seikkula and M. E. Olson, "The open dialogue approach to acute psychosis: its poetics and micropolitics," vol. 42, no. 3, pp. 403–418.

[17] N. Putman, "Introducing open dialogue training," in *Open Dialogue for Psychosis* (N. Putman and B. Martindale, eds.), pp. 9–34, Routledge.

[18] J. Aaltonen, J. Seikkula, and K. Lehtinen, "The comprehensive open-

dialogue approach in western lapland: I. the incidence of non-affective psychosis and prodromal states," vol. 3, no. 3, pp. 179–191.

[19] M. Olson, J. Seikkula, and D. Ziedonis, "The key elements of dialogic practice in open dialogue: Fidelity criteria," vol. 1.1, p. 33.

[20] J. Brown, M. Kurtti, T. Haaraniemi, E. Löhönen, and P. Vahtola, "A north–south dialogue on open dialogues in finland: The challenges and the resonances of clinical practice," vol. 36.

[21] S. M. Geller and L. S. Greenberg, *Therapeutic presence: A mindful approach to effective therapy*. Therapeutic presence: A mindful approach to effective therapy, American Psychological Association.

[22] C. Lagogianni, E. Georgaca, and D. Christoforidou, "Co-therapy in open dialogue: Transforming therapists' self in a shared space," vol. 14.

[23] M. Nosek and V. E. Meade, ""holding space": A phenomenological exploration of mindfulness-based practice with veterans who have experienced trauma," vol. 61, no. 7, pp. 47–55.

[24] S. Schubert, P. Rhodes, and N. Buus, "Transformation of professional identity: an exploration of psychologists and psychiatrists implementing open dialogue," vol. 43, no. 1, pp. 143–164.

[25] J. Piippo and J. Aaltonen, "Mental health care: trust and mistrust in different caring contexts," vol. 17, no. 21, pp. 2867–2874.

[26] J. Piippo and J. Aaltonen, "Mental health: integrated network and family-oriented model for co-operation between mental health patients, adult mental health services and social services," vol. 13, no. 7, pp. 876–885.

[27] A. C. Florence, G. Jordan, S. Yasui, D. R. Cabrini, and L. Davidson, ""it makes us realize that we have been heard": Experiences with open dialogue in vermont," vol. 92, no. 4, pp. 1771–1783.

[28] "Dialogue as a response to the psychiatrization of society? potentials of the open dialogue approach," vol. 6, pp. undefined–undefined.

[29] R. Falcone and C. Castelfranchi, "Trust and transitivity: How trust-transfer works," pp. 179–187. Publisher: Springer, Berlin, Heidelberg.

[30] C. Castelfranchi, R. Falcone, and E. Lorini, "A non-reductionist approach to trust," pp. 45–72. Publisher: Springer, London.

[31] C. Castelfranchi, R. Falcone, and F. Marzo, "Being trusted in a social network: Trust as relational capital," in *iTrust 2006* (K. Stølen, ed.), no. 3986 in LNCS, pp. 19–32, Springer.

[32] C. Castelfranchi, "Trust and reciprocity: misunderstandings," vol. 55, no. 1, pp. 45–63.

[33] C. Castelfranchi and R. Falcone, *Trust Theory: A Socio-Cognitive and Computational Model.* John Wiley & Sons.

[34] R. Falcone and A. Sapienza, "The role of trust in dependence networks: A case study," vol. 14, no. 12, p. 652. Number: 12 Publisher: Multidisciplinary Digital Publishing Institute.

[35] R. Falcone and C. Castelfranchi, "Socio-cognitive model of trust," in *Encyclopedia of Information Science and Technology* (M. KhosrowPour, ed.), pp. 2534–2538.

[36] R. Falcone and C. Castelfranchi, "The socio-cognitive dynamics of trust: Does trust create trust?," in *Trust in Cyber-societies* (R. Falcone, M. Singh, and Y.-H. Tan, eds.), vol. 2246, pp. 55–72, Springer Berlin Heidelberg.

[37] J. A. J. Seikkula, Birgitta Alakare, "Open dialogue in psychosis i: An introduction and case illustration," *Journal of Constructivist Psychology*, vol. 14, no. 4, pp. 247–265, 2001.

From Trust Modeling to Trusted AI: Enlightened by the Work of Falcone

Robin Cohen

Cheriton School of Computer Science, University of Waterloo
Waterloo, Ontario, Canada N2L 3G1
`rcohen@uwaterloo.ca`

Abstract

A Blue Sky paper at AAMAS 2019 argued that researchers studying trust modeling in multiagent systems have a lot to contribute to a topic of increasing interest in artificial intelligence, namely how to achieve trusted AI. The position of the paper was that we have an obligation to draw greater attention to the effort lasting more than twenty years now on how to judge the trustworthiness of other agents, in order for key insights to emerge on the question of engendering AI trust. In this chapter, we return to some of the central ideas of Falcone and Castelfranchi regarding how trust is defined. We argue that without very solid grounding in the notion of trust, it is challenging to reach the goal of trusted AI. We shed some light as well on the relationship between this fundamental work in modeling trust and a particular concern arising today in efforts towards responsible AI, namely the value of reasoning about mental models of users of these systems. All of this culminates in commentary on the value of continuing to enable context-specific trust modeling and in turn context-specific solutions for trusted AI...

I am indebted to Nanda Kishore Sreenivas for his assistance in preparing this chapter. I am also grateful to Liliana Ardissono and Pinar Yolum for providing detailed insights into some of their recent research for my reflection on current trends.

Towards this end, we acknowledge again the foresight of Falcone to examine the study of trust in distinct areas of application, lending support to the position that only through exploring specific contexts of use does one truly appreciate what trust modeling can contribute to trusted AI.

1 Introduction

In this chapter we explore the position that trust modeling in general and that reasoning about the specific context of use are both essential in order to design approaches for promoting trusted artificial intelligence. Explainable AI has emerged as a central concern in order to provide more confidence for users to proceed with the use of AI systems for decision making within organizations [1]. But with many possible approaches for deriving explanations, it turns out that the perspective of the user who will employ the AI system becomes very important.

We begin with commentary in a recent Blue Sky paper at AAMAS [2], which argues that returning to the area of research known as multiagent trust modeling provides a much-valued anchor for establishing trusted AI. We also discuss elements of that Blue Sky paper which reinforce the claim that users will need to see solutions mapped out in their particular context of use, attuned to their perspectives of what may be accepted as trustworthy. We highlight how previous work of Rino Falcone has shown the real value of these cornerstones for trusted AI, namely a detailed representation of the concept of trust [3] and an illustration that trust assumes distinct visions within its different context of use, for AI systems [4, 5].

We then briefly discuss current research within AI today that touches on the need for responsible AI and the avenues for achieving it. This includes reflection on the role of mental models of users [6, 7, 8, 9], the effort on achieving explainable AI [10, 11], and very recent debate on where the challenges lie in order to achieve this, given current attention on large language models, within our field of study [12, 13].

As a backdrop, we include a dedicated study of our own work over the past 20 years, noting where we observe the origins of the important themes of understanding the concept of trust and supporting distinct solutions in different contexts of use. Towards the end of the chapter we reflect on our most recent research in the field, we speculate as well on the kind of future that lies ahead, of possible benefit to society, due to concentrated efforts to not only develop AI solutions for today's problems, but to care about the ways in which these systems will be used and then ultimately accepted when developing our technological artefacts.

As we explain our current work and its adherence to our former principles what emerges is an even more resounding tribute to Rino Falcone and his seminal work, as a clear indication of our esteem for him, as he heads into an entirely new phase of his life.

2 Background

We begin with a summary of some of the work of Rino Falcone who, with coauthors, has provided our research community with essential grounding in the understanding of the concept of trust and has made clear that distinct areas of application require context-specific treatment of trust modeling.

What comes to mind first and foremost is the work co-authored with Castelfranchi from the Multiagent Systems conference of 1998 [3]. The abstract of this paper makes clear the stance of the authors that trust is both a mental state and a social attitude and relation. The descriptions that follow are grounded in a formal treatment of belief. Integrated into the model are concerns regarding credibility and conditions under which an agent would be willing to delegate. The themes that come through are embracing BDI architectures for agents and viewing trust as an inherently social relationship.

If you look at how the field of multiagent trust modeling expanded since this time, one can see consistent effort to be formal with definitions and to delve deeply into the interrelationship between trust and the inherently-social concept of reputation. Logic-based research with the work of such researchers as Vercouter and Lorini [14] and continued efforts of researchers such as Bentahar [15] highlight the respect for Castelfranchi and Falcone's committment to careful definitions and inter-relationships. By this time as well terms such as trustor and trustee, now commonplace in our subfield of multiagent systems, came into popular use, as did the distinction between trust and reputation.

Another key insight of Falcone's, moving forward from the earlier days into papers presented only in the past few years, was realizing the importance of exploring the modeling of trust in distinct areas of application. There are many different domains that are studied in Falcone's work, including ones that have become magnets for attention with artificial intelligence (robotics, ioT [16, 17, 18]). And yet more eclectic and intriguing areas of application have come into focus as well, including hydrology and weather [4] and social behaviour surrounding Covid-19 [5].

There is in fact a progression in thought here, though a subtle one, that we claim is becoming central to several of the most important subtexts in the study of AI today. If one wants to model trust of other agents, those other agents can indeed be situated in different enviroments (agents as robots, agents as iOT devices). But if one wants to comprehend whether users will accept an AI system that is used for decision making, the question then adjusts to what really constitutes the concept of trust in these settings and what causes humans to believe what is being said and done. In essence, the concept of trust itself evolves, depending on the topic of study; this is distinct from trying to transport the same well-developed characterization of trust into different environments, to appreciate the richness of its use.

This perspective became all the more apparent in the paper regarding Covid-19 [5] as what each person perceived to be the social acceptance among their peers became another important element of predicting behaviour, towards engaging users in the best possible way. Those directing the social system were judged according to distinct factors, willingness and competence. What constituted trust indeed evolved.

3 Perspective of the field of AI trust modeling

We pause briefly to clarify two of the primary threads of research in trust modeling for multiagent systems. This is the personal perspective of the author of this chapter, based on their own journey within this field of study. Our origins were in natural language processing and from here we became part of the user modeling community which formed in 1986 [19, 20]. One seminal paper that emerged, coauthored with Ardissono, demonstrated how user feedback could be essential for intelligent agents to truly assist [21]. At this point in time we realized the importance of understanding far more than just the goals of speakers in order to be of assistance when designing an AI system to work together with a user.

Some years later, papers began to surface which discussed how to help users in managing electronic commerce online, and then it was apparent that a key issue concerned how much to trust the party with which one might be entering a business partnership. The first efforts along these lines proposed modeling trust on the basis of direct experience, progressively learning and then predicting which agents would be best to consider. Our first papers in this area were with PhD student Tran [22] and integrated clever methods for modeling reputation and attempting to avoid poor choices of partners. By the time had begun to flourish, the term reputation was reserved instead for a modeling of agents based on reports from advisors, representing peers in the environment who may have had experiences themselves, to share with others when there was little evidence yet from any direct interactions of that agent. At this point our work with PhD student Zhang made clear that one needed to model the trustworthiness of advisors as well, so that the algorithms to propose which agents to accept made use of multi-level reasoning [23, 24, 25]. Imagining where incentives to honesty could be integrated, with coauthor Larson, arose as well [26]. Very interesting work was also completed with coauthors Yolum and Şensoy concerning advances for online information systems in general, grounded in trust [27].

To this date, the concept of witness-based trust modeling predominates and several new challenges have arisen, including how to mine data in order to learn which features are most important in order to personalize the modeling [28]. Other key insights that arrived within our own work were the fact that cheaters could prosper if they knew their trust was being modeled [29] and that one could still make good use of reports from peers even if they were quite subjectively different from you [30, 31]. Truly novel applications eventually came into focus in our research, including recommender systems and a role for federated learning [32, 33].This then was the backdrop of the insights which arrived just a few years ago, linking to the rising concern of whether AI systems could be more trusted, to continue their use within many organizations.

4 Motivation

The commentary in this chapter originates with the perspective of the author of this book chapter from 2019. They published a BlueSky paper at AAMAS that year entitled "Trust modeling and the contribution of trust modeling in multiagent systems". While great interest was emerging at that time in exploring issues of fairness and transparency of our AI systems, in order for organizations to rely on these systems for their decision making [34], few efforts acknowledged the role that research in multiagent trust modeling could play. Better understanding of the relative value of differing solutions for trusted AI could emerge, it was argued, if one were to leverage some of the standard metrics from the multiagent trust community. That paper also explained that current efforts for fairness, transparency and such each seemed to use distinct measures while demonstrating the success of a particular approach. Two separate use cases were showcased in the paper as well. The reason for doing so was to clarify how, for a variety of possible settings where reseachers wanted their AI systems to be embraced and accepted, trust modeling perspectives could be integrated to help to foster trusted AI. The first was that of AI vision systems applied to webpages in order to make possible greater acceptance from users with assistive needs (for example decluttering or highlighting) where individual preferences would be supported. The second was the introduction of argumentation systems for AI-directed decision making. Even then, it was clear that the context of use would matter when developing the best solutions for trusted AI.

The vision example as sketched in the paper made clear that reputation could be built during the training of the vision system, making use of some of the direct experience trust modeling first proposed to reinforce the learning in the work of Tran and Cohen [22]. The argumentation setting made the point that decision support accepted by a group would need to appreciate a nuance of opinions being influenced by others where predictions could drive the overall reasoning. This then introduced an important role for witness-based trust modeling. All this was grounded as well in important appreciation for the human needs to acquire a grounding of trust in information sources, through consideration of benevolence and integrity (not so distant from the key factors Sapienza and Falcone discovered when aiming to gain insights into the Covid-19 setting [5]).

5 Most Recent Themes in Trusted AI

Fast forward to 2024, a mere five years later, and the concept now in focus when discussing whether AI solutions can be trusted and accepted is that of large language models. What our systems have been capable of doing and the uses to which they have been put has advanced considerably and the interest in making use of massive databases and clever reasoning methods has exploded, both in the sense of seeing new opportunities and being concerned about new threats. All this is to say that trusted AI continues to be a topic of great interest and that new directions for developing appropriate solutions are emerging.

We return to expand upon the discussion of how enabling users to trust AI systems has now been thrust into a study of large language models, later in this chapter. Before doing so we explain how earlier work on reasoning about the mental models of users emerged, in order to convey the significance of this particular thread of study.

5.1 Increased interest in the mental models of users

Prior to this focus on large language models, some researchers were already devoting considerable attention to trying to tease apart the need to have transparent, understandable AI systems and the desire for these systems, once employed, to provide the most accurate and valuable output for their users (typically, organizations interested in leveraging the ability for these intelligent systems to determine the best decisions to make, given massive data that can be analyzed). One researcher who has impressed considerably in this respect is Kambhampati. Along with this students, and grounded well in the environment of AI planning, the themes that were explored led ultimately to the insight that mental models of users were an essential component of designing our AI systems to function well in the realworld.

There has in fact been a steadily increasing and encouraging trend within AI over the past 5 years or so to revisit the importance of considering the mental models of humans as part of the design of effective AI decision-making systems. The users of these systems are often those who launched the AI system to be employed within an organization and who therefore would like to know more about the reasoning of the systems in use, often conveyed in terms of explanations [35].

One of the earliest papers in this new generation of attention to this important topic is the IJCAI 2017 paper of Chakraborti et al. [6]. As clarified within the abstract, the authors suggest that generating explanations of plan-based systems simply from the perspective of the algorithms underlying the design is destined to fall short of the mark, for users. Instead, it is ideal to imagine what the user's mental model might be, to then discover how the system's reasoning may diverge from this perspective, to then expand the output to users towards a more effective trust in and acceptance of these AI systems.

In this particular paper, the authors present methods for searching a space of models and discuss why it may be important to consider more than just whether the explanations are easy to generate. a space of models and discuss the relative advantages of explanations of AI systems that are easy to compute compared to ones that are more complete. The authors leave for future work a greater examination of how to learn a human's model and also to allow for different cognitive abilities of the relevant human users.

This theme also arises in the companion work of Sreedharan et al. [8] who begin my stating that a set of possible worlds may need to be considered when trying to achieve model reconciliation. They propose introducing conditional explanations to then iterate with a user in order to bring the process closer to what that user needs to see. They reveal the primary challenge that arises, namely the computational overhead for the agent set against the communication overhead of the human. Continuing to examine trusted AI and explanation with a specific context, namely that of planning, allows the authors to develop detailed human-aware processes and enables interesting applications to specific domains such as search and rescue, for illustration. The AI system in question is in fact robots and approach is framed as determining how best to perform post-processing of actions to be taken, in order to become progressively closer to the needs of the human users.

This group of researchers has recently expanded in order to provide a deeper exploration of human mental models and this time the effort is explicitly framed as one of faciltating trust, with Zahedi as lead author [9]. The authors delve further into behavioural trust inference and its relationship to the design of AI decision-making frameworks. They refer to their formalization as a mental model based theory of trust and are invested in comparing what they develop against some competing notions of trust. While they do not select the very well known work of Castelfranchi and Falcone [3] as the comparator, selecting instead the trust scales of Muir [36], it is abundantly clear that the authors put in place a definitive anchor of what constitutes trust, in order to bring our field closer to one that can be running systems for users who will then accept these systems and continue to use them.

The discussion is then grounded in a particular approach to trust model for robots in particular, a Bayesian reasoning framework developed by Xu and Dudek [37] named OPTIMo. This enables a concrete comparison of performance in terms of the robot's ability to execute tasks as expected by the human. Soliciting user feedback as to trust enables better calibration of the outcomes. Observational studies were then included. The methodology of the authors is important, as this begins to bring the field closer to a concern raised in the AAMAS 2019 Blue Sky paper, introduced at the start of this chapter, namely the desire to be able to judge whether different solutions for trusted AI are better than competitors, in terms of some fixed metrics. While multiagent trust modeling systems have, for years, run simulations in environments such as electronic commerce, to judge whether un-trusted advice may have led to poorer outcomes, the enterprise of designing effective explainable AI systems hasn't perhaps reached the same level of comparing possible solutions. Regardless, the proposal to ground a particular method in a specific context of use, to then examine whether the outcomes are meeting user needs, is a central theme and this is what we also feel Falcone has embraced within his trust-oriented research over the past 5 years.

Zahedi et al. [9] do well to allow trust to evolve over time and to formalize appropriate levels of trust (elements which are endemic to the original model of Castelfranchi and Falcone [3] as well). What is really encouraging to see, as the logical extension of research in study-ing the concept of trust for AI systems is a detailed user study where human users provide their views, towards possible improvements of the formalism in the future. While this may not have been in focus in the 1998 paper [3], Falcone's efforts to project trust modeling into different use contexts certainly suggests that studying the acceptance of these models by the specific users within those contexts is an ex-citing step forward.

One interesting way of viewing the path through this particular set of research is as follows. Kambhampati's AAAI presidential address in 2020 [38] suggested that it may be useful, at times, to sacrifice accuracy for explainability (all in the name of achieving better accep-tance of AI Systems within various organizations). From here, there were a series of efforts exploring how to operationalize the success of AI systems (delving into applications for robotics and planning for some breadth) [39, 40, 41]. At some point the essential realization was that acceptance by users would ultimately rest on how inclined the user was to be willing to embrace the AI system [42] and this led to the observation that each user is different and thus what is needed is a representation of that user's mental model. If this could be captured with some precision, then the question was whether the match to mental model was sufficiently cohesive to achieve the desired outcome [43, 44]. Left for continued future work was continued finetuning of the representation of mental models and the affirmation of their success in acceptance of AI. Small efforts along that pathway emerged through periodic user studies, both from Kambhampati's group [45, 6] and those of other researchers [46], including my own brief examination of personalized explanation [47].

5.2 Large language models lack human insights

We are on the horizon of a particular topic of conversation when reflecting on trusted AI (the use of AI systems for decision making and the trust that users of these systems may have or may lack in these systems, when in use). The topic in question is how the challenges we face are different, now that Large Language Models (LLMs) are in predominant use. Previously, the focus was on being able to adequately explain systems that were grounded in neural networks driven big data; the reasons why certain advice would be proposed by these systems could be buried in the workings of layers of these neural networks and also being data-driven, the rationale might not have been immediately obvious, in order to provide users with appropriate clarifications. With large language models, the amounts of data have truly escalated and the transparency of the methods has declined [48]. The fact that the topic of large language models was one of the most central items in discussion at the AAAI 2024 conference is especially telling, as explained in the exposition below.

Three invited talks at the conference all provided signficant insights into LLMs. Towards the end of the conference there was a panel discussion chaired by Kevin Leyton-Brown, with panelists Sheila McIlraith, Rao Kambhampati, Charles Sutton and Christopher Manning. Earlier, Yann le Cun spoke about Machines that can Learn, Reason and Plan. And on the first day of the conference invited speaker Sarit

Kraus presented her views on the current state of explainable AI and the need to develop novel directions today in view of the use of LLMs. The talk was entitled Intelligent Agents Assisting Humans in the Super AI era.

Kraus has done some work to determine whether LLMs have some of the same cognitive biases that humans do [49]. In the end, she concluded that human-AI partnerships are essential when AI systems are enlisted to perform decision-making tasks and also that intelligent agents may help to bring the humans and the LLM systems closer in order for the partnership to flourish. One very interesting thread in her research was to go beyond scenarios where explanations of systems are derived from automated systems which determine which features are most central to the decision making (such as LIME [1]). Kraus allowed humans to take the initiative to propose certain features of the domain that they reasoned would be most important to know about, to then have the explanations that are generated revised in order to encompass these elements as well [50]. One lesson here perhaps is that if some of our newest AI models are one which may be most difficult to explain to users, we may be able to focus on avenues for bringing humans back into the loop more effectively. Certainly trust in the deployment of these super intelligent systems can be engendered if at least some control is brought back to the user. A method being explored here is known as contrastive explanations and the user studies that Kraus has conducted show promising responses from human users, preferring these kinds of explanations to other baselines. They have held some promise as well for making multiagent reinforcement learning systems more accessible to their human users [51]. The fact that the solution in this paper extends the one proposed earlier by Sreedharan et al [45] helps to reinforce the earlier message that the path forward for successful trusted AI should continue to consider the mental models of our human users.

In his invited talk at AAAI 2024, Yann le Cun characterized Large Language Models as having little commonsense, no persistent memories and limited reasoning and planning (even for the popular application of vision) [52]. He pointed to work of Valmeekam et al. [53]

that explictly studies the planning abilities of LLMs, as one relevant study. Some of le Cun's ideas for how to make LLMs more human are expressed in a paper from 2022 [54]. This is an architecture for deep learning that uses hierarchical joint embeddings to operate at even greater scales. He also points out that generative models aim to predict what will happen and are often examined when there is not enough data coming to other machine learning models. He speculates that if some of his ideas for how to make LLMs better at planning become successful, we may also see these models doing better at actually operating with common sense, allowing us to be one step closer to having these models trusted by our users.

Towards the end of the conference there was a panel discussion chaired by Kevin Leyton-Brown, with panelists Sheila McIlraith, Rao Kambhampati, Charles Sutton and Christopher Manning. Manning began his discussion by mentioning that any steps forward with LLMs should aim to be done in accordance with human preferences. Kambhampati made clear that there are inherent limitations of LLMs; he looks forward to beautiful synergies between LLMs and more symbolic approaches that can perform reasoning (because we cannot assume that the planning done by the LLMs is accurate and correct). McIlraith mentioned the value of audits, going forward and imagines there may be some external reasoner operating together with the LLMs. Sutton said he was excited about neurosymbolic systems and that this step forward may end up providing the needed promise for the future. Kambhampati also talked of the role that provenance can play, if we are able to secure this at all; if we know what has resulted from LLMs this may assist in the process of trying authenticate sources. He stressed that building trustworthy AI systems is going to be increasingly important now. The reputation of the source will matter. Sutton expressed concern that LLMs may not exercise good judgment and stressed again the importance of human input, human empathy for any of the applications to decisionm making, moving forward. All this again suggests pathways such as the one that Kraus advocated, namely greater partnerships between LLMs and human experts. Manning then encouraged us to apply caution, because human preferences are not always the best, either. Kambhampati reminded us of the age-old AI paper called Artificial Intelligence meets Natural Stupidity [55]. McIlraith reminded us that the fact that LLMs have been grounded in perception and interpreted in context is in fact good and we should seek to take advantage of whatever the introduction of LLMs can achieve. It is fair to say that the panel saw many concerns with the current introduction of LLMs, seemed to feel that LLMs will still be with us and then sought to brainstorm on what we may be able to do, for next steps. It was refreshing that trustworthiness was raised as an essential concern and that contextualizing whatever we are studying, to make progress, was advocated. As such, the central themes of this chapter may yet be in focus even for the new realities that we are facing with more powerful AI systems becoming popular.

6 Final reflection: current state, future work

We began this chapter with a chronicle of our journey into the field of multiagent trust modeling and mentioned coauthors who accompanied us on that path. We can look at where these researchers are today to gain further insights into how crucial contributions of Rino Falcone have resonated with other leading figures in the field, providing evidence of the continued legacy of his formative work.

Liliana Ardissono has devoted considerable effort working with the user modeling community, exploring interesting pathways such as the design of cultural heritage guides [56] (which can be personalized) but also continues to embrace the challenge of determining how best to model trust between intelligent agents. One recent effort examines trust in information sources based on social network data [57], and proposes a multi-faceted model which also considers what users may prefer not to have disclosed. Another thread studied as well how to provide justifications for key recommendations coming from these systems [58][1]. Expanding the vision for how trust is to be defined and modeled, and realizing the important role that context plays are both

---

[1] This helps to very nicely reveal how our AI systems can really only be trusted and accepted if we can somehow explain their reasoning to our users.

quite important, exactly the themes this chapter highlights when acknowledging the influence of Falcone.

Our retrospective also mentioned coauthor Pinar Yolum, who has recently been focusing on novel directions for addressing privacy concerns and yet the central theme of modeling trust continues to be an important component of the work. In particular, they are studying how personal assistants can facilitate user interactions with iOT devices in order for privacy to be managed [59, 60]. A critical insight of this work is that identifying the context of interactions really enables making better trust judgments for users; this is especially helpful when there may not be a large set of interactions from which to learn. Once more the central need for grounding in a definition of trust and the importance of designing solutions attuned to specific settings arise here.

In the meantime, it has also been encouraging to see former coauthors Kate Larson and Thomas Tran extending into new territory in the area of trust in AI systems; some of Larson's current research examines how to support explanations of AI systems in a way that is contextually relevant [61, 62] whereas one of Tran's directions has been to challenge recommender systems to be suitably attuned to the need for their users to obtain trust [63, 64]. It is also encouraging to see the age-old concepts of trustor and trustee, established in Falcone's earliest work [3] emerging again in Tran's exploration of how agents in multiagent systems can engender trust (a theme proposed by Sen in his Blue Sky paper [65]) [66, 67].

It is apparent that the need to model trust and to adjust its application to contexts of use continue to flourish as concerns not only for multiagent systems researchers but also within the AI community. Today's world is one in which AI systems are increasingly applied and used and where organizations that seek to integrate AI solutions are increasingly interested in better understanding how to accept these systems and continuously questioning whether their specific applications are being handled appropriately. It is up to us as AI researchers to rise to the challenge of meeting user needs when designing our systems; without this, the benefits that our systems may bring to society cannot be realized. We don't yet know what the future holds as we see the rise of powerful new AI systems that impose much greater challenges in meeting goals of transparency and explanation. But if we take it upon ourselves to examine these questions as we move forward, then we can continue the dialogue that enables our fascinating area of research to continue to advance, one day leading to a better understanding of what we may want to happen with its usage in the world at large.

References

[1] M. T. Ribeiro, S. Singh, and C. Guestrin, ""Why Should I Trust You?": Explaining the Predictions of Any Classifier," in *Proceedings of the 22nd ACM SIGKDD International Conference on Knowledge Discovery and Data Mining*, KDD '16, (New York, NY, USA), pp. 1135–1144, Association for Computing Machinery, Aug. 2016.

[2] R. Cohen, M. Schaekermann, S. Liu, and M. Cormier, "Trusted AI and the Contribution of Trust Modeling in Multiagent Systems," in *Proceedings of the 18th International Conference on Autonomous Agents and MultiAgent Systems*, AAMAS '19, (Richland, SC), pp. 1644–1648, International Foundation for Autonomous Agents and Multiagent Systems, May 2019.

[3] C. Castelfranchi and R. Falcone, "Principles of trust for MAS: cognitive anatomy, social importance, and quantification," in *Proceedings International Conference on Multi Agent Systems (Cat. No.98EX160)*, pp. 72–79, July 1998.

[4] R. Falcone and A. Sapienza, "Community dynamics in case of critical hydrogeological phenomena: Some simulated scenarios," in *TRUST workshop at AAMAS 2018*, (Stockholm, Sweden), 2018.

[5] A. Sapienza and R. Falcone, "Studying Citizens' Trust to Monitor Measures Acceptance during COVID-19 Pandemic," in *Proceedings of the 22nd International Workshop on Trust in Agent Societies*, (London, UK), May 2021.

[6] T. Chakraborti, S. Sreedharan, Y. Zhang, and S. Kambhampati, "Plan explanations as model reconciliation: moving beyond explanation as soliloquy," in *Proceedings of the 26th International Joint Conference on Artificial Intelligence*, IJCAI'17, (Melbourne, Australia), pp. 156–163, AAAI Press, Aug. 2017.

[7] T. Chakraborti, S. Sreedharan, S. Grover, and S. Kambhampati, "Plan explanations as model reconciliation: an empirical study," in *Proceedings of the 14th ACM/IEEE International Conference on Human-Robot Interaction*, HRI '19, (Daegu, Republic of Korea), pp. 258–266, IEEE Press, Jan. 2020.

[8] S. Sreedharan, T. Chakraborti, and S. Kambhampati, "Handling Model Uncertainty and Multiplicity in Explanations via Model Reconciliation," *Proceedings of the International Conference on Automated Planning and Scheduling*, vol. 28, pp. 518–526, June 2018.

[9] Z. Zahedi, S. Sreedharan, and S. Kambhampati, "A Mental Model Based Theory of Trust," Jan. 2023. arXiv:2301.12569 [cs].

[10] S. Anjomshoae, A. Najjar, D. Calvaresi, and K. Främling, "Explainable Agents and Robots: Results from a Systematic Literature Review," in *Proceedings of the 18th International Conference on Autonomous Agents and MultiAgent Systems*, AAMAS '19, (Richland, SC), pp. 1078–1088, International Foundation for Autonomous Agents and Multiagent Systems, May 2019.

[11] D. Gunning and D. W. Aha, "DARPA's Explainable Artificial Intelligence Program," *AI Magazine*, vol. 40, no. 2, pp. 44–58, 2019. _eprint: https://onlinelibrary.wiley.com/doi/pdf/10.1609/aimag.v40i2.2850.

[12] K. Valmeekam, M. Marquez, A. Olmo, S. Sreedharan, and S. Kambhampati, "Planbench: an extensible benchmark for evaluating large language models on planning and reasoning about change," in *Proceedings of the 37th International Conference on Neural Information Processing Systems*, NIPS '23, (Red Hook, NY, USA), Curran Associates Inc., 2023.

[13] L. Guan, K. Valmeekam, S. Sreedharan, and S. Kambhampati, "Leveraging pre-trained large language models to construct and utilize world models for model-based task planning," in *Proceedings of the 37th International Conference on Neural Information Processing Systems*, NIPS '23, (Red Hook, NY, USA), pp. 79081–79094, Curran Associates Inc., May 2024.

[14] A. Herzig, E. Lorini, J. F. Hübner, and L. Vercouter, "A logic of trust and reputation," *Logic Journal of the IGPL*, vol. 18, pp. 214–244, Feb. 2010.

[15] J. Bentahar, N. Drawel, and A. Sadiki, "Quantitative Group Trust: A Two-Stage Verification Approach," in *Proceedings of the 21st International Conference on Autonomous Agents and Multiagent Systems*, AAMAS '22, (Richland, SC), pp. 100–108, International Foundation

for Autonomous Agents and Multiagent Systems, May 2022.

[16] A. Sapienza and R. Falcone, "An Autonomy-Based Collaborative Trust Model for User-IoT Systems Interaction," in *Intelligent Distributed Computing XV* (L. Braubach, K. Jander, and C. Bădică, eds.), (Cham), pp. 178–187, Springer International Publishing, 2023.

[17] R. Falcone and A. Sapienza, "On the Users' Acceptance of IoT Systems: A Theoretical Approach," *Information*, vol. 9, p. 53, Mar. 2018. Number: 3 Publisher: Multidisciplinary Digital Publishing Institute.

[18] F. Cantucci and R. Falcone, "Towards trustworthiness and transparency in social human-robot interaction," in *2020 IEEE International Conference on Human-Machine Systems (ICHMS)*, pp. 1–6, Sept. 2020.

[19] R. Kass and T. Finin, "Modeling the User in Natural Language Systems," *Computational Linguistics*, vol. 14, no. 3, pp. 5–22, 1988.

[20] W. Wahlster and A. Kobsa, "User Models in Dialog Systems," in *User Models in Dialog Systems* (A. Kobsa and W. Wahlster, eds.), (Berlin, Heidelberg), pp. 4–34, Springer, 1989.

[21] L. Ardissono and R. Cohen, "On the value of user modeling for improving plan recognition," in *IJCAI95 workshop on the next generation of plan recognition systems*, (Montreal, Canada), 1995.

[22] T. Tran and R. Cohen, "Improving User Satisfaction in Agent-Based Electronic Marketplaces by Reputation Modelling and Adjustable Product Quality," in *Proceedings of the Third International Joint Conference on Autonomous Agents and Multiagent Systems - Volume 2*, AAMAS '04, (USA), pp. 828–835, IEEE Computer Society, July 2004.

[23] J. Zhang and R. Cohen, "A personalized approach to address unfair ratings in multiagent reputation systems," in *AAMAS06 workshop on Trust in Multiagent Systems*, 2006.

[24] J. Zhang and R. Cohen, "Design of a mechanism for promoting honesty in E-marketplaces," in *Proceedings of the 22nd national conference on Artificial intelligence - Volume 2*, AAAI'07, (Vancouver, British Columbia, Canada), pp. 1495–1500, AAAI Press, July 2007.

[25] J. Zhang and R. Cohen, "Evaluating the trustworthiness of advice about seller agents in e-marketplaces: A personalized approach," *Electronic Commerce Research and Applications*, vol. 7, pp. 330–340, Sept. 2008.

[26] J. Zhang, R. Cohen, and K. Larson, "Combining Trust Modeling and Mechanism Design for Promoting Honesty in E-Marketplaces," *Computational Intelligence*, vol. 28, no. 4, pp. 549–578, 2012.

[27] M. Şensoy, J. Zhang, P. Yolum, and R. Cohen, "Poyraz: Context-

Aware Service Selection Under Deception," *Computational Intelligence*, vol. 25, no. 4, pp. 335–366, 2009.

[28] A. Parmentier, R. Cohen, X. Ma, G. Sahu, and Q. Chen, "Personalized multi-faceted trust modeling to determine trust links in social media and its potential for misinformation management," *International Journal of Data Science and Analytics*, vol. 13, pp. 399–425, May 2022.

[29] R. Kerr and R. Cohen, "Smart cheaters do prosper: defeating trust and reputation systems," in *Proceedings of The 8th International Conference on Autonomous Agents and Multiagent Systems - Volume 2*, AAMAS '09, (Richland, SC), pp. 993–1000, International Foundation for Autonomous Agents and Multiagent Systems, May 2009.

[30] K. Regan, P. Poupart, and R. Cohen, "Bayesian reputation modeling in E-marketplaces sensitive to subjectivity, deception and change," in *proceedings of the 21st national conference on Artificial intelligence - Volume 2*, AAAI'06, (Boston, Massachusetts), pp. 1206–1212, AAAI Press, July 2006.

[31] N. Sardana and R. Cohen, "Demonstrating the value of credibility modeling for trust-based approaches to online message recommendation," in *2014 Twelfth Annual International Conference on Privacy, Security and Trust*, pp. 363–370, July 2014.

[32] O. A. Wahab, R. Cohen, J. Bentahar, H. Otrok, A. Mourad, and G. Rjoub, "An endorsement-based trust bootstrapping approach for newcomer cloud services," *Information Sciences*, vol. 527, pp. 159–175, July 2020.

[33] G. Rjoub, O. A. Wahab, J. Bentahar, R. Cohen, and A. S. Bataineh, "Trust-Augmented Deep Reinforcement Learning for Federated Learning Client Selection," *Information Systems Frontiers*, July 2022.

[34] C. Dwork, N. Immorlica, A. T. Kalai, and M. Leiserson, "Decoupled Classifiers for Group-Fair and Efficient Machine Learning," in *Proceedings of the 1st Conference on Fairness, Accountability and Transparency*, pp. 119–133, PMLR, Jan. 2018. ISSN: 2640-3498.

[35] T. Miller, "Explanation in artificial intelligence: Insights from the social sciences," *Artificial Intelligence*, vol. 267, pp. 1–38, Feb. 2019.

[36] B. M. Muir, "Trust in automation. i: Theoretical issues in the study of trust and human intervention in automated systems," *Ergonomics*, vol. 37, pp. 1905–1922, 1994.

[37] A. Xu and G. Dudek, "OPTIMo: Online Probabilistic Trust Inference Model for Asymmetric Human-Robot Collaborations," in *Proceedings*

of the Tenth Annual ACM/IEEE International Conference on Human-Robot Interaction, HRI '15, (New York, NY, USA), pp. 221–228, Association for Computing Machinery, Mar. 2015.

[38] S. Kambhampati, "Challenges of Human-Aware AI Systems: AAAI Presidential Address," *AI Magazine*, vol. 41, pp. 3–17, Sept. 2020. Number: 3.

[39] S. Sengupta, Z. Zahedi, and S. Kambhampati, "To monitor or not: Observing robot's behavior based on a game-theoretic model of trust," in *Proceedings of AAMAS 2019 TRUST Workshop*, 2019.

[40] Z. Zahedi, M. Verma, S. Sreedharan, and S. Kambhampati, "Trust-aware planning: Modeling trust evolution in iterated human-robot interaction," *HRI 2023 - Proceedings of the 2023 ACM/IEEE International Conference on Human-Robot Interaction*, pp. 281–289, Mar. 2023. Publisher: IEEE Computer Society.

[41] S. Sreedharan, A. Kulkarni, D. E. Smith, and S. Kambhampati, "A Unifying Bayesian Formulation of Measures of Interpretability in Human-AI Interaction," *Proceedings of the 30th International Joint Conference on Artificial Intelligence, IJCAI 2021*, pp. 4602–4610, 2021. Publisher: International Joint Conferences on Artificial Intelligence.

[42] U. Soni, S. Sreedharan, and S. Kambhampati, "Not all users are the same: Providing personalized explanations for sequential decision making problems," *IEEE/RSJ International Conference on Intelligent Robots and Systems, IROS 2021*, pp. 6240–6247, 2021. Publisher: Institute of Electrical and Electronics Engineers Inc.

[43] S. Sreedharan, T. Chakraborti, and S. Kambhampati, "Foundations of explanations as model reconciliation," *Artificial Intelligence*, vol. 301, p. 103558, Dec. 2021.

[44] T. Chakraborti, S. Sreedharan, and S. Kambhampati, "The emerging landscape of explainable automated planning & decision making," in *Proceedings of the Twenty-Ninth International Joint Conference on Artificial Intelligence*, IJCAI'20, (Yokohama, Yokohama, Japan), pp. 4803–4811, Jan. 2021.

[45] S. Sreedharan, U. Soni, M. Verma, S. Srivastava, and S. Kambhampati, "Bridging the gap: Providing post-hoc symbolic explanations for sequential decision-making problems with inscrutable representations," in *International Conference on Learning Representations*, 2022.

[46] M. Martijn, C. Conati, and K. Verbert, ""Knowing me, knowing you": personalized explanations for a music recommender system," *User Mod-*

eling and User-Adapted Interaction, vol. 32, pp. 215–252, Apr. 2022.

[47] O. Chambers, R. Cohen, M. R. Grossman, and Q. Chen, "Creating a User Model to Support User-specific Explanations of AI Systems," in *Adjunct Proceedings of the 30th ACM Conference on User Modeling, Adaptation and Personalization*, UMAP '22 Adjunct, (New York, NY, USA), pp. 163–166, Association for Computing Machinery, July 2022.

[48] Q. V. Liao and J. Wortman Vaughan, "AI Transparency in the Age of LLMs: A Human-Centered Research Roadmap," *Harvard Data Science Review*, May 31 2024. https://hdsr.mitpress.mit.edu/pub/aelql9qy.

[49] J. Shaki, S. Kraus, and M. Wooldridge, "Cognitive Effects in Large Language Models," in *ECAI 2023*, pp. 2105–2112, IOS Press, 2023.

[50] P. Zehtabi, A. Pozanco, A. Bolch, D. Borrajo, and S. Kraus, "Contrastive Explanations of Centralized Multi-agent Optimization Solutions," *Proceedings of the International Conference on Automated Planning and Scheduling*, vol. 34, pp. 671–679, May 2024.

[51] K. Boggess, S. Kraus, and L. Feng, "Explainable multi-agent reinforcement learning for temporal queries," in *Proceedings of the Thirty-Second International Joint Conference on Artificial Intelligence*, IJCAI '23, (Macao, P.R.China), pp. 55–63, Aug. 2023.

[52] Y. LeCun, "Objective-driven AI: towards machines that can learn, reason and plan," 2024. AAAI 2024.

[53] K. Valmeekam, M. Marquez, S. Sreedharan, and S. Kambhampati, "On the planning abilities of large language models: a critical investigation," in *Proceedings of the 37th International Conference on Neural Information Processing Systems*, NIPS '23, (Red Hook, NY, USA), pp. 75993–76005, Curran Associates Inc., May 2024.

[54] Y. LeCun, "A Path Towards Autonomous Machine Intelligence Version 0.9.2, 2022-06-27," 2022.

[55] D. McDermott, "Artificial intelligence meets natural stupidity," *SIGART Bull.*, pp. 4–9, Apr. 1976.

[56] N. Mauro, A. Geninatti Cossatin, E. Cravero, L. Ardissono, G. Magnano, and M. Giardino, "Exploring Semantically Interlaced Cultural Heritage Narratives," in *Proceedings of the 33rd ACM Conference on Hypertext and Social Media*, HT '22, (New York, NY, USA), pp. 192–197, Association for Computing Machinery, June 2022.

[57] L. Ardissono and N. Mauro, "A compositional model of multi-faceted trust for personalized item recommendation," *Expert Systems with Applications*, vol. 140, p. 112880, Feb. 2020.

[58] N. Mauro, Z. F. Hu, and L. Ardissono, "Justification of recommender systems results: a service-based approach," *User Modeling and User-Adapted Interaction*, vol. 33, pp. 643–685, July 2023.

[59] N. Kökciyan and P. Yolum, "TURP: Managing Trust for Regulating Privacy in Internet of Things," *IEEE Internet Computing*, vol. 24, pp. 9–16, Nov. 2020. Conference Name: IEEE Internet Computing.

[60] N. Kökciyan and P. Yolum, "Taking Situation-Based Privacy Decisions: Privacy Assistants Working with Humans," in *Proceedings of the Thirty-First International Joint Conference on Artificial Intelligence*, (Vienna, Austria), pp. 703–709, International Joint Conferences on Artificial Intelligence Organization, July 2022.

[61] M. Pafla, K. Larson, and M. Hancock, "Unraveling the Dilemma of AI Errors: Exploring the Effectiveness of Human and Machine Explanations for Large Language Models," in *Proceedings of the CHI Conference on Human Factors in Computing Systems*, CHI '24, (New York, NY, USA), pp. 1–20, Association for Computing Machinery, May 2024.

[62] M. Pafla, M. Hancock, K. Larson, and J. Hoey, "The need for a common ground: Ontological guidelines for a mutual human-ai theory of mind," in *Workshop on Theory of Mind in Human-AI Interaction at CHI 2024*, (Hawaii, USA), 2024.

[63] R. Zhang, T. Tran, and Y. Mao, "Recommender systems from "words of few mouths"," in *Proceedings of the Twenty-Second international joint conference on Artificial Intelligence - Volume Volume Three*, IJCAI'11, (Barcelona, Catalonia, Spain), pp. 2379–2384, AAAI Press, July 2011.

[64] R. Zhang and T. Tran, "An information gain-based approach for recommending useful product reviews," *Knowledge and Information Systems*, vol. 26, pp. 419–434, Mar. 2011.

[65] S. Sen, "A comprehensive approach to trust management," in *Proceedings of the 2013 international conference on Autonomous agents and multi-agent systems*, AAMAS '13, (Richland, SC), pp. 797–800, International Foundation for Autonomous Agents and Multiagent Systems, May 2013.

[66] A. Aref and T. Tran, "RLTE: A Reinforcement Learning Based Trust Establishment Model," in *2015 IEEE Trustcom/BigDataSE/ISPA*, vol. 1, pp. 694–701, Aug. 2015.

[67] A. Aref and T. T. Tran, "A trust establishment model in multi-agent systems," in *Incentive and Trust in E-Communities, Papers from the 2015 AAAI Workshop, Austin, Texas, USA, January 25, 2015* (Z. Noo-

rian, J. Zhang, S. Marsh, and C. D. Jensen, eds.), vol. WS-15-08 of *AAAI Technical Report*, AAAI Press, 2015.

T^3: for a Multidimensional, Integrated and Dynamic Theory of Trust

CRISTIANO CASTELFRANCHI
Institute of Cognitive Sciences and Technologies, National Research Council of Italy (ISTC-CNR), 00196 Rome, Italy
`cristiano.castelfranchi@istc.cnr.it`

Abstract

It is no coincidence that the name and symbol that Rino wanted to give to his research group is T^3: the Trust multiplied by itself. Not only because there were 3 T ("**T**rust **T**heory and **T**echnology Group") and to center it on the vision of the relationship between "theory" and "technologies": and to also underline our perspective towards IT, AI, Robotics,.. That is:

- that on the one hand technologies must be based on theories, explanations, models;

- that on the other hand technology will contribute with its formal and operational tools to the redefinition, clarification of concepts and models - such as the modern concept of "purpose" in Cognitive Sciences, inspired by Cybernetics - and therefore to theory, not only to its implementation [1]. This is a merit of "the sciences of the artificial" in their impact on the sciences of man (Herbert Simon). Furthermore, Technology (AI, robotics, etc.) will also provide "experimental", simulation tools for model verification.

But above all this multiplication of T (Trust) has to underline the multifactorial nature of trust: its multiple faces and dimensions which also interact. This framing to counter simplistic or reductionist approaches, such as reducing trust to the estimate of the probability of the event/act and desired outcome

[2]. I will deal with various aspects of this complexity, and multidimensionality, and dynamics of the Trust; and I apologize if I go in too many directions and often give mere ideas and intuitions. But a homage is not only to the past and what has already been done but also to the future.

1 The faces of the reliability of Y

T^3's research was dedicated, for example, to the different faces or rather components of the "reliability" of Y (trustee), under two fundamental aspects:

1. The "Can Do": X (the trustor) must believe that Y "is/will be able" to do the action (A) and achieve the purpose. And this has two faces: powers internal to Y and powers external to Y (conditions, circumstances, means, access, ..). That is, that Y has/will have competence, know-how, skills, .. and opportunities.

2. The "Will Do": X must believe that Y has/will have the intention and therefore the motivation and perseverance relating to his doing (or inducing Z to do) the action (A) that X is counting on. To rely on Y (and expose oneself to the risk not only of failure or betrayal but also of taking advantage of his openness, of his non-"distrust"[1]) X must trust and count on Y's "reliability". But what makes Y "trustworthy"? And how much"? (see later).

1.1 Trust as Mind-reading: Y's Motivations

In domain (2) there are e.g. the reasons that X attributes to Y and therefore "for what" she thinks he will do it:

- Is he a guy who keeps his commitments for consistency and seriousness, or for the image/reputation he wants to have? OR:

---

[1]Lucio Anneo Seneca "To trust an evil person is to give him the means to do harm"

- With me he keeps his commitments, he cares about me: he cares
  about the relationship, relationship with me (social, work, ..):
  does he love me? he's my friend? he takes care of me, etc. etc.
  OR:

- Because he cares about having gratitude, recognition; or put me
  in debt; or does he care about esteem? OR:

- It's a relationship of mutual exchange or cooperation and he
  cares a lot about what he expects from me, and depends on me,
  he has no alternatives and it doesn't suit him not to do his part
  because he knows he won't receive what I owe him give me. OR:

- It's for moral reasons, social rules (e.g. good manners: I'm sure
  he will give us the place or precedence reserved for us); maybe
  because it has that moral value, it's right for him; or it's simply
  a "duty" and he does what he has to. OR:

- It is because he fears sanctions, punishments (practical, judicial,
  or legal) and even if he does not agree or would not like to do
  so, he "obeys" out of fear of authority (institutional or social).
  If he could, he wouldn't. So:

  - If there is supervision, surveillance (and sanctions) or at
    least Y believes so? Then X can trust Y to do it[2]. Note
    that X trusts Y because X trusts Z, the "authority": a
    trust establishes another trust from agent to agent: a kind
    of trust transfers; one of the important social dynamics of
    the trust (see Section 6).

As can be seen, trust is a complex evaluation of Y's mind (Mind
reading/ascription) (of what he knows or believes, what motivates
him, what he feels or would feel) to predict (and explain) his conduct,
and to influence it (Mind Reading function); anything but an estimate of the probability of the event based on frequency or learning.

---

[2]Or is A against a moral or ideological principle of Y, and therefore he will
violate that task even at the risk of sanctions?

However, it is true that there is a "degree" of trust, a strength of his: quantitative dimension (see below)

1.2   Trust as Ascription of Power to Y

In domain (1) to trust Y about his action A:

- for his own purpose (to which I rely mine as part of the result: I exploit his autonomous action) or

- for my purpose adopted by me (he does it for me)

I must believe that Y has the "Power of" [3] S/ to do the action that realizes S i.e. that he has the competence/knowledge about:

- how to reach S (plan and necessary actions);

- what A produces;

- how A has to be performed; that he has the required skills: he is/will be able to do it;

- awareness and self-esteem (self confidence, sense of competence): knowing that you know how and that you have the skills[3].

For this reason the Trust is a positive "evaluation" (value judgement) of Y, an appreciation: because it is an attribution/recognition of "powers". Evaluating something positively means recognizing its powers and therefore that it is worth/useful, can (make you) achieve goals. "Power gives Value" both for others and for oneself. Negative evaluation, lack of value means that the person being evaluated does not have the expected powers (merits, qualities, skills), is devoid of them and therefore "defective", "inadequate", "inferior", "useless" (the center of shame). There is another type of "negative evaluation" which

---

[3]You don't really have power if you don't know you have it, that you "can"! Otherwise that power wouldn't be accessed, it cannot be "disposed of"; it's not there. One of the fundamental steps of Empowerment is to give you the experience of the "sense of competence"; making you aware that you can do it, are capable; making you trust yourself and rely on yourself (and therefore become autonomous).

does not attribute a lack of power but perverse powers: that is, Y has
the power to harm, to do damage; it is not inadequate, useless but
rather dangerous (the center of guilt). And this aspect of evaluation
is also relevant in Trust and towards Y: shouldn't Y be trusted as he
is incapable, inept, lacking willpower? Or as a profiteer, exploiter,
deceiver, etc.?[4]

1.3 Trust about the world/circumstances

Again in domain (I) trust involves beliefs/evaluations not only about
Y but also:

- that there are/will be the conditions to execute A (or for the
  executed A to have its result: Y calls and the telephone works);
  both physical conditions (and means) and social conditions: relationships with others necessary to perform A successfully, or
  prohibitions and permissions (i.e. socially conferred powers).

- Not only are Y's "internal" powers needed (cognitive, motivational qualities, abilities,...) but also "external powers": conditions that give Y the possibility to successfully perform A.
  Practical conditions or social relationships.

Therefore one cannot simply say or model that "X trusts Y"; trust has
an *aboutness*: about what? And it has a context/surrounding. The
TRUST predicate has at least 4 arguments: X (trustor), Y (trustee),
G or O (the goal/object), and C in what context:

$$TRUST\%(X, Y, G, C) \tag{1}$$

There is also a degree (%) strength of Trust (see below)

---

[4]It could also be said that there is a paradoxical negative Trust: that in a
certain sense I "trust" that Y can harm me, hurt me (not achieve but frustrate my
goals); I evaluate him negatively but not as "mistrust" in the sense of "insufficient
positive trust" but as "distrust" "fear". Common sense does not accept this possible
and paradoxical meaning of "trust"; trust is intrinsically a positive expectation, a
good one, i.e. one which achieves purposes, not harm. But obviously if my aim is
precisely to be harmed or to harm Z, I might trust, evaluate and rely on Y for his
perverse power to harm me or Z.

These different factors and faces also interact and can interfere with each other. For example: If Y knows or believes himself capable (in that area) then he will be non-uncertain ("If I want, I'll do it!"), and persistent. If, however, he doubts his skills or abilities, he will be more hesitant to undertake or will more easily give up. As mentioned, to fully have Power I must know I have it; if I don't know/understand/feel it I lose it, I don't acquire it.

2 Forms, Layers and Dynamics of Trust

2.1 Forms

Trust can also be or become automatic, habitual, non-deliberative. First of all trust in objects or tools – which in our theory is a perfect form of "trust". Trust is not only social: what matters is that Y, the trusted object or agent has "purposes" (not necessarily mental, intentions, but also "functions: it is "made for", "serves") and this is the purpose that the user entrusts to it. I may not trust a chair, in the sense that I am not sure that it will hold me; since the task of a chair is to support those who sit. An example of habitual and unconscious trust is our implicit trust in the legs or the floor when walking, or the implicit trust in automatically pushing the pedal to brake. It is not a true evaluative "judgment", but a presupposition. But also social trust can be automatic, habitual, non-deliberative; e.g. in the child-mother relationship (if a problematic, avoidant attachment has not been created) or the trust in the traffic policeman who I don't know in asking him if you can park or enter them.

Trust in System 1 - Trust in my brake pedal or in the floor where I walk is a habit, a habitual expectation to count on Y because it worked, it didn't create problems, and therefore it is a practice not a reasoned decision, based on evaluations. I expect (trust) that it will happen again as always (e.g. that the bus stops, that he opens the door, lets me get on). Also expectation (EXP) in an Anticipatory Classifiers, which is the model of automatic conduct (non-deliberate and intentional) due to instrumental conditioning, to reinforcement learning, represents a type of trust, albeit implicit, presupposed[4]. It is in fact a "positive expectation", of success, of achieving goals; and rather certain, confirmed; not a simple "hope". When Skinner's pigeon goes to peck on the button that he has learned that unloads the food (or when I brake the car at a red light) it expects that the food will come out (and I expect the car to slow down and stop) and counts on this. And this belief/expectation has its own *strength*: it is strengthened and maintained if the expected result comes true, it is weakened if the result doesn't happen, if the EXP fails, it is disproved. The agent counts on this when acting, he relies on his Exp.

Trust as Feeling and Disposition - Trust can also be a "feeling" that one feels towards Y or others; and is sometimes simplistically identified with this [5]; while in our view there is also a cold-minded trust: pure judgments, evaluations and decisions.

Inspired confidence - There is a trust that Y "inspires" in X ("He inspires trust in me"); merely felt ("I feel that I can trust"). A trust not based on cognitive evaluations, facts, reasoning, but evoked unconsciously or consciously. Not "arguable" (where I tell you and we can discuss the reasons why I trust). Based on the "reasons of the heart" not of the Reason[5]. However, we need to have reasons for our behaviors and preferences, to *explain* them to ourselves and feel guided by our true choices and arguments; and therefore we can invent them in a false way, we "tell it to ourselves" in a certain way but in reality we are acted upon by evocations and pulses.

Dispositional, generic trust - The trustfulness or distrust of X as a generalized disposition. This disposition obviously depends on various factors: X's personality, learning/experience, culture (ethnic or family) and education.

2.2   Levels/layers of Trust

Returning to the more complex and cognitive trust, at play in the deliberative system and in intentional conduct, in our model it is multi-level and stratified (see Figure 1).

---

[5]Pascal: "the heart has its reasons that reason cannot know".

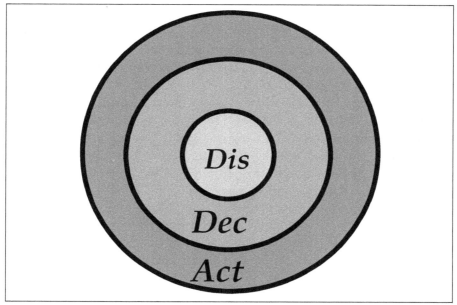

Figure 1: Our layered model of trust: disposition, decision, action

From trust such as **beliefs**, evaluations, expectations on Y, to an active trust: in deciding and acting. There is a **decision** to trust and an **act** of entrusting oneself, delegating to the other, counting on Y to reach G. But from this evaluation, and social choice/decision and act, a **social relationship** is born: trust as a relationship between X and Y. **Trust relationship and social network** - And in fact the "social network" of trust is the determining structure of social relationships and cooperation. A superlayer of the Dependency network. The network of Power differences (who is capable or has the means to do what, for a purpose) and therefore the network of Dependence that emerges (who needs whom to achieve their goal or the common goal) is then filtered by awareness and trust (see Section 3)

2.3 Degree/Strength of Trust

To these various aspects of Trust we must add the degree/strength of trust (%): do I trust 100%, enough (70%), so-so (50%), very little

(20%)? How much?

If I partially trust (50%, 20% etc.) it doesn't mean that the rest
be mistrust, suspicion. We clarified with Rino that "lack of trust"
should not be confused with "mistrust", i.e. a *negative* evaluation and
trust; the belief, expectation that A/G will NOT happen because I
evaluate Y as not capable, with adverse circumstances, not credible,
reliable, etc. It's just a *lack* of Trust. I can trust 50% and "not
trust" 50% but not due to 50% distrust but due to a gap of ignorance,
uncertainty, lack and waiting for data, sources, an amount that I
don't know; distrust might only be 20%. If I am an optimist I will
play the unknown, the gap in my favor (my heuristic is: "unless I
have evidence, data against it, why should I assume that it will go
badly?"): think positively! If I am a pessimist or a prudent one I play
against it: "unless I have evidence, data against that, why should I
assume that it is okay? Better to be foresight".

The % degree of confidence/T derives from:

- From the degrees of its components and dimensions: how much
  do you care about that purpose? How likely is it that Y knows
  this or how convinced or uncertain is Y? How skilled or prepared
  is Y at this? How honest, motivated, persistent, cares about me,
  etc etc etc.

- From the *strength/degree* of certainty of what I assume about
  him, of my *beliefs*: how sure am I that Y...

Belief Strength and Trust in Belief - Actually beliefs have a
quantitative dimension, a subjective degree of certainty, of reliability.
In a sense it's the subject's "trust" in that belief (see Section 4 on Epistemic Trust). X may be 100% sure about the competence/willingness
of Y, or quite sure, or not so sure, or doubtful about. This has nothing
to do with the other factor: the degree/level of Y's "competence" or
"willingness". The two quantities/factors are completely independent
on one the other, and all combinations are possible. Thus when we
say that "X doesn't trust so much Y" do we refer to the certainty of
X's beliefs about or to the content of that beliefs: Y's "qualities"?

3 Trust manages Dependence Network and Social Cooperation

3.1 Dependence and Networks of Power

This is indeed one of the central themes we are modeling with Rino. Dependence is a relationship between X and Y due to a difference in power: more precisely if X Depends on Y for its purpose Sx and the relative action A, there is

- lack of "Power of" S/A on the part of X;

- "Power of" S/Az by Y.

Y has the Power that X lacks (and is useful to X). But in society (and this is the function of society) these lacks and diversity of Powers are a resource: they create power, they lead to the acquisition of power (for our purposes) thanks to the *mediation* of others: I can achieve my purpose thanks to others . The difference (lack) of powers and their complementarity is a multiplication of individual and collective powers thanks to the "social division of labor" and cooperation. As Durkeim explained, it makes us increasingly "dependent" on others: explosion of social competition, skills, techniques, tools, products and therefore also explosion of *infinite* new purposes and needs achievable by combining and differentiating powers ("generative" dynamics typically of individual and collective purposes): maximum distributed dependence and maximum potential and combination. This is not just for "exchange": where everyone satisfies their personal purpose thanks to another; but also for *common* purposes, for true cooperation. Humans must combine their powers (internal and external) to build the necessary "co-power", collective/cooperative power to achieve complex goals. But this interaction and multiplication of powers presupposes two steps and mediations:

- **Mental mediation**: To determine the interaction network one must move from Objective dependence (not known by the subject) to Subjective dependence: the subject knows he is dependent and therefore wants the action of the other.

- **Trust Mediation and Trust Network**: But dependence
  awareness is not enough because to move on to interaction and
  count on the power of the other I must also have trust in him
  (skills and reliability).

Network upon network: emerging multilayer of sociality [6, 7]. It
is true that the Trust exposes us to risks and dangers but it also
gives enormous opportunities to individuals and the community. All
this creates and gives new powers; individual as in exchange (I give
you my power if you give me yours) or collective: co-power associating our complementary powers for a common objective. Also in this
sense the Trust is the glue of society and a precious resource. Only
the Trust can translate the network of dependence into a network of
exchange/cooperation and multiplication of powers.

> **Powers of Agents ==> Relationships / Dependency
> Network ==> Power over others ==> Subjective De-
> pendence and Powers ==> Potential Exchange or Co-
> Power ==> Network of Trust ==> Network of Ex-
> changes or Cooperations**

3.2 Trust as Capital

Trust is a true "capital", an accumulated and investable value and
power; but it has two faces: one competitive, the other collective/co-
operative.

- There is a *competitive Trust-capital* between individuals or social subjects (companies, schools, etc.). Accumulating positive
  evaluations, esteem and a reputation about reliability is a comparative positioning that favors us in the competition with our
  possible competitors. Will those who depend on me or others
  and who can exchange with me choose me as their partner?
  Why should they prefer me? Because they attribute more pow-
  er/value to me (Trust about ability, competence, means) that
  they need but also because they consider me more reliable, less
  risky. It is a capital also because when they turn to me they

will offer me their power (exchange) and therefore I will increase my powers, the goals that I can achieve through social intermediation. Accumulation and concentration of power in the social network partly mediated by the Trust. Given this competitive advantage and means of accumulation it can obviously become useful to deceive, to make people believe that I have those skills, means, competence and that I am honest and reliable, and that I am convenient. Fundamental chapter on deception and its usefulness ("mercator mendax").

- But the Trust is also a social capital, a collective one for the community and its participants. It is a "common", a common good which as such must be established and defended.

  - Through rules, norms, sanctions, surveillance, authority (one of the functions of the law: reliability and safety in interactions and protection/reduction of the risk to which trust exposes); but also

  - as purely interpersonal and as an experience that we have towards others in "our" community: built with family and collective education and customs, "values", shared culture.

Exposing everyone and each other to risk in a certain sense reduces everyone's risk (also through spontaneous collective surveillance and blame and exclusion) and multiplies cooperative possibilities. And that everyone behaves in a certain way in fact pushes them to conform (see work of Cristina Bicchieri, for instance [8]), and reduces the probability of exceptions. This advantage does not only apply to individuals but to groups and subcommunities: internal trust in teams, companies, teams, parties, ...

3.2.1 Why pass on your powers?

The real problem of the circulation and exchange of powers is not because Y depends on X and needs the powers of the other to achieve his goals, and turns to the other with this end.

The real problem is why X (on which Y depends) should spend her resources and powers (skills, abilities, time, goods, effort...) for the other to his advantage, i.e. to achieve his goals (needs, desires, interests, requests). It is irrational "economically". In our vocabulary: why does X "adopt" (adopt and pursue) Y's Goals? There are three main reasons/types of Goal Adoption:

a) *Self-interested, utilitarian adoption*: Where X expects a return advantage, a result that is evidently convenient compared to the sacrifice/cost; and she does it for her own benefit. Nothing irrational. The basis of every form of barter, exchange: "Do ut Des": I give/do something for you if and so that you do something for me. There is reciprocal/bilateral dependence of X on Y ands Y on X, but ach for its own purpose. It's an egoistic sociality. In this type of relationship it is rational to deceive: once I have had what I need it is irrational to give you what I promised.

b) *Cooperation*: I call "cooperation" in strict sense between X and Y when X and Y have a common/shared goal, and act together (by assigning roles/tasks to each other, and relying on the other action). In this kind of bilateral dependence their power are complementary for achieving that goal; a "co-power" is needed; the contribution of both of them. In this situation to deceive the other and do not do as promised/expected is irrational since the goal cannot be achieved. To defeat is self-defeating!

c) *Altruism*: X adopts the goal/need of Y just for Y's good; not for any return, advantage of herself. Even if there was some advantage, pleasure for her, X is not motivated by that, but by Y's advantage[6]

---

[6]The problem discussed in psychology (see for example literature in Wikipedia) is whether true altruism exists, or there is only "pseudo-altruism"; since there is always some good outcome for X: gratitude, social approval, self-approval, moral value, God's reward, relief from compassion for Y, etc. But as Seneca explained, we should not confuse "what I expect" with "what I do it for": not all the things I expect "motivate" me, they are my "motivation". Without this distinction the

3.2.2 The Needed Image and Trust for Receiving Help

What kind of image does Y have to present to enter into these adoption relationships (A) (B) in which X does something for him? An image that gives Trust. In fact, for selfish/exchange-based help (A) I must give the image that I have of those powers (skills, abilities, means, resources,..) that you need; It's me you need and better than others: I'm your right exchange partner. And you can trust me; I won't cheat you (because in this relationship it would be economically "rational" to cheat you). The first aspect of image and trust in (B) is similar: "I have the powers you need" because they are complementary to yours and necessary to achieve the common goal ("co-power"). I am your right cooperation partner. The second aspect of trust, however, is different in that in this mutual dependence it is clear that I don't screw you, since - as mentioned - to defeat would be self-defeating: not achieving the goal. Paradoxical is the image that Y must present in (C) to induce altruistic help, not motivated by any return advantage for X. It is an image of lack of power: of neediness, difficulty, impossibility, suffering (current or potential), and therefore also of *lack of value*, of *inferiority*. An image that arouses compassion, pity and which can therefore also be "humiliating", inferiorising. And in fact, paradoxically, this beautiful human empathic and moral impulse to help can be used as an insult, to hurt and humiliate you: "You cause pity!" "You inspire compassion!". This humiliating aspect of pity/compassion (and of the help motivated by it) is so true that Y can give up the help he needs, and even hide his neediness and suffering so as not to arouse pity[7]!

---

discussion about altruism or pseudo-altruism is useless.

[7]Hence the Italian motto "Better to be envied than pitied". How is it possible that a hostile emotion towards you, which hopes for your harm (envy) is better than a pro-social emotion, which is activated for your good? The answer lies in value: envy recognizes your value and superiority: something enviable that you have and the envious person does not have and would like. While pity sees your lack of power/worth. Envy gives you value while punishment takes it away. [9, 10]

4 Epistemic trust, in believing

Confidence in believing/belief, i.e. subjective certainty, derives from
the sources of my belief (direct experience with Y; what Z told me or
what is written on the WEB; how sure I am of my own reasoning[11])
and from their credibility, reliability. This is also a form of *transfer*
of the Trust (see Section 5) More exactly:

a) The more I trust the source, the more I rely on it, and therefore
   I believe it and trust that belief. For this reason we keep (and
   must keep) memory of the sources of the knowledge we acquire:
   to be able to feedback, readjust or confirm its "credibility" if
   the information received turns out to be incorrect or valid. But
   the "discredit" also depends on the *attribution* of the error (as
   mentioned in our book): *internal* or *external* attribution? (e.g.
   there was a disturbance in the net connection; there was a sab-
   otage, ...; it's me who didn't understand, ...). Is this wrong
   information due to intention or error? to ignorance (general in
   the domain or on that topic?) or to malice? etc. Furthermore:

   • I must take into account any *convergent* or *divergent sour-
     ces*, and their credibility/weight/impact. Converging sour-
     ces increase the credibility/trustworthiness of the belief.
     Combining - as mentioned - the credibility of the source
     with the degree of certainty communicated to me (see be-
     low). Divergent sources can give rise to different heuristics:
     combinatorial, or privileging a source chosen by an opti-
     mistic or pessimistic/prudent attitude (depending on the
     "risks" I run by relying on that belief), ..

   • The new candidated belief is not necessarily accepted:
     – I can suspend the decision, remaining in "doubt" and
       waiting for evidences;
     – the candidated belief must at least (minimal plausibil-
       ity) not create a contradiction with my other beliefs :
       how many beliefs would it dismantle and I should re-
       vise? We will resist the more we have to revise ("belief

revision" costs);

- possibly (better plausibility) not only must the new belief not force me to revise by creating contradictions, but it must be integrated, be coherent and consequent to what I already believe (be explained or explain other beliefs);

- if there is a contradiction, which one will win? The most credible, most integrated package: where the sources are not only external communication (other agents) but also my "perception" and my "reasoning".

b) It also counts how much the source says he believes it: "It is certain that..." "I believe that" "I suppose that" "Maybe..."... How do these weights combine? (see Sections 2.3 and 4)

c) Confidence in a source may not already be in memory but must be calculated/derived: by transitivity or similarity; for reputation and recommendation; verification of previous information from the same source, etc.

True Trust – Trust in beliefs is a real form of "trust" because it is a bet, you count on them and you accept/take a risk. In what sense do you run a risk by doing "reliance" on a belief? We decide, we pursue a goal, ... on the basis of beliefs and how certain we are about current circumstances or predictions; and we invest and do not give up based on this security: the more we believe in it, the more we risk, invest (and sacrifice alternatives, exposing ourselves to failure or regret). From confidence in believing to confidence in choosing, in doing, in taking risks. These are other cases/examples of connections between types of trust and transfer from one trust to another (see Section 6): this is from belief to action, but also the one mentioned from trust in sources to trust in coming to believe.

5 Institutional Sociality and Institutional Trust

5.1 "Count As"; act "As If"

A crucial form of Trust is the implicit reliance on the complicity of others in the conduct of "conventional" effects of the As-if and Count As. In the representation, collective staging, and creation of an "institutional" reality.

Society as representation. A crucial dimension of human sociality is staging, knowing how to take on and play one's role/part in Scripts. We must act "As if" we believe, understand, want, share what we should (that others prescribe or expect us to) believe, understand, share, want, ... This "symbolic play" (the infancy beginning of social recitation) works; has the desired effects. The "normative" ideal is that the agent introjects the Norm/Value as such [12, 13], and follows it ("obeys") because it is a norm and that's it . It is not necessary for him to understand its usefulness or purpose, or whether others obey or not, or whether there are advantages in obeying or punishments/sanctions in violating it... He "must"! And this works even if in reality in many cases it is not so; he acts "as if" he really conforms to a norm as his imperative; and instead it is now only a habit, a ritual, an automatism (Syst 1) or due to an ongoing threat (to avoid getting a fine), or to do a favor to, or to be approved, or for a prize etc. etc. In any case the norm works collectively and those people are assumed to behave deontically, "As if" [14]. The same is true for a thousand faces of mutual expectation and cooperation, and of trust; e.g. competence: he is a doctor, he will behave like a doctor, he will have medical competence, ..; he is a policeman.....

The "symbols", masks of power, *give* power; this is what the crown, the scepter, a palace, the throne are for... or the toga, the uniform, the cassock, the alb, ... Not just practical purposes but stigmergic communication for the construction of that role/power through its "recognition" (in both senses of the term). The same goes for the dress or the hair or the gait..."for a woman" or "for a man". They are needed so that they "value" as such; that is, recognizing and treating them as such, *makes* them such: the king should be/act like a king, the judge like a judge, a man like a man. And in experiencing that sign/act/person "as if" I rely on the fact that others do so (because the recognition must be collective, shared; it is not the effect of an individual act) and the others count on me; and the king, the policeman rely/trust in me because it is my task (and they implicitly prescribe it to me). Completely implicit, presupposed, unaware trust/reliance. Let us reflect further on this specific form of "power" which establishes a form of trust and which is based on this trust.

5.2 Institutional and Role Powers

Some powers of the individual and of her/his action are not personal/individual but given to the individual by the institution or organization [15]. We call them "institutional" powers. The subject has and disposes of such power only "as" member of the group/organization and player of that role (and s/he is playing such role, not for her/his personal goals/interests). For example *as* policemen you gets/has the powers of....; *as* medical doctor you has the powers of... Notice that these powers are not only "instruments", conditions you have access to, at your disposal for that performance, but also deontic powers: you has the permission, authorization to do that, to exercise those powers. Moreover not only you have the permission but you also have – in that role – the obligation to do that, to exercise your power. And this "permission" and "obligations" are not just "on paper" (legal) but social: interpersonal beliefs, expectations, prescription. We "recognize" (and so give!) such rights and obligations.

As just said, those powers and your role has to be recognized in order the people ascribe them to you, submit to them, ask to you that use since know *to be dependent* on them and on you. And recognize your practical action and its result as "counting as" (institutional action and goal), and that object as "counting as" that "institutional" instrument (and symbol)[8].

---

[8]It's a crucial case of spontaneous submission and remission/tranfer of powers: Y renounces to, delegates her/his powers: the very crucial (for sociality) "servitude

5.3   "Institutional" Trust

This has important consequences on *"institutional" Trust*: the trust
we have on you "as" policemen, doctor, ecc. (as role player). In
fact the powers given to you as role-player (preparation and training;
access to specific information, or tools; etc.) not only give you *the
first face* of trustworthiness and of our trust in you: your *competence*.
Not only we think/expect that you will able and in condition (has the
Power Of); but the fact that you have obligations, norms, role-tasks
gives you also the other face of trust: your willingness. We can also
expect that you will do/intend to/are committed to do that task for
which we count on you and delegate to you.

Institutional and Role Dependence - Institutional and role
powers creates an "institutional and role Dependence". Our dependence on you (as role player) in our role of addressees of your role:
as citizens or as prisoners toward you as policemen; as drivers towards you as traffic policeman; as sick or your patients towards you
as hospital doctor or as our doctor.

5.4   Trust from Role

My *institutional Dependence* on you (and thus the need to Delegate
to you some need of mine) creates an Institutional form of Trust: the
trust I have on you not as individual/person (that I do not know)
but I derive from your membership and role. You inherit the Trust I
have in that group/institution/role. And I trust you "as" role player
(policemen, doctor, teacher, ..) and as for your functions/tasks.

Personal Trust - This trust in you "as"/"in" that role should not
be confused with the "personal" "individual" Trust: the trust I have
specifically on you as a person. By default I can ascribe to you the
trustworthiness that you inherit as "member" of that group/role (§
6), but I might have specific information (like reputation), or previous
experience, or personal signs (of your attitude, character, conduct,..)
that give me a specific evaluation of you. And this might increase or

---

volontaire" (Étienne de La Boétie).

reduce my trust in you even on your role-playing, on your institutional task (for competence or for willingness)

5.5   Will Bots be "institutional" minds?

An interesting question about a possibility of a "true" humanoid social interaction with humans on the part of artificial Robots and Agents is this: This presupposition on the mind of the other consistent with the prescribed and executed conduct will also apply in the mixed Hybrid Society of humans and Artificial Intelligences? We will necessarily use Mind Reading or rather Mind Ascription to interact with them. We will spontaneously and necessarily assume that they act a given way (in the world and towards us or among themselves) because they want S, they believe B, they feel E.

We will attribute to them the mind that appears from their conduct and what they recite, for example. as emotions:

- he smiles → he is affectionate, he likes me, he wants to help me..;

- stern face → he is examining whether I did as/what I was supposed to; etc.

We will attribute not only thoughts, purposes but also feelings and emotions. Which in reality aren't there; given that robots (or Bots) do not have a real "body" and do not "feel" anything. And this illusion - when performed in the appropriate circumstances - works even if fake; it creates good interaction, acceptance, confidence, and effectiveness of the robot etc. Moreover the interaction with and reading of conduct and Mind will be based on common sense language/concepts; an indispensable tool for social interaction with humans and humanoids.

But the question is: will this also apply to their recognition and establishment of "institutional" powers of things, acts, and people that "count as"? Will they give us - by conforming - these powers and will they be able to take them on too? Will they understand that: "arresting" is not a kidnapping; that "giving an order" is not a request or advice; that "giving money" can be "pay" or "alms" or "corruption" or "restitution" etc.?

6 A Trust dynamic: Trust Transferts

There are other examples we have studied of transfert of trust (in addition to that examined in §1(i) of trust in the vigilance of the authority and therefore in the obedience of Y). Trust does not derive (as in many models) from direct interaction and experience with Y, but from many other possible sources. (e.g. from trust towards Z, from class categories):

- Transfert based on evaluation and reputation: Z says that Y is reliable and I trust Z as a serious and competent evaluator in that domain so I trust Y[9];

- I trust Z and I see a lot of analogy/similarity between Z and Y and therefore I trust Y by analogy; or

- I constructed a group, a membership category and generalized the trust to this "class" or set C, and since Y is a member of C I instantiate the trust on him[16].

- Another example of transfert is that the confidence attitude and trust (taking the risk) of Y towards X can induce X to trust (positive evaluation and trust) of Y.

- Even: If he/they trust Y Y acquires self-esteem, changes his evaluations and expectations; and this perhaps makes him more reliable: competent (sense of competence), confident, determined and persistent, and available. Trust creates reliability.

- One form of transfert from self-confidence to confidence in Y is as follows: when X trusts Y he also has some trust in X.

---

[9]Or by emulation without Z giving me his judgement: an act/experience of trust from Z towards Y is transferred to me: "If Y trusts him it means that he can be trusted, he is reliable". This is a case/example in which it is false that any subjection to the power of others (interpersonal or collective or institutional) necessarily arouses rejection, and an impulse or desire for rebellion (Foucault). Submitting can be a need/impulse or choice of the individual for a thousand reasons: from attachment, to hierarchy, to delegation, to group organization and action, to norms and institutions. (Note 7)

Actually trust in others and delegating implies and presupposes a certain self-confidence. I must first of all trust what I believe/-know/estimate about my problem (goal and plan) and what I believe about Y. But trust (certainty and reliance) on beliefs is not just trust in the sources of belief and that I remember well; but it implies that I trust enough - perhaps implicitly and habitually - in myself in terms of understanding and reasoning ability: I don't believe everything that comes to me but I evaluate it for credibility, coherence with what I know, non-contradiction, etc. If I believe it I *can* believe it; and take a risk on this.

Even more this applies to my evaluation of Y; first basis of trust in Y. Do I consider myself a good evaluator of Y's competence in that area, and of his abilities? Can I trust my assessment and judgment? And the same for my judgment on the reliability, intentionality, persistence and loyalty of Y; or even that Y has no advantage in deceiving me and intends to do so? Am I naive or can I trust my expectation?

Another implicit (but sometimes explicit and reflected) self-esteem in entrusting my purpose (need, desire, project) to another is the belief that I am able to understand the appropriate purpose for me, what I need. And also to be a good planner and problem solver and therefore to delegate/ask for the right thing for my purpose.

And on this I can be wrong and Y can (for my interest and for a deeper collaboration) counter me and correct me; and if I trust (at this point more in him than in me) I accept: I change purpose or plan.

Or Y can directly do something different from what I expected/delegated, that is (in our terminology) *Over-help* (more than what was requested and therefore something not required) or *Critical-help*: not doing what I requested but one thing different by violating/correcting my plan either because it cannot be executed in that context or because it is wrong for my purpose [17]. And I can either accept because I understand the solution and why he did it, or out of radical trust in Y ("if he decided like this it means that it was better/necessary"). Or I can refuse/protest: either because I don't accept initiative/autonomy from Y or also because I trust myself more and I believe that my (requested) solution was/is the right one ("No, you don't understand..."

"No it's better as I wanted").

References

[1] Castelfranchi, C. (2020, September). For a Science-oriented, Socially Responsible, and Self-aware AI: beyond ethical issues. In 2020 IEEE International Conference on Human-Machine Systems (ICHMS) (pp. 1-4). IEEE.

[2] Castelfranchi, C., & Falcone, R. (2000, January). Trust is much more than subjective probability: Mental components and sources of trust. In Proceedings of the 33rd annual Hawaii international conference on system sciences (pp. 10-pp). IEEE.

[3] Castelfranchi, C. The Micro-Macro Constitution of Power, ProtoSociology, An International Journal of Interdisciplinary Research Double Vol. 18-19, 2003 Understanding the Social II – Philosophy of Sociality, Edited by Raimo Tuomela, Gerhard Preyer, and Georg Peter.

[4] Castelfranchi, C. (2022). Purposiveness of Human Behavior. Integrating Behaviorist and Cognitivist Processes/Models. Croatian Journal of Philosophy, 22(66), 401-414.

[5] Castelfranchi, C., & Falcone, R. (2010). Trust theory: A socio-cognitive and computational model. John Wiley & Sons.

[6] Falcone R., Castelfranchi C., (2010) From Dependence Networks to Trust Networks, in Falcone, R., Barber, S., SabaterMir, J., Singh, M., (Editors) Proceedings of the Trust in Agent Societies Workshop at AAMAS 2009, Budapest, May 11- 16, pp.13-26; 2009. 2010

[7] Falcone R., Castelfranchi C. (2022), Grounding Human Machine Interdependence through Dependence and Trust Networks: Basic Elements for Extended Sociality, Frontiers in Physycs., Volume 10, Article 946095 (September 2022), DOI: 10.3389/fphy.2022.946095

[8] Bicchieri, C., Xiao, E., & Muldoon, R. (2011). Trustworthiness is a social norm, but trusting is not. Politics, Philosophy & Economics, 10(2), 170-187.

[9] Castelfranchi, C. e Poggi, I. The mental state of shame: implications for adolescence, Proceedings of Tenth Biennial Meetings of ISSBD, Jyvaskyla , Finland, July 1989.

[10] Miceli, M., Castelfranchi, C. (2007) The envious mind. Cognition and Emotion, 22, 449-79.

[11] Villata S., Falcone R., Da Costa Pereira C., Castelfranchi C., Tetta-manzi A., Paglieri F. Comunicazione e fiducia: un modello ad agenti su qualità delle informazioni e valutazione delle fonti. In: Sistemi Intelligenti, vol. 24 (3) pp. 559 - 579. Argomentazione, processi cognitivi e nuove tecnologie. F. Paglieri (ed.). Il Mulino, 2012.

[12] Conte, R e Castelfranchi, C. (1990) Issues on the mental representation of norms. Proceedings of the 2nd International Congress for Research on Activity Theory. Lahti, Finland, 1990.

[13] Andrighetto G., Castelfranchi C. (2013) Introduction to chapter III: Norms. pp. 169 - 170. S. Ossowski (ed.). Springer, 2013

[14] Castelfranchi, C., Tummolini, L. (2003). Positive and Negative Expectations and the Deontic Nature of Social Conventions, in Proceedings of the 9th International Conference of Artificial Intelligence and Law (ICAIL 2003), ACM Press: 119-125.

[15] Tummolini, L. Castelfranchi, C. (2006) The cognitive and behavioral mediation of institutions: Towards an account of institutional actions. Cognitive Systems Research 7(2-3): 307-323 (2006)

[16] Falcone, R., Piunti, M., Venanzi, M., & Castelfranchi, C. (2013). From manifesta to krypta: The relevance of categories for trusting others. ACM Transactions on Intelligent Systems and Technology (TIST), 4(2), 1-24.

[17] Falcone, R., & Castelfranchi, C. (1997, January). On behalf of..": levels of help, levels of delegation and their conflicts. In 4th ModelAge workshop.

Note: In this contribution to honor Falcone, I have decided to provide only the references to the work of our group. The international community will forgive me, but my choice is motivated by the desire to create a heartfelt tribute to my research group.

Appendix

Rino Falcone is the true international reference in studies on trust. As a small final tribute I would like to list his very numerous and well-known publications on Trust but they are too many! Let me just say that some of them have a very high citation impact: for example these near 1000:

> Trust theory: A socio-cognitive and computational
> model
> C Castelfranchi, R Falcone John Wiley & Sons, 885 cita-
> tions
> Social trust: A cognitive approach
> R Falcone, C Castelfranchi Trust and deception in virtual
> societies, 55-90 875 citations

Not to mention the conferences and WSs of which he was Chair;
or the many talks, lessons, and courses he held on Trust at a national
and international level. As well as interviews and participation in
political debates on the topic.

PART

SOCIO-POLITICAL CONTRIBUTION

LETTERA A RINO FALCONE IN OCCASIONE DEL SUO PENSIONAMENTO

LUCIO BIANCO

Professore emerito presso l'Università degli Studi di Roma Tor Vergata, già Presidente del Consiglio Nazionale delle Ricerche (CNR)

Caro Rino

Sono restato molto meravigliato nell'apprendere che il prossimo luglio andrai in pensione. Viste le tue performance sportive, anche recenti, non immaginavo che fossi già arrivato all'età in cui, per Legge, bisogna cedere formalmente il passo.

Per l'occasione i tuoi colleghi e collaboratori hanno opportunamente deciso di festeggiarti progettando la stesura di un libro in tuo onore. Hanno contattato anche me sollecitandomi un ricordo della nostra frequentazione. Ed eccomi qui a scriverti, con vero piacere, questa lettera aperta in cui provo a ricordare le occasioni che hanno determinato i nostri rapporti.

La prima che mi viene in mente, che risale, se la memoria non mi inganna, a più di trentacinque anni fa, è stata di tipo ludico: il calcetto del venerdì pomeriggio all'Istituto S. Maria di viale Manzoni.

Questa tradizione è durata qualche decennio per noi dello IASI. Per te, purtroppo, si è interrotta presto quando l'infortunio al ginocchio ti ha costretto a rinunciare.

Questo non ha impedito che il nostro rapporto continuasse, soprattutto durante il periodo della mia presidenza del CNR, quando era in discussione la ristrutturazione degli organi di ricerca.

Ricordo bene come, partendo dall'Istituto di Psicologia, costituito a Roma nel 1950 e rilanciato da Raffaello Misiti negli anni '70, si è arrivati progressivamente a immaginare l'Istituto di Scienze e Tecnologie della Cognizione in cui sono confluiti altri organi del CNR operanti in sedi territoriali diverse da Roma.

La configurazione del nuovo Istituto ha così ampliato ancora di più il suo carattere interdisciplinare, già caratteristica specifica del vecchio Istituto di Psicologia e vocazione storica del CNR. Il periodo della tua direzione non ha solo confermato questo dato, ma ha introdotto ulteriori elementi innovativi, in particolare nel rapporto con le Neuroscienze.

Non posso poi non menzionare la tua vicinanza nel periodo burrascoso della mia convivenza istituzionale con il ministro Moratti, culminato nel tentativo di commissariare il CNR. Mi opposi a quel tentativo anche perché sentivo il sostegno di gran parte dei ricercatori che, come te, cercavano di opporti a una deriva autoritaria nel governo della ricerca pubblica.

I nostri rapporto sono proseguiti anche dopo la mia uscita di scena perchè, periodicamente, mi hai informato delle vicende del CNR postmorattiano e dei tentativi da te fatti per invidere positivamente nella politica dell'Ente.

Anche la tua battaglia, non andata purtroppo a buon fine, per essere eletto nel Consiglio di amministrazione andava in questa direzione.

In questo lungo periodo di conoscenza ho apprezzato la tua serietà e dirittura morale e nel momento in cui stai per diventare un pensionato "attivo", come io lo sono ormai da anni, ho voluto partecipare all'iniziativa dei tuoi amici e colleghi con questo breve ricordo.

Certo che il pensionamento non interromperà il tuo impegno nella ricerca dell'Istituto, ti auguro un futuro di buona salute e fecondo di iniziative appaganti.

Con immutata stima e il solito affetto,

Lucio Bianco

Licio

Omaggio a Rino Falcone

Gaetano Manfredi

Sindaco di Napoli, già Ministro dell'Università e della Ricerca

È con grande piacere e profonda stima che mi unisco a questo tributo per celebrare il lavoro e la carriera di Rino Falcone, un ricercatore il cui contributo è stato illuminante negli anni, per tanti giovani ricercatori. Rino ha avuto un ruolo rilevante nelle scienze cognitive del nostro Paese, ha diretto per oltre 9 anni l'Istituto di Scienze e Tecnologie della Cognizione del CNR ed ha contribuito in particolare allo studio dei modelli cognitivi delle interazioni sociali, nel rapporto tra fiducia ed Intelligenza Artificiale su cui ha operato con lavori ampiamente citati nella comunità scientifica internazionale.

Ho avuto l'onore di conoscere Rino durante il mio mandato come Ministro dell'Università e della Ricerca, quando è stato mio consigliere per le questioni attinenti al rapporto tra ricerca scientifica e società. La sua competenza, passione e dedizione hanno aiutato l'azione del mio mandato in un periodo in cui l'impatto della ricerca scientifica sulla società era al centro del dibattito per la gestione della pandemia di COVID-19.

Uno degli aspetti più significativi del contributo di Rino durante il periodo in cui collaboravamo strettamente è stato il suo approccio al rapporto tra scienza e società, in particolare in quel periodo. Rino ha saputo evidenziare come la crisi sanitaria globale abbia posto in primo piano l'importanza della scienza e della ricerca, rendendo evidente la necessità di un dialogo aperto e costruttivo tra il mondo scientifico ed i cittadini. Lavorò a lungo a numerose iniziative, tra cui la proposta di una conferenza nazionale su "Scienza e Società" che avrebbe mirato a esplorare e rafforzare questo rapporto, mettendo in luce l'importanza della trasparenza nella ricerca, dell'efficacia della comunicazione e della fiducia nel processo scientifico.

Rino era convinto, ed io con lui, di come la pandemia avesse squarciato il velo che separava la scienza dalla vita quotidiana, mostrando in modo palpabile l'importanza della ricerca scientifica per affrontare le sfide globali. Ha anche evidenziato con il suo lavoro le difficoltà e le complessità del progresso scientifico, ricordando a tutti noi che il percorso della scienza è spesso irto di dubbi, incertezze e discussioni, ma che proprio queste caratteristiche ne costituiscono la forza e il motore di innovazione.

Il lavoro di Rino Falcone è stato caratterizzato da una profonda riflessione sulle dinamiche della ricerca e sulla necessità di una maggiore sensibilizzazione della società verso il valore della scienza. Ha posto domande fondamentali su come la politica e l'amministrazione della cosa pubblica debbano supportare a tutti i livelli la ricerca e su come gli scienziati possano svolgere un ruolo proattivo nel promuovere una maggiore comprensione e apprezzamento della scienza tra i cittadini.

Concludo questo tributo esprimendo la mia gratitudine e ammirazione per Rino Falcone. La sua carriera è un esempio di eccellenza e dedizione, e il suo impegno nel promuovere un rapporto più proficuo tra scienza e società è una fonte di ispirazione per tutti noi. Auguro a Rino ogni bene per il suo futuro, con la certezza che continuerà a contribuire al progresso della conoscenza con la stessa passione e determinazione che ha sempre dimostrato. Gaetano Manfredi

IL MIO CAMMINO CON RINO

FRANCESCO LENCI

Già Dirigente Ricerca CNR e Direttore Istituto Biofisica (IBF) CNR Membro del Council delle Conferenze Pugwash (Premio Nobel per la Pace nel 1995, `https://pugwash.org/pugwash-council/`*)*

`francesco.lenci@ibf.cnr.it`

Ho conosciuto Rino agli inizi degli anni 2000, quando si seppe che la Signora Moratti aveva messo mano alla "sua" riforma. A quei tempi io ero ancora Direttore dell'Istituto di Biofisica del CNR di Pisa. Ho cercato sulla rete e su miei vecchi documenti riferimenti precisi e siti affidabili. Purtroppo con pochissimo successo. Quanto segue, quindi, è quanto riesco a ricordare di anni ricchi di speranze (quasi tutte deluse), di impegni coraggiosi, di momenti bellissimi e di momenti tristi, di regali importanti che la vita mi ha fatto. Prego Rino e quanti hanno vissuto gli eventi che racconterò di perdonarmi per inesattezze, confusioni, dimenticanze.

Dunque: conobbi di persona Rino in un'assemblea nell'Aula Magna del CNR, alla quale parteciparono centinaia di ricercatori e Colleghi del CNR. Naturalmente c'era anche Carlo Bernardini, mio amico fraterno e maestro di scienza e di vita, che con Rino istaurò subito un legame affettivo e intellettuale profondo e ricco. Nel 2002 venne fondato l'*Osservatorio sulla Ricerca*, coordinato da Rino, una rete di ricercatori e scienziati italiani (di cui facevano parte tra gli altri, Giulio Peruzzi, Carlo Bernardini, Marcello Buiatti) che promuoveva il valore della ricerca nel Paese. Pensavamo che questa "promozione" si esplicasse sia monitorando le politiche che si applicano al settore, sia promuovendo iniziative anche di natura propositiva sul settore stesso.

Quando il testo della Legge Moratti fu noto, Rino organizzò diverse riunioni al Dipartimento di Fisica di Roma La Sapienza con lo scopo di proporre modifiche e miglioramenti al testo originale. Tra quanti si dedicarono a questo lavoro ricordo Giorgio Salvini - già Ministro MURST nel Governo Dini - Giulio Peruzzi, Gianna Cioni....e molte/i altre/i. Nel febbraio del 2004 Rino, Carlo Bernardini ed io fummo ricevuti dal Commissario Adriano De Maio (`https://www.cnr.it/it/news/4009/riordino-del-cnr-nomina-commissario`). Mi sembra di ricordare ci abbia ascoltato: niente di più.

Quando nel 2006 Fabio Mussi divenne Ministro nel Governo Prodi chiese a Rino di diventare suo consigliere proprio per le questioni attinenti alla ricerca. In questo ruolo ha contribuito ad avviare un percorso di autonomia della ricerca dal potere politico (rafforzando al contempo gli elementi di responsabilità): ne sono testimonianza una legge delega sull'autonomia statutaria degli Enti di Ricerca, il metodo del Search Committee per la nomina dei loro Presidenti, l'introduzione di un metodo di valutazione del livello dei maggiori Paesi sviluppati, l'unificazione dei vari strumenti progettuali in un unico fondo. E ancora, alcune iniziative per il rilancio europeo della ricerca nazionale come la costituzione di gruppi di esperti ministeriali per il supporto della ricerca italiana in Europa, o la costituzione di una *roadmap* nazionale per le grandi infrastrutture di ricerca di livello pan-europeo.

Il 9 Luglio 2008 una delegazione dell'Osservatorio sulla Ricerca fu ricevuta dal Presidente della Repubblica, Giorgio Napolitano. Questo il comunicato stampa preparato da Rino

"Il Presidente Napolitano: "piena disponibilità a sostenere la ricerca nazionale".

Oggi mercoledì 9 luglio 2008 alle ore 11.30 si è tenuto al Quirinale un incontro tra il Presidente della Repubblica Giorgio Napolitano e un gruppo di 20 scienziati italiani, guidati dal premio Nobel Rita Levi Montalcini, che hanno consegnato direttamente nelle mani del Presidente due appelli sottoscritti complessivamente da 3000 firme all'interno della comunità scientifica nazionale. Dati gli intenti comuni e la volontà di rappresentare il massimo di coesione della comunità scientifica, le due delegazioni hanno deciso di fondersi in un'unica che avrà come suo simbolico capo-delegazione il premio Nobel Rita Levi-Montalcini.

I due appelli che fanno riferimento alle difficoltà che il nostro Paese trova nel valorizzare la ricerca e nell'individuare le corrette forme del suo sostegno, sono stati illustrati al Presidente da Rita LEVI-MONTALCINI, Paolo BIANCO, Elena CATTANEO, Rino FALCONE, Giulio PERUZZI, Caterina PETRILLO. Il Direttore della Società di sondaggi SWG, Maurizio PESSATO, ha poi brevemente presentato un'inchiesta, commissionata qualche mese fa dall'Osservatorio sulla Ricerca, sulla percezione da parte dell'opinione pubblica delle questioni relative alla ricerca.

Nella replica, il Presidente Napolitano ha innanzitutto sottolineato il valore della comunità scientifica che più di altre manifesta dedizione per il proprio lavoro e ottempera al bene della collettività in modo disinteressato e particolarmente qualificato. Ha riconosciuto l'importanza dei temi sollevati nei due appelli e ha garantito il massimo di impegno per sensibilizzare tanto la classe politica quanto l'opinione pubblica. Ha ricordato di aver già nella fase iniziale del suo settennato rivolto le proprie attenzioni al settore e di essere consapevole delle problematiche in particolare di finanziamento e di scarsa valorizzazione del capitale umano.

Ha quindi chiesto di conservare un rapporto privilegiato con la comunità scientifica di cui sente egli stesso il bisogno di farsi interprete e intermediario. A tal proposito ha chiesto di elaborare una sintesi dell'analisi e delle proposte presentategli nella giornata per poter meglio svolgere quel ruolo di persuasione morale verso i diretti responsabili delle politiche della ricerca: in particolare i Ministri dell'Economia, dello Sviluppo Economico e dell'Istruzione Università e Ricerca.

Sul sito dell'"Osservatorio sulla Ricerca" (http://www.
osservatorio-ricerca.it) saranno disponibili da do-
mani i testi dei vari interventi della delegazione.

Osservatorio sulla Ricerca. La delegazione era costituita
da: Rita LEVI-MONTALCINI (capo-delegazione), Enrico
BELLONE, Giorgio BERNARDI, Paolo BIANCO (spea-
ker appello su "peer review"), Giovanni Fabrizio BIGNA-
MI, Marcello BUIATTI, Nicola CABIBBO, Elena CAT-
TANEO, Rino FALCONE (speaker appello Osservatorio
sulla Ricerca), Stefano FANTONI, Sergio FERRARI, Pie-
tro GRECO, Francesco LENCI, Lucio LUZZATTO, Pier-
mannuccio MANNUCCI, Elisa MOLINARI, Giulio PE-
RUZZI, Maurizio PESSATO, Caterina PETRILLO, Set-
timo TERMINI, Carlo UMILTA'.

A seguito di quell'incontro, Rino inviò al Presidente Napolitano questo
documento

Signor Presidente della Repubblica,

corrispondendo alla Sua richiesta di fornirLe una sintesi
contenente urgenze e priorità relative al settore della ricer-
ca e dell'alta formazione nazionale, Le inviamo il breve
documento che segue.

Il documento fa riferimento ai deficit strutturali cui Le ac-
cennammo nell'udienza dello scorso 9 luglio e che – dob-
biamo dirlo – rischiano di essere significativamente aggra-
vati da alcuni provvedimenti governativi che stanno per
concludere il loro iter in questi giorni (in particolare il
DL, 25 giugno 2008, n. 112). Alleghiamo anche, per com-
pletezza, i testi dei nostri interventi durante l'udienza del
9 luglio scorso.

Lei fece cenno nel Suo intervento, Signor Presidente, alle
caratteristiche particolari del Suo ruolo rispetto ad altre
istituzioni. Per noi e per tutta la comunità scientifica ital-
iana sapere che Ella opererà sempre a difesa della cultura

e del progresso del nostro Paese è certo un importante incoraggiamento.

Con stima e cordialità,

Rino Falcone, Giulio Peruzzi, Enrico Bellone, Giorgio Bernardi, Paolo Bianco, Marcello Buiatti, Nicola Cabibbo, Elena Cattaneo, Stefano Fantoni, Sergio Ferrari, Pietro Greco, Francesco Lenci, Rita Levi Montalcini, Lucio Luzzatto, Piermannuccio Mannucci, Elisa Molinari, Caterina Petrillo, Settimo Termini, Carlo Umiltà

(le firme sono di coloro che salirono al Quirinale lo scorso 9 luglio. Lo sostanza del documento è certamente condiviso dai firmatari degli appelli che Le inviammo)

P.S.: nel Paese, in particolare presso università e centri di ricerca, sta montando un forte clima di preoccupazione e di sconcerto relativo alle potenziali conseguenze del sopra citato DL 112. Ci è sembrato utile per questo riassumerLe in una pagina l'elenco delle prese di posizione critiche a noi note fino alla data del 28 luglio tanto di organismi ufficiali quanto di organismi spontaneamente costituitisi per iniziative di protesta.

RISORSE UMANE E FINANZIARIE:

come tutti gli indicatori mostrano (percentuale di PIL investito, quantità di finanziamento pubblico e privato, numero di ricercatori su popolazione attiva, export di alta tecnologia, qualificazione personale nelle aziende private, etc.) le risorse umane e finanziarie nel settore sono del tutto inadeguate rispetto alla ricchezza economica che il nostro Paese esprime. Alcuni fattori indicano una deriva declinante: riduzione dell'export di prodotti con contenuto tecnologico avanzato (partendo oltretutto da livelli già molto modesti), livello dell'indice di competitività nettamente in discesa e più basso degli altri Paesi. La ricetta che i governi europei a Lisbona nel 2000 avevano predisposto riguardava proprio una massiccia inversione di tendenza

421

*nell'investimento in ricerca e alta formazione (con l'obiet-
tivo di un 3% di media europea nel 2010). L'Italia ha
fino ad oggi marcatamente disatteso quell'accordo (ne sono
testimonianza i report di aggiornamento dell'agenda Lisbona ogni anno redatti dalla Commissione Europea). Con
il decreto legge n°112 del 25/6/08 viene introdotta una
riduzione di risorse finanziarie (che porterà, in 4 anni,
a circa 450 milioni di euro in meno per il fondo funzionamento ordinario delle università). Non solo, lo stesso decreto si preoccupa anche delle risorse umane, introducendo
la norma di quasi blocco totale del turn-over (80%) fino al
2011, con l'aggravante che il recupero di risorse che in
tal modo si otterrà non è destinato a essere ridistribuito
o reinvestito nelle università e nella ricerca in generale.
Risorse vengono infine sottratte anche alla crescita professionale degli universitari. Di fatto si inverte del tutto
la logica europea per un serio e crescente investimento (di
risorse umane e finanziarie) in ricerca e università.*

*PRIORITA': Sarebbe di grande importanza ridurre al mas-
simo i tagli del DL 112. Sarebbe inoltre necessario predisporre un significativo piano di assunzione di (giovani)
ricercatori per riportare il settore a livelli adeguati, preve-
dendo anche una ragionevole progressione di carriera per
i più meritevoli.*

AUTONOMIA E RUOLO PUBBLICO PER RICERCA E ALTA FORMAZIONE:

*l'autonomia e il ruolo pubblico delle università italiane,
anche se tra molte contraddizioni ed alcune evidenti stor-
ture e malcostumi (primo fra tutti: un reclutamento non
sempre basato sul merito), hanno garantito al Paese un
livello alto, qualificato e aperto di ricerca e formazione.
In tal modo dando anche seguito al dettato della Carta
Costituzionale (articolo 33). Nel DL n°112 viene inserita
una norma che concede la possibilità alle università italiane di trasformarsi in fondazioni private. Sono del tutto*

evidenti i rischi per l'autonomia degli atenei e dei docenti oltre che per quei settori e ambiti di ricerca che non sono appetibili sul piano economico. Di fatto il combinato disposto – taglio delle risorse e possibilità di trasformazione in fondazione privata – rischia di modificare il sistema universitario nazionale in un sistema di formazione estremamente debole e con accessi differenziati in base al censo. Citiamo una parte dell'articolo che lo storico Franco Cardini ha scritto per il Secolo d'Italia il 16 luglio 2008 e che rende perfettamente conto dei rischi cui andremo incontro: "Il passaggio dall'Università alla Fondazione è in un certo senso epocale: sarà il passaggio da una concezione culturale e comunitaria a una patrimoniale e privatistica del sapere; da una mediocre e magari, perché no?, scalcinata Università di tutti, a una (forse) buona e (certo) più costosa università per i ricchi. Privatizzandosi, alcune università potranno salvarsi: ma in questo modo andrà una volta per tutte a farsi benedire il diritto allo studio: o meglio lo studio come diritto." Sul versante dell'autonomia va sottolineata l'importanza di rendere almeno parzialmente autonome dalla politica le nomine dei Presidenti degli Enti pubblici di ricerca utilizzando metodi (del tipo dei comitati di alta consulenza introdotti dal precedente Governo) che coinvolgano scienziati nazionali ed internazionali di chiara fama. Il recente caso (18 luglio 2008) di avvio del commissariamento dell'ASI da parte del Governo (con la nomina di un commissario ufficialmente legato ad un organismo beneficiario dell'agenzia in oggetto) richiama l'urgenza delle tutele qui indicate.

PRIORITA': Si auspica che le scelte riguardanti una prospettiva di privatizzazione delle università e degli enti di ricerca siano oggetto di specifici interventi legislativi sentite le comunità scientifiche. Sarebbe opportuno rendere operativa la legge delega sugli EPR dello scorso 2007 in cui si prescrive l'introduzione dei Comitati di alta con-

sulenza (Search Committees) per la nomina dei membri governativi dei CdA.

CRITERI E PROCEDURE PER ALLOCAZIONE RI-SORSE (ex post/ ex ante):
L'erogazione dei finanziamenti pubblici avviene spesso attraverso metodi non guidati da criteri di trasparenza. Questo vale per il finanziamento di soggetti tanto pubblici che privati.

PRIORITA': Fare in modo che tutti gli investimenti pubblici vengano sottoposti a procedure trasparenti, uniformi e basate sul merito. Dovrebbero essere sviluppate norme attraverso le quali i finanziamenti pubblici, nazionali e regionali, possano essere assegnati in base ad un sistema di valutazione da parte dei pari secondo standard internazionali (peer review). In particolare un ruolo importante potrebbe essere svolto dalla costituenda agenzia per la valutazione dell'Università e della Ricerca (ANVUR), per garantire che regole e procedure di peer review siano uniformi per tutti i programmi di finanziamento e che vengano effettuate valutazioni in modo indipendente, terzo e competente.

EUROPEIZZAZIONE:
lo Spazio Europeo della Ricerca rappresenta lo strumento portante per l'integrazione e la convergenza verso una società della conoscenza, pacifica e partecipata. I parametri di valutazione dell'attività di ricerca, i meccanismi di reclutamento e di sviluppo di carriera e le risorse economiche minime da destinare alla ricerca di base dovrebbero costantemente essere riferiti a quelli fissati nell'ambito dell' Unione Europea. L'Italia sta contribuendo alla costruzione dell'ERA con i suoi scienziati e con le competenze di cui è capace. Serve però uno sforzo maggiore in particolare nell'approntare strumenti organizzativi e politiche che favoriscano e supportino il coinvolgimento dei nostri

ricercatori nell'integrazione europea. Alcuni segnali recentissimi (blocco della roadmap grandi infrastrutture, blocco progetti Eranet, etc.) sembrano contraddire significativamente questo approccio.

PRIORITA': Politiche di sostegno alle iniziative europee. Non solo fondi ma anche sostegno operativo ed efficace. Impegno per la realizzazione di una roadmap nazionale di grandi infrastrutture di ricerca di livello pan-europeo da realizzare nel nostro territorio nei prossimi 10-20 anni.

Mi sembra che risulti chiara la straordinaria capacità di Rino a promuovere e coordinare iniziative di alto valore politico e culturale. Negli anni successivi, fino al 2014, Rino coordinò l'impegno di molti di noi, come sempre *in primis* di Carlo Bernardini, per la formulazione e la diffusione del **Manifesto Per un'Europa di Progresso**.

Il mondo è in rapida trasformazione. Società ed economia della conoscenza hanno profondamente ridisegnato equilibri ritenuti consolidati. Aree geografiche depresse hanno conquistato, in tempi storicamente irrisori, potenziali enormi di sviluppo e crescita. Conoscenza, cultura e innovazione rappresentano più che mai il traino decisivo verso il futuro.

All'opposto l'Occidente, e alcuni aspetti del suo modello di sviluppo, sono entrati in una crisi profonda. L'Europa, in particolare, risulta investita da gravissimi e apparentemente irrisolubili problemi: disoccupazione, crisi del tessuto produttivo, riduzione sostanziale del welfare. A pochi anni dalla sua formale consacrazione, con la nascita ufficiale della moneta comune, l'Europa rischia di deflagrare come sogno di una comunità di cittadine e cittadini che avevano ambito ad una nuova Nazione comune: più ampia non solo geograficamente, quanto nello spazio dei diritti, dei valori e delle opportunità. Lo storico americano Walter Laqueur ha parlato della "fine del sogno europeo".

Le responsabilità sono diverse e distribuite e investono certamente l'eccessiva timidezza nel processo di costituzione politica del soggetto europeo: la responsabilità di presentare questo orizzonte politico, culturale e sociale con le sole fattezze della severità dei "conti in ordine". L'Europa dei mercanti e dei banchieri, della restrizione e del rigore: una sorta di gendarme che impone limiti spesso insensati, piuttosto che sostegno nell'ampliare prospettive di visuale sugli sviluppi del futuro.

Proprio a causa di ciò, assistiamo, in corrispondenza della crisi, ad un'impressionante crescita di egoismi locali, di particolarismi e di veri e propri nazionalismi. Fenomeni spesso intenzionalmente organizzati per sfruttare malesseri veri, e reali stati di sofferenza, ma che rischiano di produrre reazioni esattamente opposte a quanto oggi servirebbe alle popolazioni d'Europa.

Come scienziate e scienziati di questo continente - consapevoli che esiste un nesso inscindibile tra scienza e democrazia - sentiamo quindi la necessità di metterci in gioco. Di ribadire che il processo di costruzione degli Stati Uniti d'Europa è la più importante opportunità che ci è concessa dalla Storia. Che società ed economia della conoscenza -essenziali per il processo di reale evoluzione civile, pacifica, economica e culturale- si alimentano di comunità coese e collaborative, di comunicazioni intense e produttive e di uno spirito critico che permei strati sempre più vasti della società.

L'unica risposta possibile alla crisi incombente è allora la costruzione dell'Europa dei popoli, di un'Europa di Progresso! Realizzata sulla base dei principi di libertà, democrazia, conoscenza e solidarietà.

Nutriamo la stessa speranza con cui Albert Einstein e Georg Friedrich Nicolai nel "Manifesto agli Europei" del 1914 richiamarono alla ragione i popoli europei contro la sventura della guerra, e con cui Altiero Spinelli, Eugenio Colorni ed Ernesto Rossi ispirarono l'idea d'Europa nel loro "Manifesto di Ventotene" del 1943. Le stesse idee che ebbero indipendentemente fautori illustri anche in tutti i Paesi d'Europa.

Vogliamo riprendere ed estendere all'Europa lo spirito che nel 1839 portò gli scienziati italiani a organizzare la loro prima riunione e a inaugurare il Risorgimento di una nazione divisa.

Vogliamo organizzare a Pisa la "Prima riunione degli scienziati Europei" e proporvi di firmare questo appello che è il nostro "Manifesto per un' Europa di Progresso".

Tra i promotori e primi firmatari: Ugo AMALDI, Carlo Bernardini Marcello BUIATTI, Cristiano CASTELFRANCHI, Vincenzo CAVASINNI, Remo CESERANI, Emilia CHIANCONE, Tullio DE MAURO, Rino FALCONE, Stefano FANTONI, Fabiola GIANOTTI, Pietro GRECO, Francesco LENCI, Lamberto MAFFEI, Giorgio PARISI, Giulio PERUZZI, Giorgio SALVINI, Vittorio SILVESTRINI, Settimo TERMINI, Guido TONELLI

Potrei aggiungere pagine dedicate a ricordare la bellezza umana di Rino, la ricchezza del suo pensiero, l'instancabilità del suo impegno, la straordinaria simpatia e capacità di non prendersi mai troppo sul serio. Incontrare Rino e lavorare con lui è stato per me un regalo della vita. Il suo ultimo regalo è del 25 Aprile di quest'anno: un whatsapp con Bella ciao cantata da lui.

Grazie Rino. Di tutto.

Francesco Lenci

427

Rino Falcone, Champion of the Academic Community

Daniele Archibugi

Institute for Research on Population and Social Policies (IRPPS),
CNR, Rome, and Birkbeck Business School, University of London
daniele.archibugi@cnr.it

1 The Boundaries of an Academic Community

Scholars of all times and races have always had a strong desire to stay close to each other. In ancient times, the learned sought mutual proximity because they were culturally isolated from the rest of the population, primarily because very few inhabitants could read and write. Meeting among scholars and exchanging knowledge was a great luxury often worth traveling long distances. These gatherings were not only convivial but also served to fuel heated debates and even wars between rival groups. Nevertheless, nothing would have been worse than ignoring each other. An international program of academic exchanges that began in 1987 and has involved millions of young people, the European Community Action Scheme for the Mobility of University Students, was aptly named after Erasmus of Rotterdam, a humanist who travelled extensively across Europe to spread his opinions.

Over time, the community of scholars has progressively expanded, leading to increasingly marked specialization. Unlike during the Enlightenment, today's members of the republic of letters are no longer capable of making significant contributions in all fields of knowledge and often find it difficult to fully read and understand articles published in authoritative and generalist journals such as "Nature" and

"Science". Scientific conferences are now fragmented and dedicated to individual disciplines and, despite this, are enormous. Gatherings of a medical society can easily reach twenty thousand participants, those of an economists' association ten thousand, and even within each discipline, theories, methods, and languages are increasingly diversified. It seems, in short, that the scientific community is increasingly approaching a Tower of Babel.

Yet, despite this progressive fragmentation, the academic community continues to preserve institutions and ways of communication that are very similar. The physicist and the psychologist do very different work, but they sit on the same academic Senate, publish their results in scientific journals with a similar peer review system, obtain positions based on public competitions where the committees are appointed with similar criteria, and receive comparable salaries.

Its unity becomes indispensable when it must justify its role in society. If once scientists had to ingratiate and convince only their patrons — think, for example, of Galileo dedicating the four moons of Jupiter he discovered to the Medici family — today they must convince a much wider community. In a democratic system, we are all accountable to someone, and science has become a profession that no longer arises from idleness.

The similarity in institutional forms thus allows us to speak of an academic community despite the fragmentation of languages. But for this academic community to exist, there must be scientists and scholars who, in addition to managing their own laboratories, have the time and desire to address the problems of everyone. Without these figures, the academic community risks disintegrating or being governed by outsiders who may not understand what is needed to conduct research. Rino Falcone is one of those who, alongside his research activities, has consistently dedicated himself to the governance of the academic community.

The modern "patrons" of the academic community are represented by businesses and the public sector. Businesses, which today finance more than half of the Research and Development conducted in Western countries, demand and obtain that their investment in knowledge generates new products, processes, and services from which to extract profits. But even businesses have long learned that investment in science and technology can produce unexpected results and that these are capitalized only in the long term.

The public sector, which promotes a significant portion of the investment in Research and Development, operates slightly differently. While some of this spending is aimed at specific objectives, there is always another portion intended for the advancement of knowledge, based on the awareness that it is unpredictable what the future use of today's discoveries and inventions will be.

Based on these premises, I wish to testify that Rino Falcone has been more than just a valuable member of the academic community. He has been one of those colleagues who has played the role of a conductor, ensuring adequate funding for scientific research.

2 Funding Scientific Research

As is well known, Italy's investment in R&D is much lower than that of our economic, cultural, and commercial partners. The reasons for this underfunding date back to time immemorial but have proven incredibly persistent over the decades. Other countries with little scientific and technological vocation have managed to reverse course; Italy has not. One reason is associated with the nature of Italian capitalism: we have had a limited number of large companies capable of financing adequate industrial research. Not only that, but many of these companies — think of Olivetti, Montecatini, and even Fiat — have not been adequately valued and protected. Many of the companies belonging to state-owned holdings, which in the past supported industrial research in Italy, have been privatized, resulting in a drastic downsizing of research and innovation laboratories, especially when acquired by multinational companies from other countries.

People have often wondered why our country has not managed to climb back up and invest more. It has often been said that it was not the public sector's role to finance industrial research, but the connection between public and private investments has been lost. In other nations, it was precisely public investment that triggered private investment since the existence of excellent public universities and research centres induced companies to invest in innovative projects. This is what happened in Sweden, Finland, and South Korea, today among the nations with the highest R&D investment rates, where less than a third of the expenditure is supported by the public sector. A third, however, that is strategic in convincing companies to invest their resources and, where possible, locate their laboratories near universities and public centres. In the absence of a strong public component in the innovation systems of nations, it becomes much less convenient for companies to place their laboratories and innovation centres in the country.

Beyond his scientific work, Rino has devoted a substantial part of his energies to ensuring that research was adequately funded. Rino repeatedly fought to preserve and progressively increase public investment in R&D. This was, for example, his main commitment when he collaborated with Fabio Mussi, MP, at the Ministry of Universities and Research from 2006 to 2008. One of the effects of the second Romano Prodi government during those two years was to finally reverse the trend of reducing R&D spending in Italy at the beginning of the new millennium.

As his colleague at the National Research Council (CNR), I am grateful to Rino for his tenacity in defending the institution's budget whenever there was a risk of cuts in the Financial Law reducing already scarce resources. One of the last occasions was in 2018 when there was a risk that the CNR would go into default due to lack of funding. Rino was the promoter of an appeal to President Sergio Mattarella and managed to create a broad coalition of intellectuals — both within and outside the CNR — gathering more than 3,000 endorsements. The core idea of the appeal was that without further funding, already committed resources, starting with salaries for CNR staff, would be wasted. Rino pointed out one of the many paradoxes of Italian public spending: some amounts, starting with public employees' salaries, were not compressible, but if researchers were not given the tools to conduct research, what was the benefit to society of paying them a salary? It was like giving an army rifles but no bullets.

The mobilization was successful. But we must pause for a moment to consider the reasons for that success. First, thanks to his personal authority, Rino managed to bring together colleagues from diverse scientific disciplines, political orientations, and personal attitudes. It took an ecumenical spirit that is uncommon to find in someone with such a long-standing secular political militancy. Secondly, Rino had the patience and ability to listen to the needs of many different colleagues and understand their reasons.

But if that mobilization was so successful, it was also because it was not perceived as an action by a corporation demanding more resources for itself, but as that of a community eager to serve the entire society and offer in return for the resources received the fruits of knowledge. It is not always easy to convey to the public the benefits offered by science. We had recent experience with COVID when, for the first time, scientists were listened to as rarely happened in the past. On this occasion, Rino had the opportunity to connect one of his main research themes — trust — with a specific event, that represented by COVID.

3   Autonomy from Political Power

The relationship between science and power has always been particularly complicated. Scientists need resources to do their work, and power is never disinterested: in exchange for these resources, those in power often exercise control over the governance of the academic community. On one hand, this control is legitimate in a democratic system to ensure that scientific research is not conducted detached from social needs.

However, there must be scholars willing to abandon their own work a little to devote themselves to organizing the academic community. These are service jobs, often unrecognized, that do not extend one's individual curriculum with additional publications. Those who dedicate themselves to this often pay a price in terms of career and personal engagement. Rino Falcone has done so generously and selflessly throughout his academic life. These are sufficient reasons to pay him our sincere gratitude.

THE INTEGRATION OF HUMANISTIC AND SCIENTIFIC CULTURES: A CRITICAL APPROACH TO AI DEVELOPMENT

LUIGI NICOLAIS

Professore emerito presso l'Università di Napoli Federico II, già presidente del Consiglio Nazionale delle Ricerche (CNR), già Ministro della Repubblica Italiana

In a context as impoverished as today's, speaking of two dichotomic worlds, the humanistic and the scientific, manifests as a striking example of cultural regression that precludes the possibility of understanding how critical, integrated, and interdisciplinary spirit is increasingly necessary. A contraposition born between the late 19th century and the first half of the 20th century between soft sciences and hard sciences, still accentuated today by the observation that scientists are inclined to move towards the future, while humanists look rather to the past. With the Industrial Revolution, the geometrization of the world emerges, already present in Plato with his motto: "God is always a geometer," where quantitative precision triumphs, the victory of quantity over quality, and the rejection of the world of approximation. However, as we all know, not everything that can be known is measurable. It would be more logical, as Rovelli has emphasized, to speak of "scientific inculture" as well as a deficit of classical culture for the opposing field. Both are necessary, and their separation represents an anachronistic intellectual misunderstanding. As Albert Einstein maintained, scientists and philosophers need each other. Cultural phenomena, including humanistic ones, can be studied with a scientific method and thus be validated, as many theories suggest. Humanistic and scientific cultures are integral parts of economic and human development, and it is necessary to rediscover in Europe the microcosm of Italian humanism, shaped by a transverse cultural hybridism that the entire open world recognizes and appreciates in its complexity.

The rise of artificial intelligence (AI) has further emphasized the importance of integrating humanistic and scientific perspectives. AI has the potential to revolutionize various aspects of our lives, from healthcare to education, and its development requires a deep understanding of both humanistic and scientific cultures. How many times have we asked ourselves: is it really true? This approach is applicable to both social sciences and scientific disciplines, seemingly distant but characterized by a common denominator that allows us to connect them in a coherent amalgam: critical thinking. Like dwarves on the shoulders of giants, let us think of ancient Greece, the emblematic and eclectic figure of the philosopher-scientist Anaximander, or in our Italian Renaissance, the eclectic Galileo Galilei or Pico della Mirandola.

I have met many young people in my career as a professor and scientist, and I have always supported the idea of fostering critical thinking in them, rather than a paradigmatic choice between hard and soft sciences. Furthermore, in recent years, the relationship between knowledge has radically changed beyond our will. Today, frontier research is based on the principle of complementarity, as Bohr defined it, observing that motion and position of a particle are complementary quantities. Complementarity, therefore, is not contradiction but attention to putting things together. The integration of AI in these research endeavors requires a deep understanding of both humanistic and scientific cultures, as well as the ability to critically evaluate the information generated by AI systems.

In this context, critical thinking is more crucial than ever, as AI can both enhance and distort our understanding of the world. AI can perform tasks with precision and speed, but it lacks the nuanced understanding and flexibility of human thinking. Therefore, it is essential to develop critical thinking skills to evaluate the information generated by AI systems and to ensure that AI is used responsibly and ethically.

A shining example is the modern scientist who no longer lives isolated from society in their laboratory but whose curiosity-driven approach must be increasingly supported by open and global confrontation among peers, public and private institutions, and citizens. An interaction between humanistic and scientific disciplines is also discovered in a specific applied field, that of research projects managed by Departments according to new ministerial evaluation guidelines.

This reflection on the necessity of thinking about the integration of knowledge and the unity of culture allows me to recall, within this collection of contributions, Rino Falcone, a renowned physicist and researcher who has made significant contributions to the field of artificial intelligence, which has given great value to this holistic view of knowledge.

His work has been characterized by a deep understanding of the intricacies of cognitive embodied systems, multi-agent systems, and agent theory. As a colleague and friend, I had the privilege of collaborating with Rino on several projects, and I can attest his exceptional expertise and dedication to his work. Rino's research has been marked by a relentless pursuit of innovation and a commitment to pushing the boundaries of what is possible in the field of AI.

Benedetto Croce, a prominent Italian philosopher, has also addressed the issue of the separation between humanistic and scientific cultures. In his work, he emphasizes the importance of understanding that both types of knowledge are necessary and complementary. He argues that the separation between the two is a product of the way we categorize and classify knowledge, rather than an inherent characteristic of the knowledge itself. This perspective is particularly relevant in today's context, where AI is increasingly integrated into various aspects of our lives, and it is essential to ensure that these systems are developed with a deep understanding of both humanistic and scientific cultures.

Rino Falcone's work has been characterized by a deep understanding of the intricacies of cognitive embodied systems, multi-agent systems, and agent theory, and his expertise in natural language processing, plan recognition, and multi-agent systems has been instrumental in advancing the field of AI. His research has been recognized through numerous publications and presentations at international conferences, and he has been a key figure in several major research projects, including the European project Mind RACES, which aimed to develop cognitive embodied systems. In today's context, where AI is increasingly integrated into various aspects of our lives, Rino's contributions are crucial in ensuring that these systems are developed with a deep understanding of humanistic culture and the complexities of human behavior. His work serves as a shining example of the importance of integrating humanistic and scientific cultures in the development of AI.

Why, as Steve Jobs emphasized in 2011, "in the future, there will be a marriage between technology and liberal arts, that is, the humanistic knowledge of the past will give us the result that makes a song rise in our hearts." The integration of humanistic and scientific cultures is crucial for this marriage to succeed, and it is essential that we foster critical thinking skills to evaluate the information generated by AI systems and to ensure that AI is used responsibly and ethically.

For these characteristics and peculiarities and his professionalism, it is for these reasons that I nominated him to be the head director of the Institute of Cognitive Sciences and Technologies (ISTC) of the National Research Council (CNR).

Rino Falcone places great importance on the relationship between science and society, which must be carefully curated and followed. This relationship is complex and continuously evolving, with various facets. According to Rino

Falcone, it is necessary to operate so that the substantial reevaluation that science is acquiring, especially post-COVID, and its ethical valorization, because ethics are the fundamental pillars on which it operates and develops its action for social progress, becomes increasingly intensely the property of the community. Therefore, a primary task falls to politics, but not only. Scientists must also feel this as their duty. Rino Falcone is a great example of inter-professionalism, similar to inter- disciplinarity, a work at the border where contaminations and reciprocal influences between various activities result in the correct development of each field.

Scienza, società e democrazia

Lucia Votano

INFN, Laboratori Nazionali di Frascati, Via Enrico Fermi 54, 00044 Frascati, Italy

`lucia.votano@lnf.infn.it`

Sono onorata e felice di poter contribuire alla pubblicazione di un libro collettaneo in onore di Rino Falcone in occasione del suo pensionamento che mi auguro non ne interrompa la brillante attività di ricerca. I miei specifici interessi nell'ambito della fisica astroparticellare, diversi da quelli praticati da Rino, mi consentono di poter dare testimonianza solo di alcuni aspetti della sua multiforme personalità, e cioè del suo grande interesse al tema del rapporto tra scienza e società e dell'impegno a favore di una migliore organizzazione politica e sociale della ricerca nel nostro Paese. Sicuramente il percorso scientifico di Rino nel campo delle cosiddette tecnologie intelligenti può aver favorito e acuito l'interesse a osservare e misurare l'impatto della scienza sulla società. Basti pensare ai progressi delle nuove tecnologie che si stanno sviluppando intorno all'intelligenza artificiale, uno dei temi di studio dell'Istituto di Scienze e Tecnologie della Cognizione (ISTC) del CNR che Rino ha diretto, le quali ci offrono opportunità senza precedenti, ma presentano anche alcuni rischi finora non immaginabili. Massimizzarne le opportunità in termini economici e sociali attraverso la ricerca libera e al contempo cercare di prevedere e minimizzare i rischi connessi richiede agli scienziati, sia in ambito pubblico sia privato, l'esercizio di un forte senso di responsabilità sociale. Queste tecnologie, ormai sulla bocca di tutti, infatti, stanno contribuendo a modificare rapidamente e in misura crescente il mondo e la società in cui siamo immersi e al contempo la nostra mente, il nostro modo di pensare e di conoscere.

Non è ovviamente l'unico esempio, molti altri sono gli sviluppi scientifici e tecnologici che hanno un forte impatto sociale. L'innovazione generata dalla ricerca scientifica genera sempre più velocemente nuovi prodotti o processi, che possono tuttavia in alcuni casi presentare potenziali rischi per la salute dell'uomo o dell'ambiente o sollevare rilevanti temi di natura etica e sociale. D'altra parte, mai come oggi nella storia dell'uomo si è investito così tanto in ricerca e sviluppo e così estensivamente in tutto il pianeta da occidente ad oriente. Mai la scienza è stata così determinante per lo sviluppo culturale, economico e sociale della società, anche se ha sempre avuto un ruolo importante in tutte le civiltà, dall'epoca ellenistica, passando per la cosiddetta *Rivoluzione Scientifica del Seicento* e proseguendo oltre. Sappiamo che a partire approssimativamente dalla seconda metà del secolo scorso siamo entrati in un'era in cui l'organizzazione del lavoro dei ricercatori si è modificata profondamente. Viviamo ormai in un'era della scienza che gli esperti chiamano post-accademica, in cui la dipendenza reciproca tra scienza e società è un dato di fatto ineliminabile. I motivi di questo cambiamento sono molteplici. Intanto la ricerca scientifica è ormai diventata un'impresa collettiva, i ricercatori agiscono all'interno di università e istituti di ricerca pubblici e privati, che mettono a disposizione le ingenti risorse economiche e l'organizzazione tecnica e amministrativa necessarie a realizzare grandi e complessi apparati sperimentali e ad analizzare l'enorme mole di dati raccolti. Gli scienziati sono diventati forza-lavoro al pari degli altri, dipendenti quindi che devono rendere conto dei finanziamenti che hanno ricevuto e dei risultati ottenuti. È quindi naturale che sentano sempre di più, se non altro per ragioni pratiche legate ai finanziamenti, l'esigenza di un consenso sociale intorno ai temi della loro ricerca. Inoltre, Il prorompere dell'innovazione tecnologia nella nostra vita quotidiana ha fatto emergere nei cittadini una maggiore consapevolezza dei rischi reali o potenziali ad essa associati e si richiede alla ricerca scientifica di essere sempre rispettosa dei fondamentali principi etici e delle libertà individuali e al contempo dell'ecosistema. È naturale quindi, e così deve essere in una società democratica, che sempre più spesso le decisioni che riguardano il lavoro degli scienziati, nei settori di ricerca a maggiore impatto sulla vita dei cittadini, siano influenzate, quando non determinate, anche da "non esperti": politici, imprese, gruppi di pressione, stampa, tv e media, nonché il grande pubblico. L'obiettivo di conciliare il rigore, la libertà di ricerca e l'eccellenza scientifica con le giuste istanze che arrivano dalla società moderna non è per nulla facile da raggiungere e rappresenta una sfida per la comunità scientifica. Lo straordinario sviluppo della conoscenza scientifica e la sua accentuata specializzazione hanno generato, infatti, un divario cognitivo e linguistico tra specialisti e popolazione, che esclude buona parte dei cittadini da una consapevole discussione su argomenti di interesse della scienza. Anche se la facilità di accesso alle informazioni sul web può dare l'illusione che questo divario tra scienza e cittadini si sia attenuato, è vero il contrario. Abbiamo l'illusione di poter conoscere tutto, ma difficilmente siamo in possesso delle competenze per affrontare con spirito critico qualsiasi tipo di informazione. Internet o i social media ci hanno abituato a considerare la nostra opinione sempre pertinente e rilevante, anche quando non lo è. È pur vero che non possiamo pretendere che tutti i cittadini abbiano una qualche conoscenza delle discipline scientifiche e dei loro lessici, ma l'essenza del problema, a mio avviso, consiste nel fatto che pochissimi conoscono il senso profondo della stessa parola scienza. Parlo di cosa significhi fare scienza, di capire su cosa si fonda la sua affidabilità, ma anche i suoi limiti e il senso della sua transitorietà temporale. Il passaggio all'era post-accademica della scienza ha portato anche una rivoluzione nella comunicazione della scienza che negli anni ha subito una profonda trasformazione. Se c'è stato un tempo nel passato in cui la divulgazione scientifica era considerata un aspetto marginale del lavoro dei ricercatori, portata avanti su base volontaria e neanche troppo ben vista per l'avanzamento della carriera, oggi ci siamo convinti che il sapere scientifico sia inseparabile dalla sua condivisione. Inoltre, si è capito che la comunicazione non va intesa più come trasferimento unidirezionale dagli esperti al grande pubblico, ma si deve puntare al coinvolgimento, alla comunicazione partecipativa e al dialogo. La comunicazione della scienza è divenuta quindi una scienza essa stessa, ovviamente multidisciplinare.

Rino Falcone ha ai miei occhi il merito di essere stato da sempre un attento e profondo osservatore e un grande esperto di tutti questi temi. Egli si è speso molto nella promozione della ricerca scientifica attraverso il coordinamento di osservatori delle politiche che negli anni sono state applicate e promuovendo specifiche iniziative nel settore. Le competenze acquisite lo hanno meritatamente portato a essere chiamato come consulente di due ministri dell'università e ricerca e a dare quindi significativi contributi all'assetto normativo della ricerca pubblica in Italia. Ho conosciuto Rino per merito di un comune amico, Pietro Greco, un finissimo intellettuale di cultura enciclopedica, scrittore, giornalista e maestro di divulgazione scientifica, uno studioso profondo del rapporto scienza e società nella storia, uomo mite e sempre gentile, prematuramente scomparso alla fine del 2000. Di recente a Forlì osservavamo con Roberto Caporesi, un altro suo amico e Presidente dell'Associazione Nuova Civiltà delle Macchine, come l'inarrestabile attivismo di Pietro abbia tra l'altro creato una vasta e molto precipua comunità di persone, anche professionalmente molte diverse, che hanno avuto la fortuna di apprezzarlo, di frequentarlo e di collaborare a vario titolo con lui, e con cui è facile e naturale sentirsi in sintonia. Una comunità per certi versi anche virtuale di amici di Pietro tra cui esiste una condivisione di valori e visioni del mondo. È questa una delle tante eredità intellettuali e umane che Pietro ci ha lasciato. Non a caso quindi, pur non frequentandoci assiduamente, è stato possibile per me stabilire un solido rapporto di stima ed amicizia con Rino.

Sono diverse le occasioni in cui ci siamo incrociati ma mi piace ricordare in particolare la stesura e le presentazioni del volume collettaneo *La scienza tra etica e politica. L' eredità di Carlo Bernardini e le prospettive future*, dell'editore Dedalo, pubblicato nel 2020 a cura di Rino Falcone, Pietro Greco e Giulio Peruzzi. Un libro che, come scritto nella prefazione, era certamente una commemorazione del grande fisico scomparso nel 2018, ma che prendeva spunto dalla sua militanza civile di intellettuale, dalla sua attenzione ai temi sociali e politici, per guardare e riflettere sul futuro. Si era in piena pandemia da Covid19 e la prima presentazione del libro si poté fare solo da remoto in un webinar cui aveva partecipato anche il Ministro protempore dell'Università e della ricerca, Gaetano Manfredi.

In quell'occasione il discorso introduttivo fu tenuto da Rino. Egli anzitutto amaramente ricordò come si venisse da decenni in cui "*la visione politica di indirizzo del Paese non ha mai considerato, se non in marginali esperienze e singoli protagonisti, la scienza, la ricerca e l'innovazione tecnologica avanzata come il carburante essenziale del motore strategico della Nazione.*" E poi, parlando della percezione pubblica della scienza, Rino mise in evidenza come "*si sia stabilita una diffusa considerazione nell'opinione pubblica della competenza come bene separato dalla conoscenza. E come quest'ultima (ossia la conoscenza) sia stata sostanzialmente assimilata alla informazione. In questa visione semplificata: essere informati corrisponde anche ad avere acquisito conoscenza. Allo stesso modo e in tutti i domini.*" In sostanza mise in evidenza alcune tra le difficoltà nel lungo e tortuoso cammino verso il pieno esercizio della cittadinanza scientifica, cioè del diritto al controllo democratico della risorsa conoscenza. Se è vero che l'economia e la società della conoscenza sono diventati il motore principale del benessere delle nazioni e dei cittadini, una disparità di accesso all'alta formazione e alla ricerca non può che generare, come peraltro osserviamo, forti diseguaglianze economiche e sociali tra le nazioni e tra i cittadini all'interno di un singolo stato. Inoltre, se si vuole garantire ai tanti stakeholder di partecipare in modo informato e consapevole alle decisioni che riguardano le innovazioni scientifiche e tecnologiche a forte impatto sulla società, è necessaria una maggiore diffusione della cultura scientifica almeno come metodo di ragionamento. All'epoca delle presentazioni del libro in commemorazione di Bernardini, l'urgenza del contagio da Covid19 aveva riportato la scienza in primo piano su tutti i media e il grande pubblico stava riscoprendo il valore e la necessità della ricerca scientifica, in quel caso in funzione salvifica: in sostanza ci si rivolgeva alla scienza con rinnovata fiducia. Rino acutamente però già osservava come errori di comunicazione e una scarsa conoscenza del significato del metodo scientifico e di come la ricerca procede, potevano rendere effimero questa positiva inversione di tendenza nell'atteggiamento e attenzione verso la scienza. Purtroppo, dobbiamo ammettere oggi che le sue preoccupazioni erano fondate. Il timore oggi dei ricercatori è che dopo il

445

2026, quando si sarà esaurito l'effetto positivo del PNRR, gli investimenti nella ricerca, ma anche l'attenzione e la consapevolezza del valore sociale ed economico della scienza, in mancanza di nuovi incentivi strutturali potrebbero tornare ai livelli pre-Covid.

Le sempre più strette connessioni che si sono instaurate in epoca moderna tra scienza, società e democrazia hanno profondamente trasformato la figura dello scienziato e alcune delle modalità stesse del fare ricerca. Rino Falcone è stato un precursore che ha da sempre capito la necessità di mettersi in gioco per favorire il progredire dell'intera società attraverso la scienza. È su questo terreno di condivisione di valori, di riflessioni sui temi riguardanti la ricerca scientifica che la nostra amicizia si è sviluppata sempre più nel tempo. Auguri Rino per la tua nuova vita!

LIFE AS DIRECTORS

CORRADO BONIFAZI

*Institute for Research on Population and Social Policies
(IRPPS-CNR), Rome, Italy*

corrado.bonifazi@irpps.cnr.it

When I became Director of IRPPS in July 2014, Rino Falcone had already been leading ISTC for a few years. In fact, he had started in June 2011 as an acting Director, replacing Cristiano Castelfranchi, in October 2013 he was appointed Director as the winner of the selection procedure and then, at the end of his mandate, he would continue for another two years again as acting. In the four years of my direction, Falcone was therefore a constant presence, he was a real point of reference and comparison for all the directors. At that moment, our two institutes were among the largest of the current Department of Human and Social Sciences, Cultural Heritage (DSU) of the CNR, in whose acronym, who knows for what strategic plan, neither the social sciences nor those of cultural heritage are considered. The main offices of the two institutes in Rome were close at the time, one in via San Martino della Battaglia and the other in via Palestro, the latter premises were shared by the IRPPS and part of ISTC. Dimensions and spatial location therefore created a context conducive to developing a relationship which was however nourished thanks to the sharing of other factors: from the generational one, to the same idea of scientific research and the role of the CNR, to having carried out one's career at within the CNR (even if with differently jagged paths), to very close political visions and also to the same football faith (which certainly didn't hurt!).

It should then be considered that the relationship between the two instituts could count on other factors. First of all, the presence among the researchers of the IRP, first, and of the IRPPS, later, of Maura

Misiti, daughter of Raffaello who was the founder and Director of the Institute of Psychology, forerunner of the current ISTC, until to his untimely death. The relationships of friendship, esteem and affection that continued to bind Maura to a large part of the scientific community of ISTC, including Rino Falcone, could only play in favor of a strengthening of relationships when the need arose. In this sense, the Migrations Project, of what was then called the Department of Cultural Identity (DIC), was, as far as I am concerned, a first opportunity for discussion and collaboration with ISTC, through the person of the Director of the Institute Cristiano Castelfranchi with whom I shared the coordination of one of the sectors into which the activities of that research were divided.

The favorable context soon had the opportunity to reveal itself and become active. In the summer of 2014, in fact, a clear fracture was created between the Director of the DSU at the time and the Directors of all the institutes over the methods of distribution of the few new researcher positions assigned with the ordinary fund of that year to the DIC. A situation which had the positive effect of stimulating closer collaboration between all the institute directors. Another factor contributed to strengthening the collaborative relationship between me and Rino in the same period: the executive eviction issued on 7 June 2014 by the Court of Rome for the premises in via Palestro where the IRPPS and part of ISTC had their headquarters. News that struck both of us like a bolt from the blue, given that we were not aware of the dispute even though it had been underway for some time, leading to a Court order. A decidedly tangled and difficult situation, with a visit from the bailiff and the setting up of the execution of the eviction with an expected intervention by the public force.

A positive solution to a decidedly grotesque problem was only achieved in February 2015, but those troubled months served to strengthen our collaboration and lay the foundations for a relationship of mutual esteem and friendship that continued in the following years, especially in the period in which we shared the direction of the institute. After all, we are two Boomers, two kids from the seventies who grew up in a period in which political and social commitment was taken for granted and influenced most personal choices. From here also derives, in my opinion, the lofty and important idea of the role of public research and the specific function of the CNR that we shared. While working and directing a CNR institute has both given us the awareness that the optimal functioning of our institution must rely on maintaining the balance between research groups, institutes and central bodies, a specificity that cannot always be grasped by those coming from other experience. And some choices made by Governments and by the CNR itself seem to me to have largely confirmed this. These are the lines along which, based on my personal experience, Rino Falcone's action within the CNR has always moved with great commitment and excellent results. The CNR has extraordinary capabilities and demonstrates this by achieving results of notable scientific level, but it also has great structural difficulties. Not the least of which is changing too often and above all doing so without always evaluating the effects of previous reforms with due attention. Greater caution and a more in-depth knowledge of the organization would perhaps have led on some occasions to different choices and more targeted and precise interventions.

Dedica a Rino Falcone

Nando Dalla Chiesa
*Presidente onorario dell'associazione Libera, già Professore
ordinario presso l'Università degli Studi di Milano*

Era la tarda primavera del 2006. Quando venni nominato sottosegretario al Ministero dell'Università e della Ricerca nel secondo governo Prodi mi interrogai su cento cose. Totalmente inesperto degli "arcana" ministeriali, mi chiesi quali strumenti avrei avuto a disposizione per rendere quell'incarico così altisonante utile al Paese. Che tipologie di umanità avrei scoperto? Quali affinità vere avrei trovato? Quali "leggi bronzee" avrei dovuto sfidare e combattere per dare un senso alla mia nuova posizione? Domande difficili, perfino impervie con un governo che poteva contare su una maggioranza risicatissima, spuntato quasi per miracolo da una notte in cui i voti del centrosinistra (l'"Unione") sembravano destinati a farsi progressivamente divorare dai minuti che passavano. Finché con una fortunata dichiarazione il segretario dei Democratici di Sinistra Piero Fassino, in permanente collegamento con i seggi ancora non scrutinati, aveva dato l'annuncio irreversibile: abbiamo vinto noi. 24mila voti di scarto. Lì tutto si chiuse, ma lì tutto restò aperto. Quanto sarebbe durato un governo che si reggeva sui senatori a vita e sugli eletti all'estero? Fu in quell'aria di precarietà che mi guardai intorno. Il ministro era Fabio Mussi, persona colta e riottosa, pronta a uscire dal suo partito, i Democratici di Sinistra, avviato a fondare il Partito democratico con la Margherita. L'altro sottosegretario era Luciano Modica, un matematico prestigioso, grande conoscitore dei meandri dell'accademia. C'era anche un capo di gabinetto di cui ho scelto di dimenticare nome e cognome, dopo avere riposto in lui la fiducia che sempre tradisce gli ingenui. E c'era un consigliere del ministro, con l'aria del giovane manager politico. In realtà era un ricercatore del Cnr, che sapeva il fatto suo, che si presentava alle riunioni senza mai un briciolo di presunzione, avendo studiato le carte, lasciando agli altri il primato metodologico spettante loro in forza dell'investitura politica. Fu forse questo il tratto, temperamentale e culturale, che mi colpì di lui. La modestia dei toni, l'orientamento a costruire soluzioni pragmatiche e al tempo stesso lo spessore ideale che traspariva dai suoi ragionamenti. Che non fosse finito in quella posizione per strategie opportunistiche o per spartizione clientelare mi apparve subito evidente. Era un intellettuale di partito, è vero. Ma nitido nel pensiero, con una tensione morale non sbandierata e che però faceva spesso capolino nei suoi ragionamenti. Era lui, lo avrete immaginato: Rino Falcone. Ossia un intellettuale da me distante per predilezioni disciplinari, la fisica contro la sociologia, il Cnr contro Scienze politiche; ma anche per riferimento politico, i DS invece della Margherita, e, come città di riferimento, Roma invece di Milano. Materia che talora conta poco ma che nei fatti potrebbe anche contare molto. Ne avevo sentito parlare come di persona degna di stima. Ne feci esperienza diretta. Di lui mi piacque quella combinazione, rara per un intellettuale italiano, di disincanto politico (inteso come assenza di faziosità) e di passione morale. Mi coinvolse in alcune iniziative che non avevano a che fare con le nostre specifiche funzioni ministeriali ma avevano a che fare con il senso più ampio del nostro agire politico: discussioni, progetti, rapporti da costruire. Discutevamo dei problemi dell'organizzazione degli studi universitari in Italia e delle politiche culturali del Comune di Roma, ma anche dei problemi incandescenti della giustizia, in un raro clima di serenità riflessiva. Un giorno, benché ci conoscessimo da poco, mi ritrovai a pensare che come compagno di conversazione rappresentava in fondo un felice modello. Cultura, attenzione, rispetto, ironia: che altro chiedere di più? Siamo rimasti in contatto anche dopo, lui nel suo ruolo di prestigio di direttore dell'Istituto di scienze e tecnologie della cognizione del Cnr, di nuovo consigliere del Ministro (Gaetano Manfredi), premuroso nello spedirmi appelli e osservazioni puntuali, mai banali, sui temi un giorno comuni, e io purtroppo incapace di reggere il suo registro, essendomi ormai gettato sui nuovi campi da me aperti nell'insegnamento universitario e nell'impegno civile. Sempre di più mi sono reso conto che quella che Rino aveva per la ricerca e la scienza era passione autentica, non vocazione da carriera. Per questo, nonostante non ci incontriamo da più di dieci anni, porto di lui un ricordo che mi fa star bene. Quello di una vera affinità incontrata quasi per caso. Uno di quei ricordi che ci stendono le rughe e ci fanno sorridere gli occhi al solo rievocarli. Esattamente come mi sta succedendo ora, nel chiudere questo amarcord.

Il coraggio di scegliere science-based solutions
The courage to choose science-based solutions

Pierpaolo Campostrini

Direttore generale di CORILA, Consorzio per il coordinamento delle ricerche inerenti il sistema lagunare di Venezia

Nel 2006, quando Rino Falcone era consigliere tecnico di Fabio Mussi, Ministro dell'università e della ricerca scientifica del governo Prodi II, entrato in carica il 17 maggio di quell'anno, non ci conoscevamo per nulla. Era fine estate quando bussai alla sua porta, nel Ministero allora in Piazzale Kennedy all'EUR e mi presentai con la mia credenziale di ex-ricercatore CNR, prima di quella di direttore di CORILA, mio impiego del tempo ed attuale.

Forse l'approccio fu quello giusto, poichè ebbe la pazienza di ascoltarmi per un'ora buona: presentavo un tema del tutto complicato, sia dal punto di vista scientifico che da quello politico, ovvero la salvaguardia di Venezia ed il MOSE. E' bene ricordare che la prima pietra del MOSE fu salpata dal Presidente Berlusconi nel 2003, e che Massimo Cacciari, sindaco di Venezia per la terza volta nel 2005, al nuovo governo aveva chiesto (ed ottenuto) la possibilità di discutere delle proposte alternative al MOSE, secondo lui più efficiaci e meno costose.

Il Sindaco Cacciari, buon amico e della stessa area politica del Ministro Mussi, aveva chiesto pubblicamente a tutti i cittadini veneziani di presentare idee "alternative", che sarebbero state presentate al Comitato Di Indirizzo e Controllo per la Salvaguardia di Venezia (composto da 5 ministri, il presidente della Regione del Veneto, i

Sindaci di Venezia e di Chioggia e presieduto dal Presidente del Consiglio). Infatti, il Comitatone, il 20 luglio di quel 2006 stabilì che le proposte alternative del Comune sarebbero state discusse da un "tavolo tecnico" coordinato dalla Presidenza del Consiglio, cui avrebbero partecipato tutti i ministri, ovviamente Università e Ricerca incluso.

La pratica delle proposte alternative arrivò, dagli uffici del MIUR, sul tavolo del CORILA, ovvero il mio. Il nostro documento di commento alle "alternative" fu preparato dal nostro Consiglio scientifico, da me coordinato, e fu mandato al "nostro" ministero della Ricerca, che lo diffuse.

Allo stesso modo, tutti i Ministeri presenti nel Comitatone, nonchè Comune e Regione, prepararono un documento che venne spedito nei tempi stabiliti alla Presidenza del Consiglio dei ministri. In verità, i documenti "ministeriali" erano tutti verbosi e non molto chiari.

Quello che si trovò tra le mani, qualche giorno prima del nostro incontro, Rino Falcone, ricercatore scientifico prima che consigliere del Ministro, era diverso. Le conclusioni erano chiare e contraddicevano con rigore scientifico la funzionalità di tutte le "soluzioni alternative": esse erano inefficaci e in una certa misura anche dannose per l'ambiente.

Entrai nell'ufficio di Rino in Piazzale Kennedy quel fine estate, sicuro della solidità delle nostre argomentazioni, ma anche consapevole della difficile situazione in cui egli si trovava. Le dichiarazioni rilasciate dal Sindaco, che aveva chiesto ed ottenuto da Prodi un "tavolo tecnico", erano già del tutto "politiche" e difendevano un "pre-giudizio", prima ancora che il tavolo tecnico venisse convocato per confrontare le diverse risposte.

Non ero il primo veneziano che riceveva in quel fine estate. Mi confidò che da lui erano stati già i consiglieri di Cacciari, professori o già professori universitari, da tempo schierati dalla parte dei "NO MOSE", che lo avevano già inondato di apodittiche tesi, disegnando apocalittici scenari se la costruzione del MOSE non fosse stata fermata. Ascoltò il mio punto di vista, facendo domande sensate, anche se la sua formazione scientifica era diversa da quella dello studioso di idraulica o di ecologia. Si riservò di studiare le carte e mi salutò.

456

Qualche giorno dopo, mi chiamò per telefono e mi disse: "Avete ragione voi, le soluzioni proposte ai problemi devono essere basate su conoscenza e non su pregiudizio". Seguirono altre riunioni a Roma, che coinvolsero anche il Gabinetto del Ministro.

Il "tavolo tecnico" fu convocato per il 2 novembre e una seconda volta il giorno 8 a Palazzo Chigi. Rino Falcone difese la nostra presenza su quel tavolo, che assolutamente Cacciari non avrebbe voluto, e lo aveva ben detto al suo amico Ministro. Nella discussione che si aprì nella Sala Verde di Palazzo Chigi, gli esperti di CORILA vinsero ampiamente il confronto, quasi si trattasse (mi si perdoni il paragone sportivo) di una squadra di calcio di serie A contrapposta ad una di dilettanti.

Infatti, il 10 novembre, il Consiglio dei ministri approvò a maggioranza la relazione del ministro delle Infrastrutture Antonio di Pietro che riferì sulle risultanze del "tavolo tecnico". La relazione sposò in pieno le tesi di CORILA e rilevò che non erano emersi elementi nuovi che potevano indurre a modificare il progetto originario in corso di realizzazione.

Il 22 novembre il Comitatone deliberò di procedere con il Sistema MOSE, secondo il progetto approvato. I presenti videro il Ministro Mussi, rosso in volto, il quale ai margini della riunione redarguì Cacciari per la pochezza scientifica degli esperti che lui aveva presentato al tavolo tecnico.

Rino deve essere fiero di questo piccolo pezzo di storia che avvenne (anche) grazie a lui. Per una volta, il parere della comunità scientifica aveva prevalso su quello della politica, evitando conclusioni basate su una pseudo-scienza, manipolata e costretta al pregiudizio. Poteva dispiacere, poichè tale pre-giudizio proveniva da un'area alla quale Rino (ma anche il sottoscritto!) aveva dato non solo il suo voto, ma anche la sua passione civile. I ricercatori, a differenza dei medici, non hanno qualcosa che assomigli al giuramento di Ippocrate, comunque Rino si comportò come assoluto difensore dell'oggettività dei risultati scientifici, gli unici che possono essere a base delle decisioni prese da una "buona" politica, ovvero basate sulla migliore conoscenza disponibile.

Rino visitò più volte Venezia, poichè CORILA continuò ad essere un baluardo scientifico indipendente, che monitorò rigorosamente l'andamento dei cantieri di costruzione del MOSE dal punto di vista ambientale, dimostrandosi efficace sentinella per la minimizzazione del loro impatto sui dinamismi naturali lagunari.

Finchè continuò la collaborazione con il Ministero (il secondo governo Prodi durò sino al 24 gennaio 2008), Rino Falcone non mancò di sostenere il nostro operato, la nostra indipendenza ed il rigore scientifico, come valore assoluto, da mantenersi sempre anche di fronte a situazioni delicate dal punto di vista politico.

La storia della costruzione del MOSE fu ancora lunga e travagliata; alcuni aspetti di (mala)gestione amministrativa attirarono l'attenzione della magistratura: corrotti e corruttori, cattivi amministratori a vario titolo e livello vennero perseguiti e puniti dalla giustizia dello Stato. Ma il merito tecnico dell'opera non entrò in discussione. Dal 3 ottobre del 2020 il MOSE difende Venezia ed ha sinora evitato oltre 80 volte l'allagamento della città, evitando danni irreparabili al patrimonio culturale, economico e sociale.

Un po' di merito di questo risultato epocale, che tutti lo sappiano, lo ha anche il collega ed amico Rino Falcone.

In 2006, when Rino Falcone was a scientific advisor to Fabio Mussi, Minister of University and Scientific Research in the Prodi II government, which took office on May 17 of that year, we did not know each other at all. It was late summer when I knocked on his door at the Ministry, which was then located at Piazzale Kennedy in EUR, and introduced myself with my credentials as a former CNR researcher, before those of my current role as director of CORILA.

Perhaps the approach was the right one, as he had the patience to listen to me for a good hour: I was presenting a highly complicated issue, both scientifically and politically, namely the preservation of Venice and the MOSE project. It is worth noting that the first stone of MOSE was laid by President Berlusconi in 2003, and that Massimo Cacciari, Venice's mayor for the third time in 2005, had asked (and obtained) the new government to discuss "alternative proposals" to MOSE, pretended to be more effective and less costly.

Mayor Cacciari, a good friend and from the same political area as Minister Mussi, had publicly asked all Venetian citizens to submit "alternative" ideas, which would be presented to the Committee for the Preservation of Venice (composed of 5 ministers, the president of the Veneto Region, the mayors of Venice and Chioggia, and chaired by the Prime Minister). Indeed, the Committee, on July 20 of that year, decided that the municipality's alternative proposals would be discussed by a "technical table" coordinated by the Prime Minister's Office, which would involve all ministries, including of course the Ministry of University and Research.

The practice of the alternative proposals arrived, from the MIUR offices, on the CORILA table, that is, mine. Our commentary document on the "alternatives" was prepared by our scientific council, which I coordinated, and was sent to our Ministry of Research, which disseminated it.

Similarly, all the Ministries present in the Committee, as well as the Municipality and Region, prepared a document, which was sent within the established deadlines to the Prime Minister's Office. In truth, the "ministerial" documents were all verbose and not very clear, but ours.

Indeed, what Rino Falcone, a scientific researcher before becoming a ministerial advisor, had in his hands a few days before our meeting, was different. The conclusions were clear and scientifically contradicted the functionality of all the "alternative solutions": they were ineffective and to some extent even harmful to the environment.

I entered Rino's office in Piazzale Kennedy that late summer, confident in the solidity of our arguments but also aware of the difficult situation he was in. The statements made by the Mayor, who had requested and obtained from Prodi a "technical table," were already entirely "political" and defended a "prejudice," even before the technical table was convened to compare the different responses.

I was not the first Venetian to visit him that late summer. He confided that he had already been visited by Cacciari's advisors, professors or former university professors, who had long been aligned with the "NO MOSE" side and had inundated him with dogmatic theses, depicting apocalyptic scenarios if the MOSE construction was not halted.

He listened to my point of view, asking sensible questions, even though his scientific background was different from that of a hydraulic or ecology expert. He reserved the right to study the documents and said goodbye.

A few days later, he called me and said: "You are right, the proposed solutions to the problems must be based on knowledge and not on prejudice." Further meetings in Rome followed, which also involved the Minister's Cabinet.

The "technical table" was convened for November 2 and a second time on November 8 at Palazzo Chigi. Rino Falcone defended our presence at that table, which Cacciari definitely did not want, and had clearly told his friend Minister. In the discussion that opened in the Green Room of Palazzo Chigi, CORILA's experts won the comparison by a wide margin, almost as if (pardon the sports analogy) an A-league football team was facing an amateur team.

Indeed, on November 10, the Council of Ministers approved by a majority the report of the Minister of Infrastructure Antonio Di Pietro, who reported on the results of the "technical table." That Report fully embraced CORILA's positions and noted that no new elements had emerged that could lead to modifying the original project under implementation.

On November 22, the Committee decided to proceed with the MOSE System, according to the approved project. Those present saw Minister Mussi, red-faced, who, on the sidelines of the meeting, scolded Cacciari for the scientific inadequacy of the experts he had presented at the technical table.

Rino must be proud of this small piece of history that happened (also) thanks to him. For once, the opinion of the scientific community prevailed over that of politics, avoiding conclusions based on pseudo-science, manipulated and driven by prejudice. It might have been disappointing, as this prejudice came from an area to which Rino (and also I!) had not only given his vote but also his civil passion. Researchers, unlike doctors, do not have something akin to the

Hippocratic Oath, yet Rino behaved as an absolute defender of the objectivity of scientific results, the only ones that can underpin decisions made by "good" politics, based on the best available knowledge.

Rino visited Venice several times, as CORILA continued to be an independent scientific bastion, rigorously monitoring the progress of the MOSE construction sites from an environmental perspective, proving to be an effective sentinel for minimizing their impact on the lagoon's natural dynamics.

As long as the collaboration with the Ministry continued (the second Prodi government lasted until January 24, 2008), Rino Falcone did not fail to support our work, our independence, and scientific rigor as absolute values to be maintained even in politically delicate situations.

The story of the MOSE construction was still long and troubled; some aspects of (mis)administrative management drew the attention of the judiciary: corrupt individuals and bad administrators at various levels were prosecuted and punished by state justice. However, the technical merit of the project was not questioned. Since October 3, 2020, MOSE has been protecting Venice and has thus far prevented the flooding of the city more than 80 times, avoiding irreparable damage to the cultural, economic, and social heritage.

A bit of credit for this epochal result, let it be known, also belongs to the colleague and friend Rino Falcone.

SHARED EXPERIENCES FOR A PARTICIPATORY APPROACH TO SCIENTIFIC RESEARCH

MASSIMO COCCO

Istituto Nazionale di Geofisica e Vulcanologia

massimo.cocco@ingv.it

1 The social value of public scientific research

Working for public research institutions is my common ground with Rino, as we have worked in different sectors and with different professional experiences. Awareness of the role of public research in helping to address societal challenges has characterized our interactions since the beginning. Despite the value of public scientific research is often assessed in terms of technological development and economic growth, since innovation of technological and commercial significance generated by public research is used by private sector for business and economic development, we shared the recognition of the social value of public scientific research. I am a geophysicist working on earthquakes and other natural and anthropogenic phenomena affecting society and human environment. A different field of research from Rino, which facilitated the sharing of our visions and approaches to the strengthening of the value of public scientific research. Conducting research with public funding has individual and collective responsibilities, that is, as individual researchers and as members of a community of experts, which concern both the exploitation of the scientific results to ensure a science for society and the adoption of efficient, transparent, and informed policies for the governance of public scientific research.

This has been and is firmly rooted in Rino's approach to scientific research and to the role of a researcher. This was perhaps the context in which our first collaborations and interactions were consolidated and strengthened. From the beginning I admired and shared his openness to a participatory and transparent approach to research.

I don't know how much we were aware of sharing and promoting what is now called research integrity, in the sense that we did not use this term in the discussions in which we participated. However, we certainly shared and still share the meaning of this term and the importance of bringing its contents within the initiatives organized to promote public research. Research integrity indeed consists of core principles and ethical values determining the way to conduct and evaluate scientific research, which concerns not solely deontological obligations and professional standards adopted by individual researchers, but also the role of public institutions in promoting and performing scientific research. I wish to mention here a sentence written by Rino sometimes ago to emphasize the role of scientists in communicating to society: *"The integrity of science and the role of scientists is enhanced through their image in society; the more that image will account for the real benefit that science is able to produce for society, the more society will be able to attribute to it that fundamental function"*. Researchers are the key actors for the application of these principles and values, but the role of public research organizations within the national research system is equally important. This is the environment where individual researchers can contribute to the adoption of effective policies through participative processes. This is part of science policy. Promoting research integrity is key to guarantee the quality of the research and enhance the reputation and the public image of science, contributing to progress, innovation, and a science "for" and "in" society.

Research integrity in public research is also essential to foster trust and confidence in the methods used and the resulting findings. This is vital to share scientific information and knowledge as well as to propose and adopt participative approaches to research. In the framework of public research and research organizations, this has important ethical implications, which I am not going to discuss here because they are beyond the goals of this note. A key ethical aspect which is relevant for describing Rino's approach to research is impartiality for public good. This implies transparency, authoritativeness, and third-party status to defend the common good and support the value of science for society. Promoting this cultural and political framework requires the participation of scientists to the governance of public research, the adoption of participative policies as well as the support of open science. I will briefly describe my experiences with Rino on these challenges in the following of this note.

The recognition of the social value of public scientific research that I just briefly mentioned here is solely aimed at describing the cultural and political framework and vision that characterized Rino's approach to scientific research and our collaborations. I personally believe that it still represents today a valid framework to characterize innovation for science and society.

2 From awareness to participation

Public research requires targeted investments to enhance its role for innovation and society. This requires transparent investment policies, strategically oriented and finalized to the management of the knowledge system. The promotion of a participatory approach of researchers to the management of the knowledge system has motivated Rino's participation in numerous initiatives up to the creation of the "Osservatorio della Ricerca", a network of Italian scientists that Rino coordinated from 2002 to 2008, aimed at promoting the value of scientific research in Italy. The "Osservatorio della Ricerca" was a unique, stimulating, and effective initiative in which researchers/scientists from very different fields and with very different professional experiences discussed, designed, and organized activities and took actions to monitor policies for science as well as to foster participative initiatives supporting public research. I remember with pleasure the discussions during the "Osservatorio della Ricerca" meetings and the participation of established scientists and researchers from public research institutions. Among the numerous esteemed participants, I would like to mention Carlo Bernardini, if only because he was the professor of Mathematical Methods for Physics of both Rino and me. Carlo Bernardini was an inspiring person to many of us. Rino wrote an excellent article to remember Carlo as a civil servant.

There are many issues that Rino has contributed to address not only with the "Osservatorio della Ricerca", such as the autonomy of public research institutions, transparency in selection procedures for political appointments in scientific organizations (search committees), the relationship between science and society and the role of scientists in the service of our country. Some of these issues have then become political actions and decisions when Rino was the consultant of the Minister of Education, Scientific Research and University (MIUR) from 2006 and 2008. In this role, he also contributed to start the activities to establish the Italian national roadmap for pan-European Research Infrastructures fostering the submission of several proposals to the update of the European Roadmap of Research Infrastructures in 2007-2008. All this demonstrates Rino's ability to transform (shared) awareness into participative processes.

3   Open Science and research internationalization

As anticipated above, Rino contributed to support the Italian participation to pan-European Research Infrastructures that represents the beginning of the process aimed at composing and establishing the Italian National Roadmap of Research Infrastructures. This has been a fundamental tool both for the internationalization of Italian research, involving public research institutions and universities in the design and construction of pan-European research infrastructures, and for open science, promoting the sharing and access to public research products, such as publications and scientific data. This experience, which I shared first-hand with Rino, is not limited to the facilitation of the implementation of new proposals or initiatives, rather it involves the processes and policies to govern the Italian participation in these important European initiatives. This has allowed the Ministry of Education, Scientific Research and University in the following years to consolidate the mechanisms and procedures to support the Italian role in coordinating and participating in pan-European initiatives. This is another example of how Rino's vision and cultural and political approach to science has materialized in effective actions to implement participatory processes for the governance of scientific innovation.

Sharing and enabling access to scientific data and research products has been a field of activity to which I have dedicated about fifteen years of my professional life. Open science is one of the tools to support the social value of research and to contribute to the management of the knowledge system. Research products are not limited to scientific data, but they also include scientific information that is generated and used by experts and researchers, according to transparent, shared, and impartial methods, to create and share knowledge. This represents a concrete example of putting into practice those principles that inspired Rino's vision and mission.

When the Ministry of Education, Scientific Research and University decided to promote the Italian participation in pan-European research infrastructures, Rino, in his role as consultant to the Minister, adopted an efficient, transparent, and pragmatic method to explain the rules of the game by promoting the submission of Italian proposals. The idea of calling those few expert people who were already operating in this sector in the European context to participate in a steering committee was a winning one. The start of a participatory and shared pathway allowed, in the following years, to strengthen Italian participation in pan-European research infrastructures and to obtain the Italian coordination of some important initiatives. I wanted to remember this contribution by Rino because this important professional experience is not well known outside the small number of people who participated in this undertaking.

4 Challenges and prospects

Describing to others the professional and human experiences you have had with a colleague and friend is not easy. I decided to accept the invitation to contribute to this volume in honor of Rino Falcone motivated by the importance of the values and vision that I learned from Rino and shared with him. I must recognize that these principles and values still inspire or should inspire scientific innovation and the relationship between science and society in the future. Some more recent events, even tragic ones such as the Covid-19 pandemic, have rekindled and called into question some aspects, however confirming how the autonomy of scientific research, transparency in governance processes and the sharing of scientific information are all founding elements of a science for society. Remembering the experiences shared with Rino allowed me to strengthen the recognition of the importance of his role in public research. At the same time, it left me with a sort of regret in noting how the defence of these principles and values is even more difficult today. Rino teaches us that we must continue to work to affirm the cultural, ethical, and political dimension of public research in which we believe.

Un percorso nel Consiglio Nazionale delle Ricerche a servizio della Scienza e del Paese

Maria Chiara Carrozza

Presidente del Consiglio Nazionale delle Ricerche (CNR)

Sin dalla sua fondazione, il Consiglio Nazionale delle Ricerche si è posto come punto di riferimento per la collettività, profilandosi come istituzione di orientamento, promozione, valorizzazione e coordinamento della comunità scientifica nazionale e come organo di consulenza al servizio del Paese e delle istituzioni.

Il CNR, con il suo ampio capitale umano, partecipa attivamente al dialogo tra scienza e società, offrendo metodi e strumenti per analizzare il presente e identificare proposte per il futuro sulla base di una molteplicità di saperi specializzati, nonché di una preziosa e serrata dialettica multidisciplinare che coinvolge tutte le sue componenti.

Il valore della condivisione delle evidenze scientifiche con il pubblico generalista e con i decisori politici ha permeato la carriera scientifica di Rino Falcone, attuale Direttore dell'Istituto di Scienze e Tecnologie della Cognizione (ISTC) del CNR. A lui dedico questo mio breve contributo – in qualità di Presidente pro tempore del maggiore Ente pubblico di ricerca nazionale – alla vigilia del suo pensionamento, accogliendo l'invito delle persone a lui più vicine che intendono rendere merito a lunghi, intensi e produttivi anni di lavoro.

Pur non avendo mai collaborato in prima persona con Rino, riesco facilmente a cogliere ed apprezzare alcuni elementi che ne caratterizzano il lascito intellettuale: in primo luogo, la spiccata sensibilità per la ricaduta sociale e sociologica del progresso scientifico e tecnologico; l'impegno instancabile a tutela dell'autonomia della scienza; da ultimo, la sorveglianza attenta e propositiva rispetto alle politiche di orientamento e di gestione della ricerca. Nella sua lunga carriera, Rino Falcone ha ricoperto diversi incarichi istituzionali, divenendo anche consulente presso il MUR; ha coordinato svariati progetti di ricerca nazionali e internazionali, impegnandosi con serietà e zelo per la valorizzazione della ricerca e il rilancio della ricerca italiana ed europea, non senza interrogarsi criticamente sulla responsabilità degli scienziati nei confronti della collettività.

Tra le iniziative incoraggiate da Rino che hanno maggiormente suscitato il mio interesse vi è l'Osservatorio sulla Ricerca, una rete spontanea di scienziate e scienziati italiani, sorta nell'estate del 2002 con lo scopo di promuovere la ricerca nel Paese e monitorare le politiche del settore[1]. L'Osservatorio nacque al termine di un'assemblea generale della ricerca, tenutasi il 10 settembre 2002 nella Sede Centrale del CNR, come risposta della comunità scientifica alle contingenze politiche dell'epoca e, in particolare, all'iter intrapreso per la realizzazione dei decreti di riforma dei principali Enti Pubblici di Ricerca da parte dell'allora Ministro. Di questa esperienza colpiscono, soprattutto, due aspetti: da un canto, la trasversalità della partecipazione all'assemblea, che adunò dottorandi, studiosi di tutte le discipline, premi Nobel, direttori e presidenti di Enti, così come anche ricercatori provenienti dal settore privato; dall'altro canto, la proposta che ne derivò di collocare la ricerca di base all'interno di una prospettiva europea di finanziamento, oltre alla formulazione di una carta dei principi.

I sette principi enucleati rappresentano valori pienamente condivisibili, senza dubbio irrinunciabili per il nostro futuro. Tra tutti, mi sento di rimarcare il ruolo essenziale della ricerca di base, senza la quale non può esistere una ricerca applicata realmente innovativa. In effetti, la radicata distinzione tra ricerca fondamentale e ricerca applicata perde di senso qualora si adotti una prospettiva più lungimirante e di ampio spettro, che considera la scienza e l'innovazione come due poli di un continuum indivisibile. Per il nostro Paese continua ad essere urgente, poi, il bisogno di avere, accanto ad un efficiente sistema pubblico della ricerca, un sistema di ricerca privata altrettanto solido. Resta prioritario, dunque, quanto rivendicato nella carta dei principi: pubblico e privato dovrebbero svolgere entrambi i ruoli di finanziatori e di performer della ricerca. Mi preme altresì rammentare l'importanza, ribadita dall'Osservatorio, del dialogo tra scienza e società, che a mio avviso si snoda attorno alcuni concetti chiave: in primo luogo, l'importanza della ricerca finalizzata all'identificazione delle principali problematiche riguardanti la società e alla sua trasformazione (la cd. directionality); in secondo luogo, il Citizen engagement, ossia il coinvolgimento della cittadinanza nella prassi scientifica, con il duplice obiettivo di disseminare la cultura scientifica e avvalersi dei cittadini quali risorsa strategica in grado di supportare non solo l'individuazione degli obiettivi prioritari e dei temi scientifici emergenti, mediante consultazioni pubbliche, ma anche la raccolta e la divulgazione dei dati scientifici. Infine, il concetto di science for policy, vale a dire il ricorso alla scienza da parte dei decisori politici, divenuto tanto più urgente nella società attuale, caratterizzata da crisi e sfide globali sempre più complesse. A tal proposito, è auspicabile che governi e amministrazioni si avvalgano in misura sempre maggiore della competenza delle ricercatrici e dei ricercatori, del loro scientific advice, per sviluppare soluzioni e politiche sulla base dell'evidenza scientifica e di input inter- e multidisciplinari.

Un ulteriore principio fondamentale, che Rino Falcone ha cercato costantemente di tutelare durante il suo percorso al CNR, è il principio dell'autonomia della scienza, sancito dalla stessa Costituzione all'art. 33. L'autonomia delle Istituzioni e della comunità scientifica è sempre stata pensata da Rino non come indipendenza assoluta, bensì come libertà della scienza da esercitarsi al servizio della comunità. La libertà di scegliere un argomento di ricerca, di sollevare domande scientifiche, di selezionare materiali e metodologie di ricerca, di presentare pubblicamente ipotesi, risultati e ragionamenti, deve essere tutelata nella misura in cui è orientata al bene pubblico.

In questa precisa ottica Rino Falcone ha coltivato i suoi interessi di ricerca, che spaziano dall'elaborazione del linguaggio naturale al Planning, ai sistemi multi-agente (MAS) e alla teoria degli agenti. Presso l'ISTC Rino, insieme a numerosi altri colleghi, ha contribuito allo sviluppo di una vera e propria tradizione di studi, basata sulle scienze cognitive, sui "mediatori cognitivi dell'azione sociale", ovvero le rappresentazioni mentali alla base dei comportamenti sociali e dei fenomeni collettivi e istituzionali, come la cooperazione, le norme, il potere, le emozioni sociali (ammirazione, invidia, vergogna, colpa, ecc.).

Le ricerche sui modelli cognitivi dell'interazione sociale assumono molta rilevanza nel panorama scientifico e più latamente culturale del presente, attraversato dalla rivoluzione dell'Intelligenza Artificiale. Le tecnologie intelligenti pongono diverse sfide per le scienze cognitive, dal momento che possono simulare e supportare i meccanismi di base del ragionamento e della comunicazione e potrebbero ridefinire i paradigmi principali che li governano, estendendo le nostre funzioni cognitive. La modellazione cognitiva si rivela, dunque, decisiva nell'approccio allo sviluppo di sistemi artificiali intelligenti. Oggi appare necessario chiarire, a livello teorico e pratico, alcune attitudini tipicamente sviluppate dagli umani nel loro convivere, che probabilmente subiranno una forte ridefinizione se si riuscirà a trasferirle ai sistemi artificiali. Tra le attitudini sociali più significative, come l'affidabilità, l'autonomia, o la delega, Rino Falcone ha approfondito in particolare la fiducia, elaborandone una teoria generale nel volume pubblicato con Cristiano Castelfranchi, *Trust Theory, A Socio-Cognitive and Computational Model*, in cui viene formalizzato un modello organico di tale fenomeno tanto complesso e dinamico a livello cognitivo, affettivo e sociale. Ho piacere di ricordare, a questo punto, anche il progetto triennale Mind RACES: from Reactive to Anticipatory Cognitive Embodied Systems, finanziato per oltre 2,1 milioni di euro dall'Unione Europea nell'ambito del sesto programma quadro, che ha coinvolto 8 partner leader nella costruzione di sistemi cognitivi anticipatori. L'obiettivo del progetto è stato quello di indagare secondo molteplici prospettive, sia teoriche che computazionali, diversi meccanismi e architetture cognitive anticipatorie al fine di elaborare sistemi cognitivi dotati della capacità di prevedere l'esito delle loro azioni, di costruire un modello di eventi futuri, di controllare la loro percezione e attenzione anticipando gli stimoli futuri e di reagire emotivamente a possibili scenari futuri.

In quanto esperta di robotica, non posso che stimare questo filone di studi che ha un grande potenziale per il mio settore, in quanto genera conoscenze necessarie per sviluppare robot del futuro intelligenti e flessibili, capaci di vivere e interagire nella nostra quotidianità. Aspetti come il comportamento esplorativo e l'anticipazione adattiva sono imprescindibili per affrontare le sfide della robotica "embedded", in primis della robotica assistenziale e di quella umanoide. Come sottolineato più volte anche da Rino[2], la robotica umanoide costituisce un bacino prezioso cui attingere per comprendere questioni fondamentali, come la relazione mente corpo, e verificare, per esempio, se e in che misura il corpo e le sue specificità possano influenzare i processi cognitivi.

Simili interrogativi hanno animato le riflessioni di Rino Falcone lungo tutta la sua carriera, la quale, naturalmente, non si esaurisce nelle poche esperienze fin qui ricordate.

Il suo è un lascito costruttivo che stimola, nelle giovani generazioni, un atteggiamento di apertura, curiosità e tenacia: un terreno fertile in cui far germogliare nuove idee, un solco ben tracciato lungo il quale proseguire.

References

[1] Analysis, Rivista di cultura e politica scientifica, n.22003, pp. 1-7.
[2] Cognizione e Sistemi Intelligenti: il ruolo della Fiducia, Laboratorio dell'ISPS, XIX, 2022, pp. 2-15.

Nel dedalo della fiducia, seguendo Rino Falcone

Maurizio Franzini
Professore Emerito di Politica Economica presso la Facoltà di Economia della Sapienza, Università di Roma
maurizio.franzini@uniroma1.it

Alla fiducia Rino Falcone ha dedicato gran parte della sua attività di ricercatore offrendoci numerosi contributi teorici ed empirici - spesso frutto di collaborazioni con diversi coautori ed in particolare con Cristiano Castelfranchi - che si rivelano preziosissimi per orientarsi in quel dedalo di domande e di problemi che la fiducia immancabilmente suscita. Le molte prospettive dalle quali ha guardato a ciò che chiamiamo fiducia fanno tornare alla mente le parole che Robert Dahl, molti decenni fa, usò per dar conto del sottile disagio che prendeva chi si avventurava a riflettere sul potere. Dahl rilevava che sono così tanti e così diversi tra loro i significati attribuiti a quella 'cosa' che chiamiamo potere da far sorgere il fondato dubbio che quella 'cosa' non sia una cosa sola ma tante cose [1] (p. 201).

Leggendo quello che Falcone scrive della fiducia si può essere colti (a me è capitato) da un dubbio simile ed è così perché lo scandaglio che egli fa di ciò che chiamiamo fiducia - in modo assai simile a quello che fece Dahl con il potere - è approfondito quanto basta a considerare insufficiente ogni approccio tendenzialmente riduzionista al fenomeno.

Faccio qualche esempio e poi vengo a una questione specifica che mi sta, anche come economista, particolarmente a cuore.

Contrariamente a quello che si potrebbe pensare la fiducia non è necessariamente ed esclusivamente indirizzata verso uomini o donne. Essa, infatti, può avere per oggetto il funzionamento delle macchine da cui dipende il nostro benessere e la questione è sempre più rilevante oggi, nell'epoca dell'Intelligenza Artificiale Generativa, come ha chiarito Falcone in uno scritto recente [2]. Però, a ben guardare, la fiducia nelle macchine rimanda alla fiducia in chi quelle macchine ha ideato e a chi le fa funzionare e anche alla fiducia in chi ci fornisce informazioni sul loro funzionamento.

Quando la fiducia è indirizzata agli umani possono darsi due casi, tra loro assi diversi se non altro per le conseguenze che ne derivano. L'elemento di differenziazione è se la fiducia si colloca in un rapporto formale di delega (sostanzialmente regolata da norme o contratti) oppure no.

Semplificando si può dire che se non vi è delega il comportamento di colui di cui dovremmo fidarci - e che influenza il nostro benessere – è del tutto indipendente dalla nostra fiducia, di cui, peraltro, egli nemmeno viene a conoscenza. In questo caso se abbiamo margini di scelta (nel senso che possiamo in qualche modo sottrarci agli effetti dell'altrui comportamento) è cruciale la probabilità che assegniamo al comportamento per noi più favorevole. E su questo si concentrano molti approcci alla fiducia. Ciò che va sottolineato è che così intesa la fiducia incide sul comportamento di chi si fida ma non su quello di chi è gratificato della fiducia.

In presenza di delega, in particolare di delega contrattuale, la situazione appare piuttosto diversa, un punto assai rilevante – e forse paradossale – è che potrebbe esservi delega senza fiducia, almeno in una delle sue più diffuse accezioni.

Per illustrare questa possibilità farò riferimento a un cruciale rapporto di carattere economico. Quello di lavoro subordinato, imperniato sulla delega del datore di lavoro (l'impresa) al lavoratore. Questo rapporto è, agli occhi degli economisti, un tipico rapporto tra Principale (il datore di lavoro) e Agente (il lavoratore). Il risultato che il Principale vuol raggiungere (nel nostro caso il profitto) dipende dalle azioni dell'Agente (nel nostro caso determinate dal suo 'impegno' oltre che dalle sue abilità) ma quest'ultimo, secondo le tradizionali assunzioni, ha obiettivi in conflitto con quelli del Principale: vuole un salario alto e vuole impegnarsi poco, esattamente l'opposto di ciò a cui ambisce il Principale. A complicare la vita di quest'ultimo interviene il fatto che l'Agente è protetto dalla asimmetria informativa: solo l'Agente sa quale è stato il suo vero impegno, cosicché il Principale non può remunerarlo in base a quest'ultimo. Per concedere la delega al lavoratore, il datore di lavoro deve dunque fidarsi che egli terrà un comportamento conforme a quanto prevede in termini di impegno, oltre che di retribuzione, il contratto che necessariamente dovrà essere stipulato. Ma può accadere che il Principale non abbia questa fiducia eppure conceda la delega. In realtà questo è quanto prevedono i modelli economici Principale/Agente. L'idea è che il Principale ricorrerà a espedienti che rendono conveniente per l'Agente astenersi da comportamenti opportunistici, dannosi per il Principale.

Gli espedienti sono, anzitutto, sistemi di monitoraggio che limitano la possibilità di quei comportamenti da parte dell'Agente perché aumentano la probabilità che essi siano scoperti. Un esempio estremo sono i famosi braccialetti di Amazon. Peraltro, se scoperti, quei comportamenti sono sanzionati con il licenziamento che, soprattutto in un mondo con elevata disoccupazione, è una sanzione di certo temuta dal lavoratore. Il significato di tutto questo può essere così espresso: la delega non è giustificata dalla fiducia nel lavoratore ma dall'uso di una forma di potere che costringe i comportamenti convenienti per l'Agente e porta, per costrizione, se così si può dire, a servire gli interessi del Principale. Forse, con tutte le cautele del caso, si potrebbe dire che il potere (con buona pace di Dahl) è anche un mezzo per sopperire alla mancanza di fiducia e permette di avere rapporti di delega senza fiducia, se questa è riferita ai comportamenti 'spontanei' di coloro di cui dovremmo fidarci.

Ma questa delega senza fiducia non è priva di conseguenze negative per il funzionamento della società e per le caratteristiche delle relazioni sociali. Per articolare questo punto è assai utile quanto scrivono Castelfranchi e Falcone [3]. Anzitutto è importante questa loro affermazione: "crediamo che esistano anche dinamiche generative tra fiducia e rischio: la fiducia non solo espone le persone a rischi, ma aumenta i rischi e ne crea di nuovi (ad esempio, rischi di fallimento, delusione, perdita, impotenza o tradimento)". Un'affermazione assonante con quella delle filosofa Martha Nussbaum che ha scritto "La fiducia, invece, implica l'apertura alla possibilità del tradimento, ...

quindi a una forma di danno molto profonda. Significa allentare le strategie di autoprotezione con cui di solito affrontiamo la vita" e aggiunge che "vivere con fiducia comporta una profonda vulnerabilità e una certa impotenza, che può essere facilmente deviata in rabbia" [4].

Ma ancora più importante è la successiva affermazione di Castelfranchi e Falcone: "crediamo che esistano anche dinamiche generative tra fiducia e rischio: la fiducia non solo espone le persone a rischi, ma aumenta i rischi e ne crea di nuovi (ad esempio, rischi di fallimento, delusione, perdita, impotenza o tradimento). Tuttavia, la fiducia può persino ridurre i rischi perché ha un effetto sui comportamenti dei fiduciari e modifica le probabilità di successo, di reciprocità e di lealtà. In questo senso, vorremmo sostenere che la fiducia (come molti altri tipi di aspettative positive) è una profezia che si autoavvera e che può modificare radicalmente la probabilità di un risultato." Il punto è di cruciale importanza anche per i rapporti di lavoro.

Da molte analisi e studi comportamentali risulta che esiste una figura poco popolare, certamente assai meno dell'*homo economicus* con il suo egoismo auto-interessato e la sua avidità materialistica. E' l'*homo reciprocans* che calibra i suoi comportamenti su quelli che gli altri tengono nei suoi confronti, comportandosi bene con chi lo tratta bene e male. – nei limiti del possibile – con chi lo tratta male. Usa la libertà di cui dispone per decidere i propri comportamenti non esclusivamente per massimizzare il proprio benessere materiale immediato ma anche per valorizzare le buone relazioni sociali. Dunque, adotta comportamenti cooperativi che risultano favorevoli anche per il Principale. In termini economici, come provano diversi studi empirici, quando l'impresa mostra più fiducia nei confronti dei lavoratori – ad esempio evitando sistemi di monitoraggio costrittivi e lasciando loro più libertà di azione – e quando il lavoratore è più soddisfatto, a risentirne positivamente è la sua stessa produttività con ovvii effetti positivi sui ricavi e sui profitti dell'impresa.

Quindi nutrire fiducia significa stimolare reciprocità, con vantaggi – appunto – reciproci. Certo non vi è garanzia, il rischio è ineliminabile ma una cultura, anche imprenditoriale, più propensa alla fiducia può favorire la diffusione di comportamenti come quelli di cui si è appena detto con l'effetto di ridurre con il trascorrere del tempo il rischio che accompagna la fiducia.

Quindi concedere fiducia in molti contesti ha effetti trasformativi. E questo forse vale, *mutatis mutandis*, anche per la fiducia che concediamo a noi stessi. Del tennista italiano che, mentre scrivo, inaspettatamente è giunto ai quarti di finale del torneo di Wimbledon il suo maestro ha detto, commovendosi, "E' un ragazzo fragile: la tecnica l'ha sempre avuta, mancava la convinzione nei suoi mezzi'.

E nel concludere queste note vorrei far presente a Rino Falcone che siamo in molti ad avere fiducia che continuerà a guidarci nel dedalo della fiducia. E se la nostra fiducia è superflua, tanto meglio.

References

[1] Dahl, Robert A. The concept of power. In Behavioral science 2.3 (1957): 201-215.

[2] Falcone, Rino. Piattaforme e qualità delle relazioni sociali In Parole-Chiave, 1/2023, pp. 137-152.

[3] Castelfranchi C. e Falcone R. Il rischio nel fidarsi. La fiducia nel rischiare. Equilibratori cognitivi del comportamento in Sistemi intelligenti, pp. 295-312.

[4] Nussbaum Martha Anger and Forgiveness: Resentment, Generosity, and Justice in Oxford University Press

MISSIONE RICERCA, PER LO SVILUPPO DEL PAESE

DANIELA PALMA

ENEA - Agenzia nazionale per le nuove tecnologie, l'energia e lo sviluppo economico sostenibile
daniela.palma@enea.it

E' il marzo del 2009 e con un intervento su "Le sfide della ricerca" Rino Falcone prende parte al convegno "Memoria e progetto" organizzato a Napoli in occasione del quarantennale della fondazione dell'Istituto Cnr di Cibernetica "Eduardo Caianello". La conferenza, articolata su due giorni, vuole essere di grande respiro, coltivando l'ambizione di recuperare la memoria di imprese scientifiche che hanno contribuito allo straordinario sviluppo dell'Italia nel dopoguerra, e di mostrare come la perdita di quella progettualità abbia messo in crisi il futuro del paese. Ma l'argomento non è nuovo per chi, come Falcone, già da qualche anno guarda con preoccupazione a quanto la posizione dell'economia italiana in Europa sia andata arretrando, e ha ben pochi dubbi sul fatto che, in un mondo che si trasforma rapidamente sull'onda delle crescenti ricadute applicative delle conoscenze scientifiche, la tendenza in atto non possa essere invertita se non assegnando nuovamente una centralità all'investimento nell'attività di ricerca. "Memoria e progetto" si colloca così al crocevia di quella riflessione, quasi a rappresentare una sorta di "manifesto" di diversi esponenti della comunità scientifica divenuti ormai "compagni di viaggio" di Rino, in un'atmosfera di grande consuetudine e forte condivisione di vedute.

L'idea di fondo è che l'emergere di un "declino italiano" non debba essere visto come il prodotto di una sciagura ineluttabile, ma come conseguenza inevitabile della miopia che ha guidato le scelte della politica, nella credenza che il paese potesse perseguire la via di uno "sviluppo senza ricerca". Nel convegno partenopeo questa lettura della crisi italiana trapela senza esitazioni, quale frutto più che maturo di un dibattito avviato intorno alla metà del primo decennio duemila ed arricchitosi di più prospettive di analisi, con apporti anche molto specifici da versanti diversi della cultura scientifica. In tale contesto, l'Osservatorio sulla Ricerca, che Rino Falcone coordina dal 2002 e che nasce con l'obiettivo di riportare al centro dell'azione politica il tema della ricerca come valore per il paese, si mostra da subito un terreno molto fertile di elaborazione intellettuale, nutrito della convinzione che l'Italia possa ritrovare la strada della sua rinascita. L'Osservatorio richiama al suo interno personalità scientifiche che guardano all'attività di ricerca come a un grande progetto di avanzamento del benessere collettivo; un'impresa che non può essere guidata dall'interesse privato, rispondente al canone del profitto, e che per mantenere una vocazione pubblica deve essere sostenuta finanziariamente attraverso il bilancio dello Stato. In Italia questa concezione stava decadendo assai rapidamente ed era giunto il momento di rendere esplicita una posizione di netto contrasto alla deriva incalzante, iniziando ad assumere una postura più incline all'agire politico.

Facone stesso ci riporta nel vivo di quel momento nel ricordare l'incontro con il fisico Carlo Bernardini, la cui caratura di grande scienziato è indissolubilmente legata a quella del "*civil servant*", ovvero di chi intende operare al "servizio della civiltà" esercitando quella "*funzione mista in cui si incrociano scienza e politica*". In quel clima di grande sintonia cresce l'indignazione condivisa "*assieme a molti altri, davanti ad uno dei tanti scempi che il dilettantismo politico applicato alla regolazione della scienza e della ricerca può produrre (era il tempo in cui si riteneva che la riforma di un Ente di ricerca potesse farsi, affidandosi alla consulenza della Ernst&Young)*" per poi "*insieme decidere di contrastare quei piani scellerati e mobilitarsi per mobilitare. Scoprendo che alcuni valori fondamentali (relativi al ruolo della conoscenza, alla libertà di pensiero, alle equilibrate autonomie tra istituzioni) sono molto più condivisi nell'opinione pubblica, di quanto non si creda e, cosa persino più complessa, che sia possibile convincere un gruppo numeroso di intellettuali nel non dividersi nel perseguire una strategia di risposta"*. Ma ciò che di più importante scaturisce da tale mobilitazione è che le battaglie intraprese finiscono con il diventare *"elemento di stimolo e riflessione più generale"* suscitando in molti la necessità *"di continuare a mantenere uno sguardo attivo e costante oltre le [rispettive] attività scientifiche"*[1].

La discussione nell'Osservatorio sulla Ricerca si fa sempre più intensa, acquisendo maggiore complessità man mano che i segni del "declino italiano" si vanno accentuando. Non è un caso perciò che l'attenzione inizi a soffermarsi di più su come la negligenza con cui la politica ha trattato il mondo della ricerca si è riflessa sul ritardo tecnologico del paese, e non è un caso che in questa fase il terreno di confronto si vada ampliando con nuove e più articolate interlocuzioni. Su tale versante non tardano ad entrare in gioco le analisi delle dinamiche della competitività tecnologica dell'industria italiana facenti capo a un Osservatorio progettato ed avviato nella prima metà degli anni '90 alla Direzione Studi dell'Enea dall'allora direttore Sergio Ferrari. I Rapporti prodotti dall'Osservatorio Enea, che di quelle analisi erano il cardine, erano tesi infatti a mostrare come il crescente ritardo competitivo dell'Italia traducesse una debolezza del paese nella capacità di innovare e come tale debolezza fosse strutturalmente collegata alla scarsa presenza di imprese nei settori a più elevato contenuto tecnologico, determinando una cronica insufficienza dell'investimento in ricerca da parte del sistema produttivo. Ed è quindi comprensibile come, a meno di riconoscere la fallacia di quel modello produttivo, la ricerca tutta fosse irrilevante nelle scelte operate dalla politica del paese. Il punto di vista che la spesa pubblica in ricerca fosse da considerarsi uno "spreco" era in definitiva coerente con questo scenario, e le "scelte scellerate" della politica rispetto al governo e al (crescente) definanziamento degli Enti pubblici di ricerca stavano lì a testimoniarlo. Non era difficile pertanto concludere come lo stato della ricerca italiana fosse la sintesi perfetta di un modello di sviluppo nel quale il paese continuava a rispecchiarsi, incurante dell'evoluzione che stava attraversando l'economia mondiale.

---

[1] https://www.scienzainrete.it/articolo/carlo-bernardini-civil-servant/rino-falcone/2018-09-25

Il nuovo slancio che il dibattito nell'Osservatorio sulla Ricerca andava guadagnando imponeva inevitabilmente un ripensamento della politica della ricerca, con riflessioni che cogliessero il filo conduttore che collegava l'operatività degli attori pubblici alle strategie del contesto imprenditoriale. Si trattava di un passaggio cruciale, del quale sarebbe stato assai arduo ricostruire i tratti stando al di fuori di tutto quel consesso che intorno all'Osservatorio sulla Ricerca ruotava. Scriverne qui in prima persona significa aver avuto l'opportunità di seguire lo svolgersi di quella vicenda, ma al tempo stesso anche il privilegio di sperimentare una realtà che non esiterei a definire indispensabile nel percorso formativo di un ricercatore. Come ricercatrice dell'Enea impegnata nelle attività di studio ed analisi che si svolgevano nell'ambito dell'Osservatorio sulla competitività tecnologica dell'Italia, stabilire un dialogo con l'Osservatorio sulla Ricerca poteva inizialmente sembrare l'esito predestinato derivante dal semplice collegamento tra due aspetti complementari di uno stesso oggetto di indagine. Ma ciò che sarebbe diventato più chiaro nel partecipare alle occasioni di incontro che quel luogo stimolava era il fatto di leggere la ricerca scientifica innanzitutto come chiave di progresso della società civile. E d'altra parte non ne ero sorpresa, considerando che quel tipo di visione accomunava quanti tra i referenti dell'Osservatorio sulla Ricerca avevo già avuto modo di conoscere in concomitanza con la mia attività di coordinamento dello stesso Osservatorio Enea. Pietro Greco, chimico per formazione ed affermato giornalista scientifico, e Settimo Termini, fisico cibernetico dell'Università di Palermo (oltre che direttore dell'Istituto Cnr di Cibernetica di Napoli) erano il cuore di questa rete di precedenti contatti, e con Sergio Ferrari erano finiti per diventare quasi un nucleo a sé stante, con il quale gli scambi di vedute si erano fatti sempre più frequenti, soprattutto dopo l'uscita nel 2006 del quinto Rapporto dell'Osservatorio Enea. I temi lì trattati avevano catturato l'interesse di Pietro e Settimo in maniera del tutto particolare, tanto da spingerli molto presto a dedicare delle loro riflessioni al declino dell'Italia e al ruolo che su questo aveva esercitato il ritardo tecnologico del sistema produttivo, pubblicando già nel 2007 un saggio a quattro mani dagli accenti sferzanti, che non lasciava spazio a dubbi su quanto quella lettura della crisi del paese la sentissero così profondamente propria. Ma c'era di più. Fin dal titolo, "Contro il declino", lo scritto si annuncia come un pamphlet, con uno sguardo a tutto tondo sul ritardo italiano, che traspare chiaramente da quanto "dichiarato" nell'eloquente sottotitolo: "Una (modesta) proposta per un rilancio della competitività economica e dello sviluppo culturale dell'Italia". La convinzione, dunque, che quella perdurante divaricazione dell'Italia dal "paradigma della ricerca scientifica" precludeva al paese la possibilità di entrare a pieno titolo nella economia e nella società della conoscenza, pregiudicandone lo sviluppo a venire.

Sul legame tra "declino" italiano e crisi della "cultura della ricerca" la discussione era giunta a quel punto a una prima rilevante maturazione e a breve proprio lo svolgimento del convegno "Memoria e Progetto", nato sotto l'egida della stretta e intensa collaborazione tra Pietro e Settimo, ne sarebbe stata la prova più che evidente. Tra i molti e diversi interventi che avevano reso quel convegno assai partecipato da una comunità scientifica molto attiva sotto il profilo dell'impegno civile, il contributo di Rino Falcone illustrava come le questioni poste in seno all'Osservatorio sulla Ricerca si fossero fatte interpreti della complessità della crisi italiana, segnalando sempre più l'urgenza di quell'azione politica da tempo richiesta. Il valore intrinseco dell'attività di ricerca, che portava alla rivendicazione di un'autonomia della stessa, non appariva in contraddizione con la necessità di una consapevole presa in carico di un "progetto paese" che facesse della ricerca il proprio asse strategico. Bisognava superare lo sterile dirigismo politico che aveva annientato il sistema della ricerca pubblica e che l'aveva depauperata anche finanziariamente, riformulando i termini dell'intero intervento in materia di ricerca scientifica e tecnologica nella logica della programmazione delle politiche pubbliche. Nel merito Falcone ricordava come durante la legislatura degli anni 2006 – 2008, guidata dal centrosinistra, il Ministero della Ricerca fosse per la prima volta entrato nel CIPE, l'organismo interministeriale per la programmazione economica, e come in questo senso fosse stato compiuto un primo passo avanti; ed aggiungeva che sebbene questo passaggio non avesse mostrato fino in fondo la propria efficacia, nel caso in cui tale partecipazione non fosse stata intesa come operazione di pura facciata, ci si sarebbe potuta aspettare "una trasformazione culturale ed economica molto articolata e di lungo periodo"(Memoria e Progetto, (a cura di) Pietro Greco e Settimo Termini, 2010).

I tempi non erano (ancora una volta) tra i più propizi. La crisi internazionale del 2007-2008 aveva trascinato l'Europa nella recessione e la situazione dell'Italia era tra le più deteriorate, insieme a quella delle altre economie del sud dell'area. Ma la profondità della crisi italiana era andata anche al di là delle pur oggettive condizioni del quadro economico, poiché nel dare la precedenza alle urgenze imposte dal risanamento delle finanze pubbliche, conseguenti all'esplosione dei "debiti sovrani", la spesa pubblica in ricerca veniva ulteriormente penalizzata, non tenendo in alcun conto che la maggiore fragilità del paese era non poco il frutto della carenza di quel tipo di investimento protrattasi troppo a lungo nel tempo. Né aveva alcuna risonanza il fatto che, seppur in un clima generale di austerità della spesa pubblica, i maggiori paesi europei, come Francia e Germania, non solo non avessero diminuito i (già) assai più elevati finanziamenti alla ricerca, ma ne avessero addirittura incrementato la consistenza. In qualche misura la crisi internazionale era diventata per l'Italia un alibi per non affrontare gli "antichi mali" del suo sviluppo. Ma immaginare che ci fossero spazi di proposta politica per una inversione di rotta sembrava ancora possibile. Superata la fase più grave dell'emergenza creata dalla crisi internazionale, la prospettiva di un nuovo avvio di legislatura nel 2013 creava infatti qualche ragionevole aspettativa.

Bisognava ripartire dalla necessità di riconoscere alla ricerca un ruolo fondamentale nella politica economica del paese. Chiedere di porre fine al pesante sottofinanziamento della spesa pubblica in ricerca era ancora importante, ma non poteva essere più posto come condizione sufficiente. Per tale motivo doveva essere meglio sviluppata l'idea che la ricerca divenisse un asse della programmazione economica, anche al fine di predisporre il terreno sul quale avrebbero dovuto prendere forma le altrettanto necessarie politiche orientate al potenziamento tecnologico dell'apparato produttivo italiano. E l'occasione per iniziare ad esporre le linee di quel ragionamento fu quella che si offrì con il convegno "+ Sapere=Sviluppo" organizzato dalla rivista "Left" presso il Teatro Piccolo Eliseo di Roma a metà di febbraio 2013, nell'imminenza delle elezioni politiche; un evento con una ricca partecipazione di quanti tra le fila dei ricercatori pubblici, tanto dell'Università quanto degli Enti di Ricerca, intendevano far emergere la deriva (anche legislativa) che aveva investito il sistema della ricerca pubblica, aggravandone lo stato di sottofinanziamento.

"La ricerca e l'innovazione per lo sviluppo del paese" fu una breve ma densa relazione che presentai in quella sede a cura del gruppo tematico costituitosi per la circostanza in quello stesso collettivo che aveva mosso i primi passi con l'Osservatorio coordinato da Rino Falcone. L'obiettivo era quello di sottolineare la necessità per l'Italia di un "nuovo modello di sviluppo", mettendo in atto politiche sistemiche capaci di correggere strutturalmente il ritardo tecnologico del sistema produttivo e a tal fine correggendo fin da subito la tendenza a considerare il finanziamento pubblico della ricerca come fattore residuale del bilancio di spesa. In questa direzione andava la proposta di costituire una "regia" dei ministeri coinvolti nel finanziamento pubblico della ricerca, tipicamente derivato in subordine agli appostamenti di spesa dei singoli dicasteri ed assoggettato alle linee programmatiche di ciascuno di essi con modalità indipendenti. L'idea era allora quella di dare vita a una sorta di "CIPE Ricerca" in grado programmare nel suo insieme il finanziamento pubblico della spesa in ricerca, evitando (tra l'altro) specifiche carenze e/o ridondanze implicate dalla gestione separata dei bilanci ministeriali. Ma un aspetto non meno rilevante di una tale modifica del sistema di finanziamento pubblico della ricerca sarebbe stato anche quello di integrarsi con le politiche pensate per lo sviluppo del paese, a cominciare da quelle indirizzate all'industria, che nel caso dell'Italia avrebbero dovuto puntare ad un allargamento della base produttiva nei settori tecnologicamente avanzati ad elevata intensità di ricerca.

A poco più di dieci anni da quel giorno al Teatro Piccolo Eliseo è possibile (purtroppo) solo constatare la stringente attualità di quella proposta. Il recupero dell'economia italiana rispetto agli anni più critici della crisi internazionale del 2007-2008 è stato tra i più deboli in tutta Europa e tale che, poco prima che subentrasse la nuova crisi determinata dalla pandemia covid-19 all'inizio del 2020, il Pil del paese fosse l'unico insieme a quello della Grecia a collocarsi ancora al di sotto dei valori registrati nel 2007 (-4%). Il prolungarsi della crisi pandemica e gli shock geopolitici che gli sono succeduti negli ultimi due anni non hanno di certo contribuito a migliorare la situazione, e la fragilità dell'economia italiana deve ora fare i conti con una maggiore fragilità dell'Europa intera, con la rivisitazione di un patto di Stabilità votato all'austerità (sebbene in forma leggermente più mite), e con le incertezze del quadro internazionale. Ma non per questo è meno lecito richiamare l'attenzione sulla necessità di rafforzare il potenziale di sviluppo del paese, a fronte di un ritardo tecnologico tuttora molto significativo e destinato perfino a crescere se si vogliono traguardare le grandi sfide tecnologiche del presente, che muovono dal contrasto al cambiamento climatico e che sollecitano un'accelerazione della transizione dei sistemi produttivi verso un uso esteso delle fonti energetiche rinnovabili. La sfida della ricerca per l'Italia continua dunque oggi più che mai a rimanere aperta. E credo che Rino potrebbe essere altrettanto d'accordo.

Uno Zibaldone di pensieri, parole e (potenziali) azioni

Alberto Silvani

già Dirigente di Ricerca presso il Consiglio Nazionale delle Ricerche (CNR)

1 Una giustificazione per questo esercizio

Ho conosciuto Rino più di trenta anni fa. Eravamo due (all'epoca e lui più di me. . .) giovani ricercatori che operavano in due strutture "anomale" del CNR (l'Istituto di Psicologia e l'istituto sulla Ricerca e la Documentazione Scientifica); anomale in quanto non strettamente associabili a un prevalente settore disciplinare ma, allo stesso tempo fortemente rappresentative delle potenzialità dell'Ente. Un Ente troppo spesso "utilizzato" in quanto serbatoio di competenze ma non sempre "valorizzato" per le sue capacità e per i suoi contributi, in particolare rispetto alla multi e interdisciplinarietà, all'apertura di nuove aree, alla possibilità di convergenza di conoscenze e capacità rispetto a problemi nuovi e complessi, al saper fare "massa critica" su tematiche da affrontare e su soluzioni da costruire. Un Ente che, posso affermare con cognizione di causa avendoci trascorso in sedi, ruoli e responsabilità differenti un'intera vita lavorativa, avrebbe avuto bisogno di una maggiore visione di lungo periodo, una continuità nei finanziamenti, un investimento sul suo capitale umano, piuttosto che una sistematica messa in discussione di quanto c'era, di un avvio di riforme e cambiamenti, spesso modificate nel tempo prima di una verifica della loro efficacia o interpretate in base a mutamenti di organigrammi, quasi sempre eterodiretti. Dei malesseri e delle criticità del CNR abbiamo avuto modo di discutere spesso con Rino: anzi, il loro protrarsi nel tempo e il loro adattarsi ai vincoli interni ed esterni rappresentano, forse, l'immagine speculare, direi riflessa, dei "nostri" cambiamenti e di come siamo cambiati negli anni. Ma in queste brevi note non è del CNR che vorrei parlare. Tra le tante qualità di Rino, bene espresse dalla sua produttività scientifica e dai contributi forniti in ambiti spesso al confine tra più discipline, accompagnati da concreti momenti gestionali in diversi contesti operativi, non vorrei venisse dimenticata una caratteristica di cui sono stato testimone diretto. La potremmo riassumere nella capacità di stabilire contatti con i suoi interlocutori dando seguito alla sua curiosità scientifica e culturale attraverso scambi di messaggi e di opinioni che avevano - ed hanno - al centro un'attenzione al futuro e la necessità di impegnarsi in prima persona. Ingredienti basici per un buon ricercatore, tanto più se utilizzati al di fuori dei canonici ambiti del proprio corredo disciplinare. Lo strumento di analisi di queste "qualità" è abbastanza insolito e può essere individuato in una sorta di rilettura del repertorio di messaggi, auguri e saluti, mai banali in quanto ricchi di contenuti e di richiami, che, sebbene contestualizzati ai tempi in cui sono stati espressi, mantengono una leggibilità e una validità particolarmente interessante se letti in sequenza. Un'attenzione al futuro, per sua natura positiva, che non nasconde sconfitte e pessimismi ma che stimola l'interlocutore, il destinatario di questi messaggi, a utilizzare lo stesso strumento su di se. Ne sono derivati scambi di cui conservo preziosa testimonianza ma che credo potrebbero riguardare tanti altri destinatari e sarebbe interessante poterne raccogliere e potersi confrontare. Ma forse gli esempi più interessanti, la sintesi del tutto, stanno nella testa, più che nel PC, di Rino. Il piccolo contributo che propongo prova a realizzare questo esercizio sulla documentazione che ho a disposizione e a trarne commenti più generali, partendo dall'accettazione di come siamo cambiati negli anni e di come in questo percorso "non ci siamo persi di vista".

2 Una lettura per parole e passaggi chiave

Per non eccedere nell'analisi longitudinale e anche per evitare di sovrapporsi ai due periodi che hanno caratterizzato negli anni la nostra interazione per l'attività svolta, prima da uno e poi dall'altro, presso il Ministero con la funzione di Consigliere del Ministro, e che finirebbero per introdurre variabili di non facile lettura a terzi, mi limito a riprendere, in forma di parole chiave, alcuni concetti emersi negli ultimi dieci anni.

2.1 Etica e Politica

E' sempre difficile fissare una data spartiacque da cui far derivare una temporizzazione. Un po' strumentalmente vorrei citare il 2014 in quanto il suo contributo alla predisposizione e presentazione del "Manifesto per un'Europa di progresso", volto a proporre, in occasione delle elezioni europee, una visione comune di scienziati italiani verso gli "Stati Uniti d'Europa", in analogia al ruolo svolto in passato dalla scienza nella sua dimensione transnazionale, ha rappresentato un momento significativo della capacità di Rino di coagulare, anzi direi di catalizzare, energie e proposte. Gli esiti di quel percorso avviato dieci anni fa, purtroppo, non sono stati in linea con le aspettative; da qui una sorta di "chiamata alle armi" in corrispondenza della dipartita di Carlo Bernardini, già co-promotore di quell'appello e di cui peraltro Rino era stato compagno di viaggio nell'avventura dell'Osservatorio sulla Ricerca. Una perdita certamente, per i molteplici ruoli svolti da Carlo sui vari fronti, dalla scienza alla politica della scienza, dalla divulgazione alla comunicazione scientifica, ma che Rino sottolinea in particolare sul versante umano e sulla volontà di guardare al futuro, sul costruire le basi per condividere le azioni, per dare visibilità al rapporto della scienza con etica e politica. Tutti elementi contenuti sia nell'articolo pubblicato un paio di mesi dopo la morte da Scienza in rete nel 2018, sia in quello, sempre su Scienza in rete, del 2020, relativo al convegno/webinar di illustrazione del volume, curato da Rino con Pietro Greco (di cui non mi stancherò mai di rimpiangere le qualità, le conoscenze e la generosa disponibilità) e Giulio Peruzzi per Dedalo, anche a seguito di un incontro nazionale tenutosi il 6 maggio del 2019 con un ampio patrocinio: Accademia dei Lincei, Dipartimento di Fisica della Sapienza, INFN, CNR-ISTC e Unione Scienziati per il Disarmo. Praticamente tutti gli ambiti, a parte la rappresentanza politica vera e propria, a cui Bernardini aveva molto contribuito. Nell'introduzione al webinar Rino, siamo nell'anno del Covid 19, evidenzia come il ridimensionamento della scienza da parte di decisori ed opinione pubblica sia dovuto a un processo di semplificazione (conoscenza intesa come informazione; mancata verifica di attendibilità e ruolo critico dei "mediatori", e degli strumenti utilizzati, nelle relazioni con i fruitori) messo radicalmente in discussione dalla pandemia. Con una visione premonitrice Rino anticipa come questa rinnovata attenzione avrebbe potuto evolvere in varie direzioni (e così è stato...) evocando in tal senso la necessità di un attento monitoraggio.

Concetti già presenti in uno scambio di auguri per l'anno nuovo (in questo caso siamo al 2019 e quindi a metà strada tra i due articoli citati sopra). Vale la pena di riprenderli in quanto tali poiché Rino rivendica voglia e ragioni per opporsi alla semplificazione (che nasconde impostura) e alla mistificazione (che si maschera da innovazione) e all'intolleranza (che cancella diritti e civiltà). E si chiude con un appello ai "doveri" degli uomini responsabili, accompagnato da una mia risposta dubitativa circa la percezione dell'esistenza di un praticabile spazio a disposizione in quanto coperto da una sorta di festival della presunzione, dell'inganno e dell'incompetenza. Erano, giova ricordarlo, gli anni nella teorizzazione ed attuazione del cosiddetto governo giallo-verde, portatore di quei tre sostantivi contenuti nella mia risposta.

2.2   Il momento dei bilanci

La scadenza dell'arrivo del 2020 rappresenta una data che oggi associamo al fenomeno pandemia ma che all'epoca sembrava semplicemente fosse la chiusura di un decennio. E come tutte le chiusure, l'occasione per un bilancio che Rino associa a tre parole chiave tra di loro correlate: ricostruzione, riscatto e rinascita. Parole che ben si riferiscono al già citato evento a due anni dalla scomparsa di Carlo Bernardini tenutosi a Padova e che trovano conferme negli auguri di fine anno. Rino auspica forza e fiducia nella impresa comune, a cui rispondo sottolineando come lo sforzo atteso, per raggiungere i risultati, richieda un contributo individuale e una visione collettiva e come questa non sia la mera sommatoria degli individui. Parlare di bilanci, e di compiti futuri, implica anche la ricerca di un equilibrio tra i singoli e l'azione comune, proprio come Pietro Greco, all'epoca da poco scomparso, ci aveva insegnato a fare e di come questo dovesse rappresentare un testimone da raccogliere e fare proprio.

Un primo bilancio, a metà del 2021, comprendeva anche l'uscita dalla crisi pandemica, dovuta alla vaccinazione di massa su cui peraltro Rino aveva lavorato scientificamente attraverso un'indagine condotta dal suo Istituto. Le mie perplessità riguardavano la facilità con cui si era passati dal cantare Bella ciao dai balconi ai festeggiamenti per l'aperitivo ritrovato. O all'anomala inclusione dell'accreditamento dell'agricoltura biodinamica, oggetto di un pronunciamento del Parlamento senza alcuna valutazione o competenza scientifica, su cui Rino, giustamente, lamentava un oltraggio alla scienza.

2.3 La (ennesima) riforma dell'ente in un paese che cambia

Nel dibattito sulla legge finanziaria 2022 ha trovato grande evidenza l'avvio del percorso di riforma del CNR da realizzarsi attraverso il cosiddetto Piano di riorganizzazione e rilancio, già autoesplicativo nel titolo. Per quanto detto all'inizio di queste note non si vuole affrontare in questa sede l'analisi di dettaglio, peraltro contenuta nel suo articolo di Scienza in rete del novembre 21 che, giustamente, paventa il rischio per l'autonomia dell'Ente, in quanto processo sostanzialmente eterodiretto senza alcuna indicazione di obiettivi e come, dove e perchè si vuole arrivare. Basti anche ricordare la raccolta di firme sull'appello al Presidente della Repubblica. E l'anno si chiude con l'auspicio, da parte di Rino, dell'arrivo di un anno "liberato" da superstizioni e pseudoscienze (Covid docet) e di un ritorno ai conflitti reali. Senza dimenticare di confermarsi vigili e reattivi, requisiti fondamentali per vincere gli attendismi e le passività. Tutti pensieri e propositi messi in discussione dalla nuova dimensione della guerra in Europa nei primi mesi del nuovo anno, e dal ruolo e dal comportamento degli scienziati, spesso chiamati a schierarsi e, in ogni caso, parti in causa come firmatari di appelli. Tra i diversi temi su cui ci siamo confrontati in quell'anno, spesso ispirati dalla condivisione di scritti pubblicati, ne cito due, accanto al già ricordato tema dell'autonomia degli scienziati e del loro posizionamento nei confronti/scontri internazionali. Il primo riguarda il crescente peso dei sondaggi non tanto nell'anticipazione degli esiti elettorali quanto nel loro orientamento, argomento anche ripreso dal punto di Paolo Pagliaro in Otto e Mezzo in tv sulla 7. Il secondo, per introdurre una piccola nota di costume, è il condiviso entusiasmo per la vittoria nella finale della Conference League a Tirana della maggica Roma, da troppo tempo all'asciutto negli ultimi anni.... Tutto questo a significare come gli interessi e le passioni abbraccino diversi argomenti.

2.4   Gli ultimi due anni

L'arrivo del 2023, in un paese a trazione meloniana, viene affrontato sotto l'egida di tre parole chiave. Discernere, battersi e... sperare, per poter contribuire a progettare visioni di società future ritenute raggiungibili. Senza quindi farsi prendere dallo scoramento continuando ad esserci e a voler contare.

Anche se l'intero andamento di quell'anno lascia, parole di Rino, "segni dolorosi indelebili, distacchi non rimarginabili....l'animo appare devastato di fronte alla pervasiva disumanità e disperazione che i conflitti in atto propongono". Ma questa percezione non si traduce in un pessimismo, bensì in un invito a costruire sia intenzioni, sia azioni per battersi e opporsi, con gli strumenti della ragione e dell'equilibrio, dell'equità e della tolleranza, della giustizia e della pace agli inaccettabili sconquassi sociali. Il tutto per costruire percorsi e per recuperare un senso al nostro vivere.

E dopo aver citato a lungo i pensieri e le parole, nonché gli scritti, di Rino consentitemi la presunzione di chiudere con la mia risposta a quest'appello/auspicio per un anno che è ancora in corso.

Ragione, equilibrio, equità, tolleranza, giustizia, pace. Tutte parole chiave a cui aggiungere futuro e progetto. Con una volontà di condivisione, senza false semplificazioni e rispettando le diversità. Una missione difficile ma non impossibile cui vale la pena di impegnarsi e di provarci. In queste mie parole ci sta tutto il Rino che conosco e che sono fortunato di aver conosciuto.

A COMMITMENT ON ALL FRONTS

SETTIMO TERMINI

Associazione Marina Diana Mercurio
Accademia nazionale di scienze, lettere e arti di Palermo
DMI Università degli Studi di Palermo

My first contact with Rino occurred on the occasion of an article on his research activities with Cristiano Castelfranchi asked for the journal *Lettera matematica Pristem.* I wish to start from here because, in its implications, this minimal episode contains many of the aspects of both our relationship (but this - while important to me - would be marginal) and - more importantly - of Rino's way of doing things. I did not like the article. Too technical and with language not suitable for a non-specialist reader. Just the opposite of what I would have wanted (and what I was convinced I had asked him for). I told him of my dissatisfaction (in polite form, but, to myself, thinking of the delay this hiccup would entail) while also pointing out the points that needed to be changed more radically (something a bit rude, I must confess). The paper came back, in a very short time, in the very enjoyable form that you can still read[1]. An interpretive dissonance, immediately happily resolved, an example of a constructively critical attitude and, I would still like to say, the *embodiment* of the scientific way of doing things in a small happening. The attunement with Rino emerged immediately. Since then - also because of the fact that in 2002 I had returned, as director of the Institute of Cybernetics "Eduardo Caianiello," to that CNR that I had left in 1987 to go, first, to the University of Perugia and, then, to the University of Palermo - there have been many occasions to collaborate with him (and with a small group of friends formed by, among others, Carlo Bernardini, Pietro Greco, Sergio Ferrari, Sergio Bruno, Daniela Palma, Lucio Bianco) on problems and initiatives that we *then* considered *simply important.* Seen today, in perspective, they were *crucial* points to avoid the backward step that the our Country has taken (and of this perhaps, at the time, we were not aware either). I mention just two of these occasions. Rino's creation of the *Osservatorio sulla Ricerca*[2] and the drafting and dissemination of the *Manifesto per l'Europa*[3]. Regarding the first, I recall that the climate was one of having to defend the world of research, and the CNR in particular, from the deleterious effects of incomprehensible projects (which took the form of real attacks) of Minister Brichetto Moratti. The mobilization of the scientific community, thanks also to the commitment and effectiveness of the *Osservatorio*, reached levels that had never been reached before (let alone would be reached afterwards). As for the *Manifesto per l'Europa*, the idea, launched by Carlo Bernardini, had been elaborated and discussed in many meetings almost all of which took place in Rino's office at the ISTC. I still remember him, on more than one occasion, torn between discussions with us and frequent interruptions by the secretary or others. An exercise, I must say, in which he could juggle very well given how he was able to resume dialogue with us with enviable continuity.

The publication of the *Manifesto*, with the discussion that would follow, was intended to move the level of debate higher (as a general issue and to be conducted at least at the European level), and Rino effectively contributed (in tandem with Pietro Greco) to also involve people and groups, from broader cultural circles than those of the narrow academic world, who turned out to be in total agreement with the project. The Manifesto was presented on April 8, 2014 at the CNR headquarters. Carlo planned to involve colleagues, as was the spirit, from all over Europe with the ultimate goal of presenting the Manifesto itself to the European Parliament. This later did not happen because doubts emerged that this initiative that had been quite successful in Italy could equally involve many colleagues from other European countries. Underlying the debate in Italy was the presentation of the thesis of revitalizing research also for the purpose of triggering greater economic development in the high-tech sectors, in many of the other European countries this was a topic that was less felt as an element of a battle to be engaged in because, to some extent, it was already put into practice by governments (by Germany, first and foremost, then, but also by small countries such as Finland with its involvement in Nokia, which for a number of years had been at the frontier of technological innovation). Viewed in retrospect, it was a mistake not to move forward on this ambitious path. Many things are born with the support of a few but, if sensible, gain their relevance later when more favorable moments arise. And a more important trace would remain of this important initiative than just the presentation to our CNR.

To the memory of these moments must be added that of the years in which Rino was strenuously engaged, as adviser to the Minister of Research (with Pietro Greco, among others), in trying to undo some of the damage inflicted on the CNR by Fabio Pistella[4, 5, 6]. This commitment of his, temporally, lies between the two indicated above, coinciding with the two years of the second Prodi government. It is well-known history but I would like equally to recall his total commitment on two fronts. On the one hand, with analytical work, showing Minister Fabio Mussi the damage inflicted on the CNR by the "corporatist" model (but which had nothing of the positive characteristics of companies). A model that had already been imposed but not in all its articulations, and the goal was to prevent its consolidation, hoping to succeed in changing course quickly. No less demanding was his work on the other front, that of presenting and defending models of Research Institutions that would allow the scientific community to express all its great potential, recognized, by the way, internationally, as an article by Alexander King, which appeared in *Nature* in those years, had highlighted.

On a directly personal level, I am still grateful to Rino for his help in organizing a meeting in Naples[7, 8], realizing an idea that took form in dialogues with Pietro Greco, which would highlight, by initiating a frank debate, what should and could be the essential points of a turning point in the scientific policy of our country. There was (and there is) a need to trigger a true dialogue between scientific research (starting with basic research) and the productive world, according to the model outlined by Vannevar Bush to Roosevelt[9, 10]and basing ourselves, for Italy, on the decades-long hard work of analysis done by Sergio Ferrari with his ENEA Observatory[11, 12]. A model[13] that, by the way, had been implemented in Italy, for a brief and successful period, in the late 1950s and early 1960s . Serious and challenging programs and projects. Equal generosity shown by Rino in intervening in the development of a reflection on the interactions between scientific research and society starting from an occasional (but significant) event such as the 40th anniversary of the founding of the Institute of Cybernetics of the CNR[14]. I would like to quote the introductory piece of his contribution[15]:

> *Participating in this event was of particular value to me for two main reasons. The first has to do with the fact that the cybernetic approach, interdisciplinary in its intimate nature, is the choice I have made for my profession as a researcher ever since I decided to specialize in electronics-cybernetics, that was the name of the major I chose at the faculty of physics, at La Sapienza University in Rome. [...] The second reason why I consider this invitation of value is the very approach that was decided to give to the conference. From its beckoning title (memory and project), it intends to link science, research, and knowledge development to society and the potential models that are to be realized from it. Scientific research not only as a dispenser of extraordinary cognitive advances [...] but also as an extraordinary methodological tool to set - design - society and to reflect critically on the choices made and future prospects. (page 273)*

I wanted to quote these remarks of his because they carry a vision of the role of science in society in which not only we believed (along with a small but not too small core of colleagues and friends) but which we believed then (and still believe today) to be valid for countering the decline of our country. We defended the value scientific research not only in itself for its cognitive value (motivation, by the way, more than sufficient on its own) but also as a tool, if properly used, to combat economic decline. This is what we have challenged and invited political forces to act on. Political forces that never understood the game that was and is being played. Some, on ideological grounds, devastating what was there, others assisting without understanding what was happening. Only a *few* protagonists of *some* of these political forces were aware that the stakes were significant but - for various reasons - failed to be effective. With lucidity, in 2009, Rino summarized the situation:

> *many good intentions, some even notable achievements, some glaring omissions, a short-sighted acquiescence and tolerance to bureaucratic blockades and parallel lobbies. There remains this hiatus between a vision that has also shown to exist and an inability, a lack of awareness of the enterprise through which it is necessary to proceed. (page 279)*

Although in recent years I have not followed directly the changes that have taken place in the CNR, and with detachment those concerning the University, I am convinced that, today, the situation of scientific research in our country is markedly worse than it was twenty or twenty-five years ago. But this should not discourage us; on the contrary, it should be a reason for a greater commitment to understanding. Personally, I am deeply optimistic although in the long run. Indeed, I am convinced that a development of knowledge, of the understanding of how the world works and of the productive forces as broad and deep as the one that has been with us in recent decades can only ultimately defeat those who oppose a change in the existing balance of power. But mine, precisely, is a long time optimism. I am sure that Rino is even more optimistic in the sense that, more than me, he believes that with our immediate actions (which he knows very well how to devise and put into practice) we can not only shorten the time frame of what was assumed before but can, in any case, lessen the negative effects of battles that we can lose and have often lost.

All the preceding recollections refer to aspects of Rino's engagement in which I was personally involved. I would like to add our many fruitful (at least for me) discussions for many of which I cannot find an adjective to characterize them. I will call them *unusual*, far from the ideas always debated and, in this sense at least, not trivial. They *had to be unusual* because they were precisely about the future that research in Italy should have. It was *something new* that *we had to design*. Discussions on scientific issues (including specific ones) were many and varied, but they almost always started from (and inevitably tended to return to) the broader topic of their relations with scientific institutions and research policies. A partial exception can be considered the exchange, albeit only at the methodological level, that took place immediately after the publication of the book *Trust Theory*[16], a hard copy of which arrived to me, unannounced, by mail. Unexpected, therefore, as well as a pleasant surprise that allowed me a reading unmediated (and not conditioned) by previously formed expectations and that induced me to immediately put in writing some comments and inserting them in a paper[17] that I was writing at the time and in which it fit perfectly.

Our dialogues have, perhaps, been fostered by our common background as physicists lent to other new areas of research. If I mention this point, it is because I believe that the role played by physicists (as individuals and as a community) in the establishment and consolidation of activities in the field of information sciences in Italy deserves to be explored more than has been done so far because it represents (and even more so has represented in the past) an interesting (and positive) Italian *anomaly*. And understanding anomalous behavior always leads to a better understanding of the "normal" functioning of the systems in which the anomaly occurred, with obvious benefits for improving their functioning. Finally, I want to attest how useful some of Rino's writings on the latest developments in AI have been to me, even though I have not had a chance to talk to him in person, recently. His analyses, at once wide-ranging and constructively critical, enabled me to understand immediately some conceptual knots in current developments that would have taken me much longer to dissect on my own. The items quoted in the footnote show also his ability to use a clear language.

What is mentioned here does not, by any means, exhaust the spectrum of activities in which Rino was fruitfully engaged. The aspects touched upon are only a kind of documentation of the points and moments when my interaction with Rino was strongest, in a dialogue that - even when it was just us discussing - was not just between the two of us. We both know that these are complex issues that need a lot of study and insight and everyone's contribution. We have been fortunate in the past to be able to collectively explore these issues in depth with many open minded friends and colleagues. Some of the protagonists are no longer with us. We have a duty to continue the dialogue by involving others, younger ones.

In discussing with Carlo Bernardini, there often emerged from his words, an appreciation for the figure of the "civil servant" (an expression that is difficult to translate into Italian because this figure is lacking in our society, not because there is a lack of words) more than for that of " intellettuale impegnato", a term abused for some decades in Italy. Despite the difficulty of giving an Italian definition of the term, Carlo was certainly a *civil servant*. For in him the ways of doing research and analyzing the role of science in society were both taken very seriously. The same can be said of Rino. The modes of his research work and his efforts to transform scientific institutions were happily intertwined, reinforcing and refining each other. This, in my view, is one of his important legacies along with a characteristic of his - as a researcher and as a critic (and purposeful builder) of scientific institutions - which I would summarize as follows: "a radicality in the analysis and content of criticism coupled with a great flexibility about the ways to achieve goals."

I wish Rino *only* a happy birthday. I just cannot bring myself to write the phrase "happy retirement", since it is impossible for me to associate this term with him. His desire to act and intervene effectively will never *retire*. Rino will always continue to play a reflective and critical role in all the many areas in which he was, and still is, present, helping - both those who are young and those who are older than he is - to understand how this world of ours, so permeated by science and yet so fragile, is proceeding today. *Understanding*, associating with this term all the uncertainties that the scientific mindset has taught us to connect to knowledge, is the necessary premise to be able *to change what is wrong with the world.* Thank you Rino.

References

[1] R. Falcone, C. Castelfranchi. Il paradigma ad agenti: l'IA torna alle sue sfide originarie (e al successo) in Lettera matematica PRISTEM 36 (2000), pp. 21–29. Springer-Verlag Italia, Milano.

[2] R. Falcone L'Osservatorio sulla ricerca: una sfida per la comunità scientifica nazionale in Analysis, Rivista di cultura e politica scientifica N. 2/2003.

[3] Carlo Bernardini, Vittorio Bidoli, Marcello Buiatti, Vincenzo Cavasinni, Remo Cesarani, Paolo Dario, Tullio De Mauro, Rino Falcone, Sergio Ferrari, Pietro Greco, Francesco Lenci, Tommaso Maccacaro, Paolo Nannipieri, Pietro Nastasi, Giorgio Parisi, Giulio Peruzzi, Claudio Pucciani, Carlo Alberto Redi, Giorgio Salvini, Vittorio Silvestrini, Settimo Termini, Virginia Volterra. Manifesto per una Europa del Progresso in Il Manifesto

[4] Natalini, Roberto Da Vito Volterra a Fabio Pistella in Analysis, Rivista di cultura e politica scientifica N. 2/2003.

[5] Termini, Settimo Alcune osservazioni sugli Enti (pubblici) di ricerca: il loro ruolo e brevi cronache di una devastazione annunciata in La Minerva ferita, a cura di Stefano Boffo e Enrico Rebeggiani, Napoli, Liguori, 2011, pp. 37-51.

[6] Natalini, Roberto CNR: novità non ce ne sono in Lettera matematica Pristem 60, 2006.

[7] Pietro Greco e Settimo Termini Oltre il declino. Ricerca scientifica e competitività economica in Roma, Franco Muzzio editore, 2007.

[8] Rino Falcone Ricerca e sviluppo tecnologico. La svolta e le iniziative italiane oggi in Oltre il declino, pp. 101-107.

[9] Vannevar Bush Manifesto per la rinascita di una nazione. Scienza, la frontiera infinita in Torino, Boringhieri, 2013. .

[10] P. Greco, S. Termini, Contro il declino, in Torino, Codice edizioni, 2007.

[11] Sergio Ferrari Società ed Economia della conoscenza in Mnamon, 2014.

[12] Sergio Ferrari et al. L'Italia nella competizione tecnologica internazionale in Quinto rapporto ENEA, Milano, Franco Angeli, 2007.

[13] Marco Pivato Il miracolo scippato. Le quattro occasioni sprecate della scienza italiana negli anni sessanta in Donzelli, 2013.

[14] Pietro Greco, Settimo Termini, Memoria e Progetto. Idee per lo sviluppo del Mezzogiorno come modello per un Progetto Paese in Gruppo editoriale Muzzio, 2010.

[15] Rino Falcone, Le sfide della ricerca in Memoria e Progetto, pp. 273-280.

[16] C. Castelfranchi, R. Falcone. Trust Theory. A Socio-Cognitive and Computational Model In John Wiley and Sons, 2010..

[17] Termini, Settimo. On Some family resemblances of Fuzzy Set Theory and Human Sciences In R. Seising and V. Sanz (Eds.), Soft Computing in Humanities and Social Sciences, STUDFUZZ 273, 39–54, Springer-Verlag, Heidelberg 2012.

Io e Rino

Marco Pagani

*Institute of Cognitive Sciences and Technologies, National Research
Council of Italy (ISTC-CNR), 00196 Rome, Italy*
marco.pagani@istc.cnr.it

Mi è difficile scrivere di un vero amico e collega, compagno di
battaglie politiche e interne all'Ente, impareggiabile controparte in
questioni calcistiche e gastronomiche.

La consapevolezza dell'esistenza in vita di Rino è nata a Viale
Marx con il mio passaggio all'Istituto allora di Psicologia. I colleghi,
ma soprattutto le colleghe, mi narravano di questa figura affascinante,
carismatica che era già un punto fermo nelle dinamiche dell'Istituto.
In quegli anni lo percepivo più che altro come componente, con Cristiano e Rosaria, di una triade indissolubile e la nostra frequentazione
si limitava a rapidi saluti e battute nei corridoi dell'Istituto o al ris-
torante a pranzo.

Non so esattamente dire quando abbiamo cominciato ad apprez-
zarci ed avere una frequentazione più assidua dentro e fuori dall'Istituto ma sicuramente in corrispondenza o nei dintorni della sua nomina
a Direttore d'Istituto, del trasferimento dello stesso a San Martino
della Battaglia e del mio ruolo all'interno dell'Amministrazione Centrale nella formazione del Dipartimento di Scienze della Vita. E' stato
un momento molto particolare nella vita dell'Ente e le discussioni con
Rino sono state non solo produttive professionalmente ma molto profonde a livello umano. Ci siamo trovati entrambi a combattere con
un sistema ingessato che lasciava molto poco spazio alla logica ed alle
novità organizzative e soprattutto burocratiche. La cartolarizzazione
che ha svenduto il patrimonio immobiliare dell'Ente e l'istituzione dei
Dipartimenti che ha aggiunto ruoli su ruoli senza migliorare l'efficienza

e lo status economico generale sono stai un punto di svolta che abbiamo vissuto entrambi con occhio critico passando ore ed ore a discuterne.

Un altro momento aneddotico delle nostre esperienze in comune è stata la battaglia per vedere riconosciuta l'anzianità di servizio relativa agli anni trascorsi come ricercatori a tempo determinato. Sorvolando sulle questioni tecniche, abbiamo passato momenti esilaranti scambiandoci opinioni e battute varie sulle modalità con le quali l'avvocato che ci seguiva affrontava i vari passaggi legali: vista da fuori una lotta contro un muro di gomma, vista da dentro, con occhio ironico e sarcastico, una comica. Andata poi fortunatamente a buon fine: niente di meglio per considerare poi non troppo male la propria presenza e impegno all'interno dell'Ente.

Negli anni si sono invece perfezionate le nostre affinità elettive per il buon cibo, il buon vino e il buon ... calcio. Compatibilmente con i multiformi impegni della coppia e con la mia presenza a roma abbiamo sempre cercato di ricavarci tempo e spazio per delle serate insieme. Le cene con ivana e rino sono state sempre caratterizzate dagli ottimi vini portati da loro e dalle mie doti culinarie molto amplificate dalla bontà degli ingredienti utilizzati, spesso e volentieri salmone e prodotti ittici vari, vista la mia frequentazione trentennale di Svezia e dintorni. Serate inoltre dedicate con passione ad aggiornamenti a volte amari sui rispettivi stati lavorativi, a disquisizioni letterarie (ivana è nota autrice di libri di cultura e arte), a mini dibattiti politici, difendendo con passione le rispettive posizioni, convergenti o divergenti che fossero, e non ultimi amorevoli commenti sul parallelo progresso di crescita, scolastico e sociale, dei nostri rispettivi figli.

Capitolo a parte il calcio o qualunque sport fosse d'attualità al momento. Fedi calcistiche opposte ma sempre commentate con goliardica parzialità, ironia e spensierata allegria. Serate passate a distanza a vedere e commentare in tempo reale eventi sportivi di tutti i tipi ma in realtà dedicate al godimento di momenti di assoluto relax, di distacco completo dalla routine giornaliera, di condivisione di uno spirito che accomuna competenza e divertimento fine a sé stesso.

Cosa dire per concludere questa semplice e sincera testimonianza?

Alla fine siamo stati fortunati caro rino a vivere nei "tempi andati".
Abbiamo avuto la possibilità di appassionarci, partecipare attivamente a momenti in cui esisteva ancora la vera dialettica, abbiamo
attraversato mutazioni della vita individuale e collettiva che forse non
torneranno, abbiamo dato il nostro contributo con uno spirito critico,
a volte costruttivo a volte distruttivo, sia alla società sia al CNR. Ce
lo siamo potuti permettere per il contesto nel quale abbiamo vissuto e
per le condizioni storiche di questi nostri sessanta e oltre anni di vita.
Fra poco ci aspettano la panchina e i commenti ai lavori stradali:
sono sicuro che ci divertiremo e saremo capaci di commentarli con la
stessa ironia che ci ha accomunato durante tutti questi anni. Il solito
abbraccione amico mio.

Rino Falcone in quattro dimensioni

Oliviero Stock
Fondazione Bruno Kessler - FBK
stock@fbk.eu

Rino è una persona fuori dall'ordinario. Vive in uno spazio a quattro dimensioni. Pur essendo Rino fisico di formazione, la quarta dimensione qui non è il tempo, che pure sembra scorrere per lui in modo più lento rispetto a quello in cui vivono altri mortali. E anche le altre dimensioni a cui mi riferisco non sono quelle della sua collocazione nello spazio (sarebbe troppo facile: è coerentemente, stanzialmente romano). Proverò a ricordare passaggi di quello che mi ha colpito di Rino su quattro dimensioni caratteristiche, in particolare riferendomi alla mia esperienza diretta.

Comincio dalla dimensione principale per la vita di un ricercatore: quella scientifica. Conobbi Rino quando era un giovane collaboratore alla Fondazione Ugo Bordoni. Faceva parte di un gruppo coordinato da Andrea Paoloni, dedicato al riconoscimento del parlato. Io, che lavoravo all'Istituto di Psicologia del CNR e mi ero occupato di elaborazione del linguaggio naturale da metà anni Settanta, prima a Pisa e poi a Roma, ebbi la possibilità di guidare la ricerca con alcuni di loro. Fu così che, pur non specificamente esperto di riconoscimento del parlato ma piuttosto di analisi sintattica automatica, mi proposi di adattare la mia esperienza in linguistica computazionale per le esigenze specifiche del riconoscimento del parlato. Trovai in Rino, e in una tesista di nome Patrizia Insinnamo, di cui poi ho perso traccia, due ottimi giovani partner per lo sviluppo di questo lavoro. Il concetto da cui si partiva era il *chart parsing*, ovvero una tecnica di analisi di frase, che consente, nell'elaborazione dell'analisi di frase che procede da sinistra a destra, di salvare in una tabella strutture parziali già analizzate. Le frasi in lingua naturale sono spesso complessivamente ambigue, ma l'incertezza è ancora molto maggiore localmente, nel corso dell'analisi l'ambiguità è diffusissima. Con il chart parsing, anche se si segue un tentativo che poi risulta complessivamente sterile, le strutture già costruite non le si devono ricostruire, i risultati parziali potranno essere riutilizzati. Per quanto concerne il parlato, il problema è però che l'input stesso non è uniformemente chiaro, il segnale di ingresso può lasciare delle parti (parole o sequenze di parole) non bene riconoscibili. Per questo, un approccio proposto per il riconoscimento del parlato era quello di puntare su un procedimento "a isole": anziché procedere da sinistra a destra, si parte dalle zone della frase più sicuramente riconosciute e si cerca di estenderle. La nostra idea fu di introdurre il chart parsing a isole, quindi con modalità bidirezionale. Così facendo si possono predire pezzi di input incerto, sulla base dei due contesti, sinistro e destro, e senza ripetizione del lavoro di analisi. Lavorammo allo sviluppo di questa tecnica nel 1986-1987 e posso dire con orgoglio che da queste ricerche vennero le prime pubblicazioni internazionali di Rino: il suo primo articolo su atti di conferenza internazionale fu nei Proceedings di COLING 1988 [1], che si tenne a Budapest, e il suo primo articolo su rivista fu su Computer Speech and Language [2]. Io ormai avevo lasciato Roma (e successivamente, tra le cose, mi occupai della definizione delle proprietà matematiche del chart parsing bidirezionale). Rino continuò a occuparsi di chart e input malformato. Quando, qualche anno dopo cominciò a lavorare al CNR con Cristiano Castelfranchi a temi di pragmatica e in particolare al riconoscimento di piani, potè anche sfruttare questa sua esperienza precedente e importare il concetto del chart parsing nel nuovo contesto del suo lavoro di ricerca [3].

L'incontro intellettuale con Cristiano Castelfranchi fu particolarmente fortunato. Il lavoro scientifico di Rino assieme a Cristiano si indirizzò ai sistemi multiagente, tema fondamentale del vasto settore che va sotto il nome di intelligenza artificiale distribuita. In quell'area Rino e Cristiano si distinsero per il portare concetti di psicologia sociale (ricordo a questo proposito il contributo di Rosaria Conte e Maria Miceli, tristemente scomparse prematuramente) nel mondo dell'AI, concetti fondamentali che erano stati poco approfonditi precedentemente all'interno di quella comunità. Mi riferisco al concetto di delega, a quello di fiducia, a quello di controllo, a quello di autonomia, con molte implicazioni interessanti i sistemi multiagente. La ricerca ha portato tra l'altro alla pubblicazione di un importante volume di riferimento sulla teoria della fiducia [4]. Nel libro gli autori propongono un modello con diverse caratteristiche: integrato, con vari aspetti normalmente considerati separatamente e una struttura complessa; socio-cognitivo; analitico ed esplicito; multifattoriale e multidimensionale; basato su relazioni sistematiche tra fiducia e altri fenomeni (come previsioni, aspettative, rischio, reputazione ecc.) ed emozioni (come sorpresa, delusione, tradimento ecc.). Infine, un modello che non intende essere prescrittivo.

Il lavoro ha offerto un contributo importante alla comprensione degli aspetti sociali nel mondo computazionale, con varie implicazioni teoriche e molte potenzialmente applicative. Trovo che la modellizzazione ha anche offerto un motivo di riflessione su aspetti che riguardano l'integrazione in un universo dove agiscono agenti umani e agenti artificiali. In questo contesto sono di grande rilievo aspetti che vengono studiati da chi si occupa di organizzazioni sociali, ma anche aspetti relativi al singolo, alla personalità degli individui coinvolti. La prospettiva complessa permette di entrare a un livello profondo anche nel campo dell'*augmented intelligence*, in cui c'è bisogno di chiarezza su cosa si sta delegando o cosa si può delegare al sistema artificiale. Rino e Cristiano hanno saputo sviscerare molti contenuti affascinanti di questo, ponendo la loro attenzione sia sulla struttura di potere preesistente, che sta alla base delle organizzazioni e delle relazioni sociali, che su vari aspetti dinamici. Tra questi, i conflitti, ma anche i raggiri e molte forme di azione che tendono ad approfittare della fiducia riposta e travalicarne i limiti; e poi anche le attribuzioni involontarie di capacità e potere ad altri agenti, dovute a errori o alla disponibilità di insufficienti strumenti di valutazione.

Naturalmente questo tema diventa anche centrale in relazione ad aspetti giuridici ed etici. I lavori, oltre che per il loro valore nell'avanzamento della comprensione dei fenomeni e all'aspetto modellistico, forniscono anche un bagaglio di riferimenti utili, ora che finalmente la nostra società, soprattutto in Europa, sta affrontando questioni relative alla regolamentazione dell'AI. Ma essi potranno venire valorizzati per l'etica anche in una prospettiva più ambiziosa: si stanno facendo dei passi sul tema dell'etica computazionale, cioè lo sviluppare la capacità dei sistemi di ragionare e prendere decisioni su base etica. In sostanza distinguere il bene dal male.

Naturalmente parlando di fiducia ed emozioni vengono in mente i tanti problemi internazionali nel nostro mondo attuale, così travagliato. Quando si affronta il tema dei negoziati in relazione alla soluzione di un conflitto, si parla di solito di bisogni primari delle parti in conflitto, ma probabilmente maggiore attenzione dovrebbe essere posta anche su temi come fiducia ed emozioni, spesso fattori essenziali, soprattutto quali precondizioni per risolvere questioni complesse. Spingendomi in là con gli auspici, forse l'approccio di Rino e Cristiano un giorno potrà avere contribuito ad un apporto computazionale alla soluzione dei conflitti.

Vengo a un'altra dimensione essenziale dell'attività di Rino, quella organizzativa e di servizio. Anche qui desidero partire da un mio ricordo personale. Come accennavo, negli anni Ottanta sono stato ricercatore all'Istituto di Psicologia del CNR, ora Istituto di Scienze e Tecnologie della Cognizione. Il direttore dell'Istituto era Lello Misiti, una persona eccezionale e indimenticabile, che avviò l'istituto nelle scienze cognitive, integrando al suo interno varie competenze, con psicologi, filosofi, linguisti, antropologi, informatici, biologi, medici e altro ancora. Con Lello Misiti l'interdisciplinarietà divenne davvero una cosa vissuta, anzi l'essenza dell'istituto. Lello sapeva gestire un istituto così complesso, che aveva temi diversi, tra cui era stato avviato appunto il gruppo di intelligenza artificiale di cui facevo parte; ma anche tante personalità diverse, e, come succede quando vi sono vari gruppi, anche tante primedonne con esigenze "improcrastinabili". Misiti sapeva fare sentire tutti protagonisti e sapeva trasmettere il valore sociale della ricerca che si stava conducendo. Un trauma fortissimo per l'istituto fu quando Misiti morì improvvisamente nel settembre del 1986. Sono seguiti vari direttori di grande qualità scientifica e capacità, tutti colleghi che erano stati ricercatori di punta dell'istituto ai tempi di Misiti. Nel 2011 venne nominato direttore Rino, fu il primo direttore che aveva conosciuto Misiti solamente da giovane frequentatore dell'istituto. Rino è stato anche il direttore che più a lungo ha ricoperto quel ruolo nell'era post Misiti, tra periodi con incarico di "direttore facente funzione" e di incarico pieno; ma, dal mio punto di vista di osservatore lontano ma con qualche conoscenza dell'istituto, mi sembra che il punto più importante sia che Rino è stato quello che di più ha incarnato lo spirito di Misiti. Sicuramente ha esteso ancora la vocazione interdisciplinare e, pur continuando a fare ricerca di persona molto attivamente, ha dato spazio ad altri settori di ricerca e, da quanto ho sentito da colleghi a più riprese, è stato generalmente apprezzato dal personale. C'è un altro elemento fondamentale che lo fa avvicinare ancora di più a Misiti: come Lello, Rino ha inteso l'importanza dell'assumere responsabilità in funzioni strategiche del CNR. Questo è soprattutto determinante perché l'istituto, poco univocamente inquadrabile nei grandi dipartimenti del CNR, corre il continuo rischio di essere schiacciato su una dimensione sola, che verosimilmente porterebbe a stravolgerne la natura. Quasi contestualmente al primo incarico di direzione, Rino venne nominato dal Presidente del CNR Profumo a far parte di una commissione di esperti nazionali e internazionali per lo studio e la definizione del *Documento di Visione Strategica Decennale del Consiglio Nazionale delle Ricerche*. L'anno dopo venne riconfermato dal Presidente Nicolais nella commissione ristretta che avrebbe predisposto il piano. Nel 2016 entrò a fare parte della commissione per la riorganizzazione della programmazione scientifica del CNR.

Naturalmente questo, oltre a beneficiare l'istituto, ha contributo in generale all'indirizzo del CNR, un ente che ha un ruolo speciale nel panorama della ricerca italiana. L'evidenziare questo impegno di Rino, mi porta diritto in un'altra dimensione della sua attività: quella dell'attenzione alla politica nazionale della ricerca, con incarichi di rilievo e con frequenti articoli nella comunicazione pubblica. In realtà, già nel periodo 2006-2008 Rino era stato nominato dal Ministro dell'Università e Ricerca Mussi consigliere per le questioni attinenti la ricerca scientifica italiana. In quel ruolo ha tra l'altro offerto consulenza agli organi parlamentari per la definizione della legge delega sugli enti di ricerca e contribuito all'avvio della procedura per la realizzazione di un *Piano nazionale per la costituzione di Grandi Infrastrutture di Ricerca di livello pan-europeo*, che oggi è operativo.

Per quanto riguarda la comunicazione pubblica sulla ricerca, spesso ho avuto occasione di leggere un qualche articolo scritto da Rino, e il mio commento di solito è stato: speriamo i decisori politici lo leggano con attenzione. In particolare ricordo la sua vicinanza al pensiero di Carlo Bernardini (vedi [5]). Un paese come il nostro, che investe così poco in ricerca e futuro, ha un bisogno assoluto di venire risvegliato, in realtà in tutte le forme che assume la politica, anche le meno peggiori.

C'è però un'altra dimensione di Rino che voglio mettere in rilievo, anche se trascende il contesto di questo volume: non quella genericamente personale, ma specificatmente quella sportiva. Dicevo che in questo scritto volevo prendere spunto da miei ricordi diretti. Nel 1986 vi fu un'epica partita di calcio a sei, tra due gruppi in istituto, in cui i componenti della nostra squadra, quella degli intelligenti artificiali, si atteggiavano a virtuosi brasiliani. La squadra era composta da Dinoccolao (Roberto Caraceni, mio tesista, soprannominato Dinoccolo da Federico Cecconi per il suo andamento), Amadeus (Amedeo Cesta, allora giovane collaboratore), Falcon (Rino appunto; solo lo spelling lo differenziava da Falcao, grande giocatore dell'epoca), De Oliveira (sarei io), Castelo Franco (si capisce chi era), Laudaninho (Alessandro Laudanna, ricercatore dell'istituto nel settore psicolinguistico). Giocavamo contro la squadra dei neuralisti, che avevano assunto nomi russofoni, capitanati da Zamparov (Roberto Zamparelli). Arbitrava il signor Parisi di Roma (naturalmente Domenico Parisi). Si vedeva che Rino era un atleta. Trentasette anni dopo, Rino mi fece sapere che aveva da poco fatto a nuoto la traversata dello stretto di Messina, un paio di miglia con mare mosso in un'ora e otto minuti. Il nuoto, specie in acque libere e ancora di più con una sfida di questo genere, porta a pensare a cose essenziali della vita e Rino mi mandò alcune considerazioni e anche un filmato ripreso dalla barca in cui lo si vedeva procedere con forza in mezzo ai flutti, distante dalla rassicurante costa.

In risposta gli mandai queste righe:

Era lì a poca distanza
lo vedevo
e già presentivo il piacere di raggiungerlo
nell'acqua mi muovo veloce
e lo stretto è un punto ideale
il percorso è segnato
e le correnti certamente non mi turbano
però mi è sfuggito
Firmato: il pescecane

Con l'augurio che Rino possa proseguire per molto tempo a contribuire alla ricerca sui temi importanti su cui si è affermato, e possa anche continuare a essere produttivo nelle altre dimensioni.

References

[1] O. Stock, R. Falcone, and P. Insinnamo, "Island parsing and bidirectional charts," in *Proceedings of COLING-88*, pp. 636–641, 1988.

[2] O. Stock, R. Falcone, and P. Insinnamo, "Bidirectional charts: a potential technique for parsing spoken natural language sentences," *Computer Speech & Language*, vol. 3, no. 3, pp. 219–237, 1989.

[3] R. Falcone and C. Castelfranchi, "Plan recognition for explanation: a chart-based system," in *Proceedings of the Eighth International Conference on Applications of Artificial Intelligence in Engineering (AIENG-93)*, (Toulouse, France), pp. 423–438, June 1993.

[4] C. Castelfranchi and R. Falcone, *Trust theory: A socio-cognitive and computational model*. John Wiley & Sons, 2010.

[5] R. Falcone, P. Greco, and G. Peruzzi, "Introduzione," in *La scienza fra etica e politica, l'eredità di Carlo Bernardini e le prospettive future* (R. Falcone, P. Greco, and G. Peruzzi, eds.), pp. 5–24, Edizioni Dedalo, June 2020.

Dalla ricerca scientifica alla politica della ricerca: la struttura di un impegno pubblico

Paolo De Nardis

Presidente dell'Istituto di Studi Politici San Pio V, già Professore ordinario di Sociologia presso Università degli Studi di Roma La Sapienza

Conosco Rino Falcone da tempo immemorabile. Ed è una delle persone più belle che la vita possa farti incontrare. Beninteso dal punto di vista di chi scrive sono tante le affinità elettive ma ne desidero sottolineare due in modo particolare: la ricerca scientifica da un lato e l'impegno pubblico abbinato alla militanza politica dall'altro.

Entrambi questi elementi costitutivi della sua persona mi sono invero particolarmente cari e vicini e rappresentano parte costitutiva dell'insegnamento di due grandi Maestri: Carlo Bernardini e Umberto Cerroni, per i quali il binomio scienza/democrazia ha rappresentato un faro di luce indissolubile che ha illuminato il cammino dell'individuo sociale il quale a sua volta si occupi di scienza non disconnettendo mai la militanza scientifica da quella pubblica.

Parto dal primo punto: la ricerca scientifica. Ricordo Rino giovane fisico del CNR impegnato in quella che è la ricerca pura di base che diventava per lui un *habitus* intellettuale appassionato ma non solo teorico perché è sempre sfociato nell'analisi critica di come la politica si sia atteggiata e disposta nei riguardi della stessa ricerca. Insomma di come la *politica della ricerca* abbia avuto sviluppo nel nostro paese. E di come Rino sia stato particolarmente critico nei riguardi di un piglio sovente ambiguo e distratto, ondivago direi, dello sguardo della politica ufficiale nei riguardi della ricerca scientifica.

Si può notare come la sociologia della scienza "classica", di mertoniana memoria, si sia concentrata proprio sullo studio degli aspetti normativi e dell'ethos scientifico, trascurando quelli organizzativi e proprio per questo forse si può dire che si sia occupata di quella che viene comunemente chiamata "ricerca pura". Merton ed epigoni, infatti, hanno concentrato la propria analisi non soltanto sul patrimonio assiologico della scienza, ma anche sull'attribuzione di credito scientifico all'interno delle comunità professionali e sui meccanismi di scambio e di controllo formale. In tal modo è stata operata una sorta di ipostatizzazione della categoria della "ricerca pura", in una specie di dualismo pressoché indissolubile tra ricerca pura e ricerca applicata, per il tramite di una fissazione di un sistema di riferimento privilegiato quale è quello della scienza "pura", che si adotta come parametro universale ed apodittico per la docimologia della ricerca scientifica e la valutazione della "normalità" o meno della ricerca stessa e in questo lo stesso Rino in più occasioni e in varie forme è intervenuto.

Tale operazione di "distinguo" ha incoraggiato quindi l'attitudine a separare ricerca pura e Università, quale sua sede privilegiata, da un lato ricerca applicata ed altri enti extra-universitari che la possono portare avanti, dall'altro. Ciò comporta quindi l'isolamento teorico dell'Università e della c.d. "ricerca accademica" rispetto alle reali esigenze della società civile; invece occorre rilevare, come è stato notato, come le Università spesso di fatto portino avanti ricerche applicate, tecnologiche e di sviluppo, e come, dall'altra parte, spesso, le agenzie pubbliche di ricerca applicata (il CNR in Italia, il CNRS in Francia e il DFG – *Deutsche Forschung Gemeinschaft* – in Germania) o le aziende private sovente siano committenti (quando non le conducano in proprio) di ricerche che, *prima facie*, potrebbero essere connotate come "pure", in quanto non immediatamente utilizzabili nella prassi. In realtà, ne abbiamo parlato spesso con Rino, quando si deve valutare la ricerca scientifica, ci si trova di fronte a un complesso universo di variabili di carattere strutturale, individuale ed epistemologico che rendono difficile il non considerare come intrecciati tra loro i fattori sociali, quelli cognitivi e quelli organizzativi.

Ciò sta a significare, quindi, che la vecchia dicotomia aperta dalla *querelle* mertoniana non può più reggere, ma, d'altra parte, anche le visioni oltranzistiche che propongono la "salsa verde" di un'attività di ricerca tutta omogenea, rischiano di imbastardire il concetto di ricerca scientifica, che si sottrarrebbe, in questo modo, ai canoni di una rigorosa definizione. Da un punto di vista diacronico, anche il sempre maggior intreccio dei programmi di ricerca dei Paesi industrialmente avanzati a livello internazionale ha condotto negli ultimi decenni a convergere su una definizione convenzionale della ricerca scientifica.

La stessa OECD, fin dal giungo del 1963, nel *Manuale di Frascati*, puntualizzava l'esigenza di integrare la ricerca scientifico-tecnologica e di poter esprimere valutazioni sulle organizzazioni deputate alla ricerca soprattutto in merito alla loro, oggi diremmo, "produttività". Dopo il *Manuale di Frascati* altre organizzazioni che hanno proposto definizioni ufficiali della ricerca scientifica sono:

1. il CREST (*Comité pour la recherche scientifique et technologique*) che ha pubblicato il NABS (*Nomenclature pour l'Analyse et la Comparaison des budgets et programmes scientifiques*, Bruxelles, 1975);

2. l'UNESCO che ha pubblicato *Considerations on the International Standardization of Science Statistics*, Paris, 1972;

3. il Nordforsk, che ha pubblicato *Handbok för Fou Statistik* (Manuale di Statistica della ricerca e sviluppo), Stockholm, 1974.

Lo stesso *Manuale di Frascati* è stato più volte aggiornato ed è forse ancora il più idoneo a dare definizioni più comprensive e nello stesso tempo rigorose dei vari tipi di ricerca, per poter costruire una tipologia sulla base di determinati criteri. È chiaramente da notare che si possono anche adottare altri criteri e in questo caso muterebbe la morfologia tipologica. Esso propone una distinzione fondamentale tra: 1) ricerca fondamentale; 2) ricerca applicata; 3) ricerca di sviluppo sperimentale. Questi tre tipi di ricerca sono ascrivibili alla categoria più generale della "Ricerca e Sviluppo", che tende a denotare le attività coordinate tese ad accrescere il bagaglio delle conoscenze (comprese le conoscenze ascrivibili al campo delle scienze umane, culturali e sociali) e l'utilizzabilità di tali nuove conoscenze a fini applicativi; i tre

tipi di attività, si dice, possono essere condotti nella stessa organiz-
zazione/istituzione e dallo stesso personale di ricerca; e questo è un
punto importante sul piano dell'innovazione.

La ricerca fondamentale è composta di lavori sperimentali e/o teorici che si compiono per avere nuove conoscenze sui c.d. fondamenti
dei fenomeni e dei fatti osservabili, senza che a monte del programma sia prevista o sia prevedibile un'applicazione ovvero un particolare
utilizzo; perciò le proprietà, le strutture e le relazioni poste sotto di-
samina sono tese a formulare e a controllare ipotesi, leggi e teorie. I
risultati non sono oggetto di negoziazione e il flusso informativo dei
risultati si ha per il tramite di pubblicazioni o di input-output diretto tra gli organismi di ricerca o i ricercatori interessati. La ricerca
fondamentale è quindi portata avanti da ricercatori, che possono decidere essi stessi gli obiettivi della ricerca e quindi organizzarsi con
ampio margine di discrezionalità; la ricerca fondamentale può anche
essere definita "orientata", dove fosse effettivamente orientata a dare
risposte a domande provenienti da grandi ambiti di interesse generale. E questo rappresenta un punto di approdo comune con Rino; ne
abbiamo in effetti molte volte parlato proprio a partire da un'intesa
sui cardini fondamentali della ricerca di base.

Infine, *last but not least*, mi si permetta di ricordare lo spessore
particolare del lato più empatico di Rino; una persona ricca di umanità e di intelligente vivacità che ha saputo costruire con Ivana (cara e
stimata studiosa e per me compagna di lotte da sempre) un vero modello di coppia di intellettuali "organici" riferita tale espressione alla
possibilità di una riqualificazione del felice concetto gramsciano alla
luce del presente; di un presente che muove dalla fine del secolo scorso
al difficile passaggio di secolo e di millennio, per vivere i primi ormai cinque lustri di questo secolo nelle difficoltà di una tremenda crisi
economica, della pandemia, delle nuove guerre in cui il tutto si regge
solamente nella capacità di mantenere lucida l'analisi e nell'impegno
politico, soprattutto nei termini della trasformazione della più diffusa
rabbia sociale in salutare conflitto e nella possibilità di tradurre le nostre categorie analitiche nel rinnovamento della cassetta intellettuale
degli attrezzi. In tale contesto la dinamica del tandem e della coppia

culturalmente impegnata fa da sfondo esemplare anche al modello di un progredito rapporto uomo-donna nella società borghese e all'unico modo di poter concepire la famiglia come struttura affettiva aperta, nell'impegno politico, sociale e culturale e non come monade chiusa in un'atavica concezione della stessa. Quella concezione che invece si perde troppo facilmente oggigiorno in una nuova fallimentare edizione del vecchio familismo amorale così come individuato inizialmente dagli antropologi.

Vedere oggi Ivana e Rino nel loro sempreverde spessore esistenziale fa ben capire come si possa veramente vivere insieme la gioia della conoscenza nelle lotte per una società più giusta ed eguale.

Scienze Sociali e ICT: Innovazione al Servizio dell'Umanità

Salvatore Capasso

Direttore di Dipartimento, CNR-DSU, Piazzale Aldo Moro, 7 00185 Roma

salvatore.capasso@cnr.it

L'interazione tra le scienze umane e sociali e le tecnologie dell'informazione e della comunicazione (ICT) rappresenta un terreno fertile di ricerca e sviluppo, che mira a garantire che l'innovazione digitale non solo migliori la produttività delle singole imprese e del sistema economico nel suo complesso, ma contribuisca anche al benessere della società in modo equo e sostenibile. Questa impostazione di studi si basa sull'assunto che la tecnologia non possa svilupparsi isolatamente, ma debba tenere conto dei fattori umani e sociali che ne influenzano l'adozione e l'efficacia. Eppure non sempre è pacificamente accettato che le scienze dure e lo sviluppo di nuove tecnologie debbano avere un'intersezione con le scienze umane e sociali e che obiettivi espliciti di miglioramento del contesto economico-sociale siano prioritari. Questo è particolarmente vero nell'approccio utilizzato da molti ricercatori e studiosi proprio nel campo delle scienze dure.

Pur essendo un fisico di formazione e, quindi, impegnato in attività di ricerca nell'alveo delle scienze informatiche, Rino Falcone ha da sempre dato centralità ai risvolti delle sue ricerche sul piano sociale. L'idea di fondo che traspare dal suo impegno è che lo sviluppo di nuove tecnologie e la loro applicazione non possa prescindere dal risvolto che queste hanno sul benessere della collettività. In questo senso, la sua attività di ricerca ma anche la sua attività di direzione di gruppi di ricerca e dell'Istituto di Scienza e Tecnologia della Cognizione (ISTC) è stata fortemente diretta a favorire la ricerca interdisciplinare e in particolare alla commistione tra scienze dure e scienze sociali.

Ho avuto modo di interagire con Rino in numerose occasioni sia per la definizione di attività di ricerca e progettuali, sia per la risoluzione di problemi amministrativi o gestionali. In tutte le circostanze, anche di fronte a problemi di natura puramente tecnica, Rino ha sottolineato l'importanza di considerare l'impatto sociale delle soluzioni proposte, insistendo sul fatto che una tecnologia non è mai neutra, ma deve essere valutata anche rispetto alle sue conseguenze sulla vita delle persone. La sua visione si riflette chiaramente nell'interazione dell'ISTC con gli altri istituti del Dipartimento di Scienze Umane e Sociali, Patrimonio Culturale (DSU), con cui ha sempre cercato sinergie. In questa prospettiva, come direttore del DSU ho trovato un grande alleato in termini di obiettivi da raggiungere e di costruzione di un ambiente sinergico e fortemente interdisciplinare. Rino ha, infatti, promosso con entusiasmo numerosi progetti di ricerca congiunti, convinto che solo integrando discipline diverse si potessero ottenere risultati che avessero un reale impatto positivo sul benessere sociale e sulle dinamiche umane. Questa collaborazione interistituzionale ha permesso di sviluppare approcci innovativi e multidisciplinari, rafforzando il legame tra tecnologia e umanità, e dimostrando come la scienza possa e debba essere al servizio della società.

Nel corso degli ultimi decenni, il progresso tecnologico ha trasformato profondamente il modo in cui viviamo, lavoriamo e interagiamo. L'introduzione di nuove tecnologie digitali, come l'intelligenza artificiale (IA), i big data e l'internet delle cose (IoT), ha creato nuove opportunità ma anche sfide senza precedenti. Se da un lato l'informatica e le scienze ingegneristiche rappresentano la spina dorsale di queste innovazioni, dall'altro la complessità delle dinamiche umane e sociali non può essere trascurata nello sviluppo e nell'implementazione di tali tecnologie. Solo con l'integrazione delle scienze umane e sociali, il processo di progettazione tecnologica può essere efficacemente incentrato sugli utenti ed essere socialmente sostenibile.

Un elemento chiave nel successo delle nuove tecnologie basate sulla innovazione digitale è, infatti, una ricerca che nasce dall'interdisciplinarietà in senso ampio. Troppo spesso, in altri campi le tecnologie sono sviluppate in isolamento rispetto ai bisogni e alle dinamiche sociali che andranno a influenzare. Tuttavia, per creare soluzioni che rispondano realmente ai bisogni umani, è essenziale che discipline come l'informatica, l'intelligenza artificiale e l'ingegneria lavorino a stretto contatto e interagiscano con le scienze organizzative, di gestione della conoscenza e con le scienze sociali in genere. Queste ultime, infatti, forniscono una comprensione approfondita del comportamento umano, delle interazioni sociali e delle dinamiche culturali che influenzano l'adozione e l'uso delle tecnologie.

Uno degli ambiti in cui la collaborazione interdisciplinare sta emergendo con forza è la progettazione di interfacce uomo-macchina funzionali ed efficienti. Le scienze cognitive, psicologiche e sociali offrono preziosi strumenti per comprendere come le persone percepiscono e interagiscono con le macchine, migliorando così l'efficacia delle soluzioni tecnologiche proposte. Quando queste conoscenze vengono integrate nelle prime fasi di sviluppo, le tecnologie risultano più intuitive, accessibili e, soprattutto, in grado di rispondere ai bisogni reali degli utenti. In questo processo di sviluppo, risulta essenziale non solo comprendere le dinamiche dell'interazione tra uomo e macchina, ma anche garantire che tali tecnologie siano accettate e utilizzate in modo efficace. Un aspetto chiave di questa accettazione è la fiducia che le persone ripongono nelle macchine e nei sistemi intelligenti con cui interagiscono. Alla fin fine le tecnologie vanno implementate e utilizzate e la fiducia è un elemento cruciale nell'interazione tra uomo e macchina, specialmente nel contesto delle tecnologie basate sull'intelligenza artificiale. Senza fiducia, le persone saranno riluttanti ad adottare nuove tecnologie o a utilizzarle in modo efficace. È quindi fondamentale capire come costruire fiducia nei sistemi intelligenti e autonomi, affinché questi possano essere pienamente integrati nelle attività quotidiane e professionali.

La fiducia, tuttavia, non si costruisce solo attraverso l'affidabilità tecnica del sistema, ma anche attraverso la trasparenza, la comunicazione e il rispetto delle aspettative umane. Sul ruolo della fiducia nel processo di implementazione delle tecnologie digitali e sul rapporto uomo-macchina, Rino Falcone ha dedicato gran parte della sua attività di ricerca, e di questo abbiamo spesso discusso. Rino ha approfondito il ruolo cruciale della fiducia nell'interazione tra persone e sistemi autonomi, offrendo una chiara spiegazione e proponendo una teoria completa della fiducia, radicata in approcci propri delle scienze sociali abbracciando non solo gli ambiti cognitivi e culturali ma anche quelli istituzionali e normativi [1, 2, 3]. Di fatto le scienze sociali ci insegnano che la fiducia è un fenomeno relazionale che si sviluppa nel tempo, basato su esperienze pregresse e sulla percezione delle intenzioni altrui. In questo senso, l'intelligenza artificiale e le macchine devono essere progettati in modo tale da fornire non solo soluzioni tecniche avanzate, ma anche interazioni che rassicurino gli utenti sulla sicurezza, l'eticità e la coerenza dei loro comportamenti.

Uno degli approcci più promettenti è lo sviluppo di sistemi che siano in grado di adattarsi ai modelli di comportamento umani, comunicando chiaramente il proprio funzionamento e rispettando i valori e le norme sociali del contesto in cui operano. Ad esempio, i robot collaborativi (cobot) stanno trovando un crescente impiego nelle fabbriche moderne, dove lavorano fianco a fianco con gli operai. Il loro successo dipende non solo dalla loro abilità tecnica, ma anche dalla capacità di interagire in modo sicuro ed efficiente con i lavoratori umani, guadagnando la loro fiducia.

Oltre alla fiducia, le questioni etiche occupano un posto centrale nel dibattito sullo sviluppo dell'intelligenza artificiale e delle nuove tecnologie digitali. Le decisioni prese da sistemi artificiali, infatti, possono avere un impatto diretto sulla vita delle persone, sollevando interrogativi importanti su equità, trasparenza e responsabilità. Le scienze umane e sociali sono essenziali per delineare un perimetro di riferimento nell'affrontare queste sfide tenendo conto di valori quali l'equità, la giustizia sociale e il rispetto dei diritti umani.

L'implementazione di tecnologie senza una solida riflessione etica può condurre a gravi conseguenze. Si pensi agli algoritmi di intelligenza artificiale utilizzati nel settore del reclutamento del personale o nella giustizia penale. In questi ambiti, una raccolta dati sbilanciata o omissiva e modelli interpretativi e applicativi rigidi possono perpetuare pregiudizi esistenti o introdurre nuovi bias. Per evitare tali rischi, è necessario che le fasi di sviluppo e implementazione delle tecnologie includano una valutazione approfondita degli impatti etici, sociali e culturali. Questa analisi dovrebbe coinvolgere non solo ingegneri e tecnologi, ma anche esperti di etica, filosofi e sociologi, per garantire che i sistemi sviluppati siano equi e rispettosi dei valori umani.

C'è, inoltre, un'altra questione che emerge proprio dal nuovo e più profondo contenuto tecnologico delle innovazioni digitali più recenti e che esige una nuova interpretazione filosofica e giuridica. La crescente somiglianza delle macchine agli esseri umani, sia in termini di capacità che di comportamento, può portare a un fraintendimento del ruolo etico e delle responsabilità ad esse associate. Da un lato, l'evoluzione tecnologica sembra richiedere una riflessione specifica sulle responsabilità e sui comportamenti che le macchine dovrebbero adottare, soprattutto quando agiscono in autonomia. Dall'altro, sul piano teorico, c'è ancora molto da esplorare: le macchine, per loro stessa natura, non sono soggetti morali e quindi non possono essere direttamente responsabili delle loro azioni. La questione di chi dovrebbe essere ritenuto responsabile – progettisti, programmatori, utilizzatori o le stesse istituzioni che ne regolamentano l'uso – è ancora irrisolta. Questa dinamica è pericolosa, poiché si rischia di spostare la responsabilità morale e decisionale dall'essere umano alla macchina, una transizione che può apparire inconsapevole ma che ha profonde implicazioni.

L'innovazione tecnologica, per essere davvero efficace e responsabile, deve allinearsi ai principi dello sviluppo sostenibile, che va oltre la semplice crescita economica e tiene conto delle dimensioni ambientali, sociali ed etiche. In questo contesto, il ruolo delle scienze umane e sociali diventa essenziale. Le scienze sociali, ad esempio, aiutano a comprendere come la tecnologia impatta sulle comunità, le dinamiche lavorative e i comportamenti collettivi, mentre le scienze umane offrono strumenti per riflettere sulle questioni etiche, filosofiche e culturali legate all'adozione di nuove tecnologie. Integrare queste prospettive significa non solo promuovere tecnologie più accettabili dal punto di vista sociale, ma anche prevenire potenziali rischi, come la creazione di disuguaglianze o la perdita di controllo sui processi decisionali da parte dell'uomo. La sostenibilità non è solo una questione ambientale, ma riguarda anche il modo in cui le innovazioni tecnologiche possono migliorare la qualità della vita, rispettare la dignità umana e rafforzare i legami sociali. Senza una profonda comprensione delle dimensioni umane e sociali, l'innovazione rischia di essere miope e di generare più problemi di quanti ne risolva. Inoltre, le scienze umane e sociali possono fungere da ponte tra innovatori e utenti finali, facilitando il dialogo e la comprensione reciproca. In questo modo, l'innovazione può essere guidata non solo da criteri tecnici ed economici, ma anche da valori umani, rendendo i processi di sviluppo tecnologico più inclusivi, etici e orientati al bene comune.

In definitiva, lo sviluppo e l'implementazione delle nuove tecnologie digitali non possono prescindere dall'integrazione delle scienze umane e sociali. L'approccio interdisciplinare consente non solo di migliorare l'efficacia tecnica delle soluzioni proposte, ma anche di garantire che esse siano in linea con i bisogni, i valori e le aspettative della società. La fiducia, le considerazioni etiche e la sostenibilità sono elementi centrali per garantire che l'innovazione tecnologica porti benefici diffusi e duraturi. Solo attraverso una collaborazione continua tra scienze tecniche e scienze umane, sarà possibile costruire un futuro tecnologico equo e sostenibile, che risponda realmente alle esigenze delle persone e delle comunità.

Intersecare le scienze dure con le scienze umane e sociali è un'impresa che richiede una visione chiara, una capacità di guardare oltre i confini disciplinari, e un'energia straordinaria. Questo tipo di approccio multidisciplinare non è per tutti, ma è una sfida che certamente ha chi possiede la determinazione, la resistenza e il coraggio di attraversare a nuoto lo Stretto di Messina. Proprio come quella traversata, dove si combattono correnti contrarie e si richiede uno sforzo continuo, integrare discipline così diverse implica il superamento di barriere culturali, concettuali e metodologiche, richiedendo un impegno costante e la capacità di trovare punti di convergenza tra mondi apparentemente distanti. È un atto di volontà, ma anche di grande lungimiranza, per costruire un sapere che sia davvero al servizio della collettività e capace di rispondere alle complesse sfide del presente e del futuro.

References

[1] Castelfranchi, C., & Falcone, R. (2010). Trust theory: A socio-cognitive and computational model. John Wiley & Sons.

[2] Falcone, R., & Sapienza, A. (2020). Trust and autonomy for regulating the users' acceptance of IoT technologies. Intelligenza Artificiale, 14(1), 45-58.

[3] Falcone, R., & Castelfranchi, C. (2022). Grounding Human Machine Interdependence Through Dependence and Trust Networks: Basic Elements for Extended Sociality. Frontiers in Physics, 10, 946095.

Consigliere Rino Falcone e il ruolo chiave del Ministero dell'Università e della Ricerca (2006-2008)

Francesco Profumo
Professore Emerito presso il Politecnico di Torino, già Presidente del Consiglio Nazionale delle Ricerche (CNR), già Ministro della Repubblica Italiana

1 Introduzione

Nel periodo tra il 2006 e il 2008, il consigliere Rino Falcone ha giocato un ruolo centrale nelle politiche per l'Università e la Ricerca in Italia, sotto la guida del Ministro Fabio Mussi e del Presidente del Consiglio Romano Prodi. Questo fu un periodo cruciale per il nostro Paese, in cui le strategie europee, in particolare la Strategia di Lisbona e il Settimo Programma Quadro (FP7), funsero da cornice per definire gli indirizzi nazionali. Il governo italiano si trovò a confrontarsi con i principali attori europei – Francia, Germania, Regno Unito e Spagna – nel tentativo di rilanciare l'istruzione superiore e la ricerca, settori essenziali per lo sviluppo competitivo a livello internazionale. Io ero Rettore del Politecnico di Torino e ho vissuto intensamente questa fase storica per il nostro Paese.

2 La Strategia di Lisbona e il FP7: il contesto europeo

Nel 2000, la Strategia di Lisbona fu adottata con l'obiettivo di fare dell'Unione Europea "l'economia basata sulla conoscenza più dinamica e competitiva del mondo" entro il 2010. Gli obiettivi principali della strategia erano favorire la crescita economica, creare occupazione e incrementare l'innovazione. All'interno di questa visione, le Università e gli Enti di Ricerca rivestirono un ruolo cruciale, essendo considerati motori fondamentali per il raggiungimento di una società della conoscenza. Il Settimo Programma Quadro (FP7) rappresentava lo strumento finanziario più importante a disposizione dell'UE per la ricerca, con un budget complessivo di oltre 50 miliardi di euro, finalizzato a potenziare la capacità di innovazione dell'Europa. Per l'Italia, la partecipazione attiva a questi progetti era essenziale non solo per ottenere finanziamenti, ma per migliorare il posizionamento scientifico del Paese a livello internazionale.

3 Le politiche italiane per l'Università e la Ricerca

Sotto la guida di Fabio Mussi e con il determinante contributo di Rino Falcone, il Ministero dell'Università e della Ricerca adottò una serie di misure volte a migliorare la governance degli atenei e degli enti di ricerca, con l'obiettivo di renderli più autonomi e competitivi. Le azioni messe in campo si concentrarono su cinque aree principali:

1. **Autonomia**: promosse un ampliamento dell'autonomia degli atenei e degli enti di ricerca, introducendo maggiori libertà nella gestione delle risorse e nella programmazione didattica e scientifica. Questo movimento verso una maggiore indipendenza rappresentava un elemento centrale per incentivare l'innovazione, favorendo una più stretta collaborazione con il mondo imprenditoriale e l'accesso a fondi privati. Durante questo periodo, **il numero di progetti finanziati tramite partnership pubblico - private aumentò del 15%** rispetto al biennio precedente.

2. **Governance**: In ambito di governance, furono introdotte nuove regole per aumentare la trasparenza e la meritocrazia nella gestione delle istituzioni accademiche. L'obiettivo era migliorare

l'efficienza decisionale, potenziando il ruolo dei Consigli di Amministrazione e dei rettori, ma anche incentivare la partecipazione di rappresentanti del mondo produttivo all'interno dei consigli. **Il 30% degli atenei italiani introdusse membri esterni nel Consiglio di Amministrazione**, una pratica che fu pioniera per migliorare il legame con il tessuto economico locale e nazionale.

3. **Finanziamenti:** Il governo cercò di diversificare le fonti di finanziamento per l'università e la ricerca, stimolando il settore privato a contribuire maggiormente. Ciò passava anche attraverso incentivi fiscali per le aziende che investivano in ricerca e sviluppo. Nonostante un contesto economico difficile, **i finanziamenti alla ricerca universitaria aumentarono del 5% tra il 2006 e il 2008**, portando il totale a circa **7,2 miliardi di euro**, un incremento significativo rispetto agli anni precedenti.

4. **Personale:** Un'altra area strategica riguardò la gestione del personale accademico e dei ricercatori. Durante questo periodo, furono introdotti meccanismi per agevolare il ricambio generazionale. **Il numero di ricercatori a tempo determinato crebbe del 12%**, grazie all'introduzione di politiche di reclutamento più flessibili e mirate. Tuttavia, la questione del precariato accademico rimase rilevante, con oltre **40.000 ricercatori precari** che attendevano stabilizzazione.

5. **Valutazione:** Un elemento di forte innovazione fu l'introduzione di nuovi sistemi di valutazione delle Università e degli Enti di Ricerca. L'obiettivo era monitorare costantemente la qualità della didattica e della ricerca, con parametri basati su criteri internazionali. La nascita dell'ANVUR (Agenzia Nazionale di Valutazione del Sistema Universitario e della Ricerca) fu uno degli atti chiave in questa direzione. Nel periodo 2007-2008, **oltre il 70% degli atenei italiani implementò sistemi di autovalutazione** per migliorare la qualità della ricerca e dell'insegnamento.

4 Confronto con le politiche degli altri Paesi europei

Nel confronto con le politiche di altri Paesi europei, l'Italia si trovava di fronte a sfide specifiche. Francia, Germania, Regno Unito e Spagna avevano già avviato riforme significative per potenziare i loro sistemi universitari e di ricerca. La Francia, ad esempio, aveva introdotto il progetto Campus Plan, con investimenti pari a **5 miliardi di euro** per modernizzare le infrastrutture universitarie, mentre la Germania, attraverso l'iniziativa Exzellenzinitiative, stava investendo **1,9 miliardi di euro** per creare poli di eccellenza. Il Regno Unito, a sua volta, spingeva verso finanziamenti competitivi basati su criteri di eccellenza, mentre la Spagna aumentava i finanziamenti statali del **7%** annuo per promuovere la ricerca e l'innovazione. In questo scenario, l'Italia, pur muovendosi verso una maggiore competitività, si trovava ancora in ritardo rispetto agli altri Paesi. **Gli investimenti pubblici in ricerca e sviluppo in Italia rappresentavano solo lo 0,8% del PIL**, contro l'1,9% della Germania e l'1,7% del Regno Unito. Tuttavia, grazie alla partecipazione crescente ai progetti europei, il Paese iniziò a recuperare terreno, soprattutto nelle aree scientifiche di eccellenza.

5 Le azioni chiave e i risultati ottenuti

Tra le azioni più rilevanti promosse Mussi e Falcone, quelle che ebbero maggiore impatto includono l'ampliamento dell'autonomia delle università, l'introduzione di sistemi di valutazione e la promozione di politiche di internazionalizzazione. In particolare, **l'aumento del numero di progetti italiani finanziati dal FP7 fu del 18%** tra il 2007 e il 2008, portando l'Italia ad ottenere circa **2 miliardi di euro** in finanziamenti europei per la ricerca. Questo incremento rappresentò un segnale positivo per il sistema universitario italiano, che iniziava a competere con successo a livello internazionale. In termini di governance, **il 45% delle università italiane adottò nuove forme di organizzazione manageriale** per migliorare l'efficienza operativa, e oltre **il 60% degli atenei italiani partecipò a consorzi internazionali di ricerca** durante il periodo 2006-2008.

6 L'eredità delle politiche di Mussi e Falcone e l'impatto attuale

A distanza di oltre un decennio, molte delle politiche promosse da Mussi e Falcone continuano a influenzare il sistema universitario italiano. L'introduzione di criteri di valutazione rigorosi è diventata una prassi consolidata, e la crescente autonomia delle Università ha permesso loro di muoversi in modo più flessibile all'interno di un contesto globale in costante cambiamento. Allo stesso tempo, l'Italia è riuscita ad aumentare il numero di ricercatori e progetti finanziati a livello europeo, migliorando la propria presenza scientifica internazionale. Nonostante i passi avanti, permangono ancora diverse criticità: il problema del finanziamento rimane attuale, così come la gestione del precariato e la necessità di un più incisivo ricambio generazionale. Tuttavia, le politiche di Mussi e Falcone hanno tracciato una direzione che ha contribuito a rafforzare la presenza dell'Italia nell'arena scientifica europea, ponendo le basi per le sfide future. In definitiva, il periodo 2006-2008, pur non essendo esente da difficoltà, segnò un momento cruciale per le Università e gli Enti di Ricerca italiani, con scelte strategiche che ancora oggi influenzano il panorama accademico nazionale.

PART

AFTERWORDS

LE PROMOZIONI INASPETTATE: QUANDO SCIENZA E SOCIETÀ TI SORPRENDONO
UNEXPECTED PROMOTIONS: WHEN SCIENCE AND SOCIETY SURPRISE YOU

RINO FALCONE

Director of the Institute of Cognitive Sciences and Technologies,
National Research Council of Italy (ISTC-CNR), 00196 Rome, Italy
`rino.falcone@istc.cnr.it`

1 Introduzione

In questo capitolo riporterò sensazioni, percezioni e considerazioni che la sequela di contributi di questo volume, dedicatimi in occasione del pensionamento, ha scatenato in me: nella mia mente e nella mia anima, ossia nelle forme razionali e affettive del mio pensiero.

E' però bene che prima lo dica chiaro: la forzosa ed inevitabile "collocazione a riposo", mi si è paventata come un inconciliabile e prematuro traguardo. Anche con tratti di ingiustizia: perché professori universitari, medici, magistrati, etc. vengono pensionati a settant'anni e noi a sessantasette?

Abbiamo persino cercato di promuovere politicamente, con colleghi solidali come Daniele Archibugi, iniziative legislative di recupero. Nulla è stato possibile.

Ovviamente non tutti la pensano così.

Molto dipende da quanta parte del proprio essere e sentire rappresenti l'attività lavorativa che si esercita. Quale strumento di esplorazione del mondo e di se stessi essa determini. E quanto ancora ci si senta attivi nel percorso di costruzione della propria opera: nel caso particolare che mi riguarda, in riferimento alla conoscenza e alla ricerca su cui si indaga.

Esistono è vero, anche per non disperdere un patrimonio variamente qualificabile, delle "reti di recupero" di questo ruolo (associarsi al CNR, possibilità di contratti di insegnamento con università private, e così via), ma la cesura che concettualmente si determina con il pensionamento sottolinea un limite cui non ci si sente adeguati. A me questo è successo.

In realtà il mio pensionamento è arrivato durante l'incarico di direzione di ISTC-CNR e, dato che la procedura per la selezione del nuovo direttore risultava, nell'avvento del pensionamento, in itinere, il CNR mi ha proposto di proseguire in questa attività di direzione, che al momento di pubblicazione di questo volume è ancora in corso, con un contratto di diritto privato (non retribuito). Oltretutto in un periodo straordinariamente ricco di scadenze e di cambiamenti gestionali del CNR (cambio di contabilità, esercizio della VQR, etc.): con una conseguente richiesta di impegno e attività tutt'altro che riducentesi. Mi trovo quindi nella curiosa posizione di essere in pensione ma iperattivamente operativo su richiesta dell'Ente.

La generosa e affettuosa iniziativa di Alessandro, Filippo, Fabio e Luca per la stesura di questo volume rappresenta di fatto un risarcimento inatteso rispetto al profondo disagio di questo cambio di prospettiva. Iniziativa che mi ha sorpreso per alcuni aspetti e commosso per altri e di cui sono grato ben oltre quello che ho direttamente espresso loro. Così come grato ed empaticamente riconoscente sono a tutti quanti hanno offerto un loro omaggio, esprimendo spesso giudizi nei miei confronti che pur non sapendo quanto meritare, mi piace prenderne atto e considerarli sinceri.

Spenderò per questa ragione poche ma, spero, significative parole per ciascuno di loro.

2   Contributi strettamente scientifici

A **Munindar Singh** mi lega una lunga e intensa esperienza di collaborazione nei workshop su Trust che si sono ripetuti presso la conferenza di AAMAS (Autonomous Agents and Multi-Agent Systems).

E' stata questa conferenza il risultato del nuovo approccio scientifico ai sistemi di Intelligenza Artificiale. Approccio in cui si abbandonava la visione dell'intelligenza centrata su un singolo agente, sostituendola con l'interazione e la relazione tra più agenti (AI distribuita). Il workshop su trust in particolare si è focalizzato negli anni su questa proprietà/attitudine dell'intelligenza (la fiducia per l'appunto), così essenziale per l'evoluzione di sistemi efficacemente collaborativi e in grado di individuare soluzioni ed equilibri avanzati nei sistemi costituiti da agenti autonomi e cognitivi.

Nel suo bel pezzo (*Social Artificial Intelligence and the Interplay of Norms, Values, and Trustworthiness*) Munindar nel contesto di riconoscimenti al lavoro sulla fiducia che abbiamo svolto con Cristiano Castelfranchi negli anni, individua un elemento di criticità, facendo riferimento ad un articolo scritto con Cristiano e Francesca Marzo. L'obiezione che egli stesso rivela come in contraddizione con la nostra più vasta teoria, è di considerare la affidabilità come una proprietà intrinseca del fidato. In realtà, la sua critica è assolutamente corretta: ci sono elementi di dinamicità sostanziali nella affidabilità di qualunque trustee che sia capace di correlarsi con il contesto in cui opera e con il fidante con cui interagisce (oltre che con la propria storia e situazione interna contingente). Ha ragione Munindar! Ma in quell'articolo si provava a approssimare questo dinamismo e renderlo costante con quelle condizioni al contorno che si volevano rappresentare. Comunque Il pezzo di Munindar è gustoso e intelligente. L'approfondimento sulle relazioni strette tra *norme, fiducia e valori* rappresenterà certamente una delle frontiere per trasferire correttamente e proficuamente gli elementi essenziali dell'intelligenza sociale ai sistemi artificiali misti.

Un contributo di analogo interesse a quello di Singh data la relazione fiducia/norme sociali, è stato realizzato da **Mario Paolucci** e **Giulia Andrighetto** (*Trust, Reputation, and Norms: A Simulative Approach to Cognitive Artifacts*). Gli artefatti cognitivi che questi ricercatori decidono di mettere in relazione con la *fiducia* sono per l'appunto le *norme sociali* e la *reputazione*. In effetti la rilevanza di questo capitolo sta anche nel sottolineare e ricostruire il ruolo fondamentale che ISTC ha svolto in quella che essi definiscono "Theoretical Cyber-Psychology" (TCP) dove cioè la manipolazione simbolica costruisce il ragionamento. Non può restare senza risposta inoltre l'appello che viene rivolto a simulare di più e meglio la fiducia in relazione a norme e reputazione. Non solo simulare ma anche testare nel mondo reale i modelli teorici sviluppati. Un appello ad uscire con maggiore determinazione dal mondo della teoria e trasformare le acquisizioni in strumenti di indagine pratica e fruttuosa della TCP. Pur avendo attitudini al ragionamento teorico maggiori di quello alla sperimentazione pratica (su questo la condivisione con Cristiano Castelfranchi è totale), ritengo che questa valutazione non solo è corretta ma che vada perseguita con serietà e approfondimento. Grazie Mario e Giulia!

Il capitolo di **Anna Borghi** e **Luca Tummolini** (*Semantic Delegation: Understanding Concepts With A Little Help From Our Friends*) esplora in forme originali e per certi versi intriganti una dimensione dell'approccio socio-cognitivo della fiducia e della delega quando applicate al senso delle parole, ossia alla semantica. Lo sviluppo degli argomenti è coerente e particolarmente elegante, oltre che convincente. La delega (e la fiducia che ne è alla base) viene vista come un meccanismo sociale per acquisire e padroneggiare conoscenze complesse. Particolarmente attuale è come queste considerazioni possano rappresentare uno strumento per la comprensione dei modelli di intelligenza artificiale generativa (LLM). La conclusione a cui giungono Anna e Luca sembra preconizzare la necessità di un ruolo sempre più centrale nel futuro per quei concetti (fiducia, delega, autonomia) su cui, con Cristiano, stiamo da tempo elaborando teorie e ipotesi. E si apre uno spazio interessante di collaborazione con loro che consideriamo un grande regalo e una straordinaria opportunità per il futuro.

Il contributo di **Valeria Seidita** e **Antonio Chella** (*From Trust Theory to Explainable Agents: Enhancing Agent-Based Systems*) si muove nell'ambito della cosiddetta Human-Robot Teaming Interaction (HRTI), ossia quella disciplina che ambisce a sviluppare e studiare la collaborazione tra esseri umani e robot in un contesto di squadra. Valeria e Antonio si concentrano in particolare sulla necessità di dotare agenti e robot con capacità cognitive avanzate, in particolare le capacità di auto-modellare e giustificare le azioni. Riferendosi al modello della fiducia da noi sviluppato e integrandolo nel ciclo BDI, essi mostrano in questo contributo come abbiano creato agenti artificiali in grado di comprendere e spiegare le proprie azioni e di adattarsi in ambienti sconosciuti e dinamici. Inoltre, dotando questi agenti con la capacità di gestire il proprio linguaggio interno (inner speech), sembrano avviarsi verso la generazione di agenti particolarmente evoluti. Con Valeria e Antonio sembrano quindi prospettarsi promettenti possibilità di riprendere il filo di una collaborazione già molto produttiva negli anni passati.

Il capitolo dedicatomi da **Elisa Colì** (*Trusting in the Online World: From Theoretical Studies to the Application of the Socio-Cognitive Model*) riflette su una questione di grande rilevanza affrontata insieme in vari esperimenti e lavori pubblicati: che succede della fiducia, di questa nostra fondamentale attitudine sociale, quando le nostre relazioni si trasferiscono dal mondo reale a quello virtuale, per esempio, alle interazioni online? La collaborazione con Elisa è iniziata in anni recenti, immediatamente a ridosso della pandemia COVID, come lei stessa ricorda nel suo contributo. Ma è stata una splendida fucina di aggiornate conoscenze e di intensa collaborazione. Ed ancora è pregna di potenziali sviluppi anche grazie alle sue continue sollecitazioni e stimolanti idee. L'omaggio che mi dedica nell'introduzione rappresenta uno splendido esempio del rapporto che si può stabilire tra un ricercatore esperto ed un suo più giovane collaboratore. Come la forza del lavoro di ricerca possa produrre straordinari scenari di collaborazione e amicizia. Grazie Elisa!

Antonio Lieto, nel suo intervento (*Trust in the Rational and Social Bands of Cognitive Architectures*) sintetico ma efficace, sottolineando la rilevanza dei modelli di fiducia, propone alcuni indirizzi di ricerca nell'ambito delle architetture e dei modelli cognitivi. In particolare, partendo dalle bande della cognizione proposte da Alain Newell e analizzando le modalità architetturali che ne hanno visto l'applicazione, Antonio evidenzia lo scarso sviluppo dei modelli macro-cognitivi in queste architetture e preconizza l'utilità di guardare alle conoscenze acquisite sui modelli di fiducia per aggiornare nel futuro in modo più efficace ed adeguato tali architetture. Leggere questi apprezzamenti da un giovane ma già noto e affermato ricercatore, è per me un grande orgoglio.

Un tema di ricerca che ha attratto l'interesse mio e del gruppo di ricerca T^3 di ISTC e che merita considerevole attenzione nella prospettiva della nuova AI generativa, è quello relativo alle proprietà di adattamento del comportamento autonomo degli agenti alle differenti e mutevoli situazioni e ambienti. Questo tema viene per semplicità definito come "adjustable autonomy" e **Amedeo Cesta**, **Gabriella Cortellessa** e **Angelo Oddi** ne hanno preso spunto per il loro contributo (*Adjustable Autonomy in AI and Assistive Robotics: Experiences and Lessons Learned*). In questo capitolo essi presentano alcune sfide affrontate in questa prospettiva mostrando come l'appropriato livello di autonomia sia un corretto compromesso tra varie dimensioni e, quando ben esercitato/progettato, risulti determinante per l'efficacia e l'evoluzione delle interazioni tra agenti. Un particolare ringraziamento voglio dedicarlo ad Amedeo con cui abbiamo condiviso un lungo percorso di comuni esperienze e di dinamiche e dialogiche interazioni e di cui mi piace riconoscere la profonda professionalità, il forte senso di rispetto e appartenenza istituzionale e la sincera lealtà che lo caratterizza.

Uno dei grandi problemi che l'umanità si trova ad affrontare ogni volta che una rivoluzione tecnologica impatta nello sviluppo economico e sociale, riguarda la trasformazione del sistema produttivo e del mercato del lavoro. Con l'Intelligenza Artificiale e con lo sviluppo esponenziale delle sue applicazioni questa trasformazione si annuncia ancor più profonda e rapida. **Pierluigi Zoccolotti** e **Fabio Lucidi** con il loro contributo in questo volume (*How will artificial intelligence affect our lives? Some reflections in the areas of labor and health*) affrontano questo tema con la sagacia e l'autorevolezza della loro analisi, considerando alcuni dati già valutabili di questo fenomeno. L'obiettivo sullo sfondo che Pierluigi e Fabio si pongono è comunque quello di sottolineare, in coerente (e da loro richiamata) analogia con quanto mi è capitato di indicare assai spesso nella mia attività e prassi di ricercatore, come il ruolo della scienza e degli scienziati sia di stabilire un dialogo intenso con la politica e la società. Solo grazie a questo dialogo è possibile valutare per tempo alcune sfide e individuare percorsi sinergici risolutivi. Il confronto con Pierluigi e Fabio ci ha impegnati spesso in riflessioni e occasioni di confronto che hanno trasformato la nostra relazione professionale in vera e profonda amicizia. Un'amicizia che mi onora e mi gratifica.

Un tema straordinariamente importante in questa fase di sviluppo dell'AI generativa riguarda i rischi di sostituzione degli umani in una seria di attività che non sono solo di basso livello ma che riguardano alcune precipue attività intellettuali. **Fabio Paglieri** nel suo bel contributo al volume (*Trust, delegation, and cognitive diminishment: the case of generative AI*) decide di confrontarsi con questo problema per provare ad indicare se possiamo decidere di fidarci in questo senso della nuova Intelligenza Artificiale generativa e come, fino a che punto sia utile o deleterio. Affrontando il tema (fidarsi dell'AI generativa) con gli strumenti messi a disposizione dalle teorie sviluppate nel gruppo di ricerca T^3, Fabio mostra una notevole capacità e profondità di analisi. Ma dimostra, al contempo, come sia coerente e produttiva la costruzione di collaborazione scientifica e l'avanzamento condiviso di conoscenza. Il pezzo è, come spesso capita con Fabio, godibile, intelligente e acutamente argomentato. Fabio è anche tra gli editor di questo volume: il mio ringraziamento è doppio.

Il tema dell'Intelligenza Artificiale e dei suoi sviluppi e impatti è preso in considerazione anche nel contributo di **Vieri Giuliano Santucci** e **Giovanni Pezzulo** (*Harmonizing Artificial Intelligence, Minds, and Society: A Cognitive Science Perspective*). Il focus del loro capitolo è sugli approcci che la scienza può assumere per affrontare il problema fondamentale dell'integrazione dell'AI con mente e società. E' necessario coinvolgere le varie aree di ricerca della scienza cognitiva (psicologia, neuroscienze, robotica cognitiva, linguistica, intelligenza artificiale, antropologia e filosofia) per comprendere appieno e sviluppare al meglio come armonizzare L'AI con la cognizione e i soggetti collettivi costituiti nel genere umano. In tal modo è possibile focalizzare e chiarire più chiaramente questioni come autonomia, adattabilità

547

e interazione. Vieri e Giovanni sono studiosi con differente esperienza di ricerca: Vieri più giovane ma già brillante; Giovanni con intuito e valenza scientifica di altissimo e riconosciuto valore. A Giovanni mi lega una collaborazione intensa e proficua. Con Vieri alcuni tentativi non ancora completamente definiti (ma non dispero).

Jaime Simao Sichman può essere considerato a tutti gli effetti un ricercatore del nostro gruppo di ricerca in ISTC, pur essendo un professore dell'Università Brasiliana di San Paolo. Jaime ha infatti non solo svolto ricerca da dottorando presso il nostro istituto, ma ha assunto la filosofia dell'approccio socio-cognitivo all'intelligenza artificiale esprimendo la nostra tradizione nella comunità internazionale dei Multi-Agent Systems e Autonomous Agents. Nel suo contributo (*Reputation Interoperability in Multi-Agent Systems*) riassume i risultati di un certo numero di progetti di ricerca sviluppati negli ultimi 20 anni con al centro il tema della reputation, uno dei temi centrali dell'approccio socio-cognitivo alla Distributed Artificial Intelligence. Jaime è quindi non solo un valente scienziato internazionale ma una vera e propria estensione della ricerca ISTC nel mondo. Il suo unico difetto è di avere simpatie per una squadra di calcio della capitale, innominabile, di secondo piano, oltre che di squalificata reputazione. Più volte ho provato a redimerlo verso la grande e "maggica" Roma ma tutti i tentativi sono andati a vuoto. Ciononostante gli voglio bene!

Stefania Costantini e **Andrea Formisano** nel loro contributo (*Epistemic Logics Programs: a survey of motivations, recent contributions and new developments*) intendono offrire strumenti sia concettuali che pratici adatti ad indagare la semantica esistente e nuova dei programmi di logica epistemica (ELP). Secondo gli autori questi strumenti dovrebbero risultare utili sia ai ricercatori che agli utenti, facilitando un'esplorazione e una comprensione più approfondite in quest'area. La semantica degli ELP è fornita tramite metodi che caratterizzano le visioni del mondo, che possono essere diverse, dove ciascuna è un insieme di insiemi di risposte, al contrario di un singolo insieme di insiemi di risposte come nel più tradizionale Answer Set Programming (ASP). Stefania e Andrea propongono un metodo per la prototipazione rapida di un risolutore (motori di esecuzione) che fornisce le visioni del mondo di un dato programma per la classe di semantica in considerazione. Con Stefania mi lega una recente ma profonda condivisione delle problematiche scientifiche e di politica della scienza. Il mio ringraziamento per questo contributo è sincero.

I due principali ingredienti di base della attitudine mentale della fiducia sono gli aspetti motivazionali e quelli espistemici (conoscitivi). **Emiliano Lorini** nel suo contributo (*Should I Trust? On the Epistemic Foundation of Arguments*) indirizza su questi ultimi la sua indagine, analizzando il concetto di argomento che costituisce la base per costruire i fondamenti teorici di supporto alla fiducia. Emiliano è un esperto di modelli formali di interazione tra agenti cognitivi, utilizzando logica e game theory. In questo caso, per la formalizzazione del concetto di argomento utilizza una semantica di logica epistemica che si basa su credenze di base. In tal modo egli può rappresentare le ragioni per credere o non credere qualcosa. Emiliano, un po' come Jaime, nasce nella scuola di ISTC ed è legittimo considerarlo come una estensione del modello conoscitivo che ISTC ha esportato nella comunità scientifica internazionale.

Un problema controverso, affrontato anche da Cristiano e me in vari studi è quanto sia importante il comportamento trasparente per accrescere la fiducia interpersonale e sociale. **Aron Szekely** e **Luca Tummolini** (*When Transparency Shapes Trust: A Signalling Approach*) nel loro paper affrontano questo tema con arguzia e originalità, anche riferendosi ad importanti tendenze: dalle iniziative di Open Governance al movimento Open Science. Ovviamente, sostengono a ragione gli autori, il problema si sposta sui segnali che definiscono trasparenza e determinano l'affidabilità. La loro suggestione è che integrare Signalling Theory e Trust Theory possa essere il modo opportuno per affrontare seriamente il tema. Lo stimolo è attraente e gli argomenti portati assai convincenti. Sarebbe bello nascesse a breve una collaborazione con cui investigare questo affascinante problema.

La relazione tra fiducia e coordinazione e il ruolo essenziale che questa relazione può assumere per l'ingegneria dei sistemi intelligenti e in particolare di quelli cosiddetti socio-tecnici viene approfondita da **Andrea Omicini** e **Alessandro Ricci** (*Coordination and Trust in MAS Towards Intelligent Socio-technical Systems*). I due autori indagano alcune intersezioni di questi due concetti in particolare nei sistemi multi-agente e nei sistemi socio-tecnici. In particolare si focalizzano sulla gestione dell'interazione in sistemi intelligenti complessi e sulle questioni comuni di dipendenza, così come sulla sostanziale differenziazione tra punto di vista oggettivo e soggettivo. L'approccio di Andrea e Alessandro è certamente convincente. La loro previsione su un ruolo rilevante della relazione fiducia/coordinazione nella futura AI è pienamente condivisibile. L'impegno sulla ricerca può risultare alcune volte foriero di relazioni umane frutto di solidali sfide conoscitive. Con Andrea e Alessandro questo è successo.

Robert Demolombe (*Reasoning about trust in information transmitted by a sequence of agents*) sviluppa un interessante esercizio per indagare come un agente può fidarsi di una certa informazione che gli viene trasmessa e che è il risultato di un passaggio tra vari agenti, prima di arrivare a lui. L'acutezza concettuale dell'analisi si basa su alcune interessanti proprietà che Robert ha nel passato introdotto nel contesto della comunicazione dell'informazione e che rappresentano per me uno straordinario patrimonio di pulizia formale e concettuale: la *validità*, la *sincerità*, la *competenza*, la *completezza*, la *cooperatività* e la *vigilanza*. Il pezzo va gustato per la chiarezza formale mai distaccata dal reale e anzi esplicativa della articolazione concettuale. Conosco Robert fin da quando iniziai ad organizzare presso la Conferenza Internazionale degli Agenti Autonomi e dei Sistemi Multi-Agente (AAMAS Conference) il Workshop su "Trust in Agent Societies". Questo suo contributo è un grande regalo.

Il capitolo di **Filippo Cantucci** e **Alessandro Sapienza** (*Trust and Cognition in Human and Hybrid Societies*) ripercorre lavori comuni degli ultimi tempi. Con Filippo ed Alessandro abbiamo affrontato in tempi recenti nuove sfide di ricerca. In particolare si sono affacciate importanti domande, emerse con la nuova stagione dell'Intelligenza Artificiale, non solo generativa, che hanno rivelato la sostanzialità delle acquisizioni precedenti ma al contempo la necessità di sviluppare nuovi approcci e direzioni. L'acquisizione di nuovi finanziamenti progettuali e professionalità scientifiche conseguenti (penso all'arrivo di **Francesco Stella**), hanno fornito nuova linfa su temi la cui rilevanza sembra crescere e assumere rinnovato vigore nelle tendenze internazionali. Il rapporto con Filippo e Alessandro è ormai di lunga data e di comprovata efficacia. La mia riconoscenza per questo dono, fatto assieme a Fabio Paglieri e Luca Tummolini, nell'organizzare questo volume, è enorme.

La fiducia è un tema trasversale a differenti ambiti di ricerca. Il contributo di **Francesca Camilli**, **Tarek el Sehity** e **Raffaella Pocobello** (*Trust the Process: Exploring Trust as a Foundational Element in Open Dialogue*) lo affronta mettendone in evidenza il ruolo essenziale che svolge nelle crisi di salute mentale. Nel capitolo viene mostrato come ci siano evidenze che gli individui con psicosi o a rischio di essa mostrino una fiducia di base inferiore rispetto alle persone sane. Con conseguente inibizione o compromissione delle relazioni sociali. In questo senso, gli autori sostengono come una parte sostanziale dell'intervento di salute mentale dovrebbe mirare a ripristinare proprio la fiducia nelle proprie percezioni, sentimenti e decisioni. La sfida che da qualche anno gli autori stanno perseguendo è quella di enfatizzare gli ambienti terapeutici collaborativi, come il *dialogo aperto*, dove il ripristino della fiducia è attivamente promosso attraverso i suoi principi fondamentali. In questa sfida si incontrano quindi le loro indagini con il lavoro da noi svolto sulla teoria della fiducia. Infine, riferendomi alle parole conclusive di Raffaella, con cui mi accomuna un'intensa stima e un rapporto di amicizia e simpatia, sono contento di poter affermare come sia stata per me una grande opportunità quella di poter scommettere sia sulle potenzialità di questa linea di ricerca quanto sulla capacità dei ricercatori coinvolti nel perseguire gli obiettivi che si erano posti. Questo genere di successi, sono la ragione più importante per aver dedicato tempo e fatica a pratiche che altrimenti potrebbero apparire come vuote attività puramente burocratiche.

Nel suo bello e generoso contributo, **Robin Cohen** (*From Trust Modeling to Trusted AI: Enlightened by the work of Falcone*) affronta un tema centrale della accoglienza dei sistemi di Intelligenza Artificiale nella vita quotidiana di tutti noi. In particolare, sottolinea la necessità di un'AI responsabile e le vie per raggiungerla. La sua riflessione affronta il ruolo dei modelli mentali degli utenti, lo sforzo per rendere "spiegabile" il comportamento dei sistemi che la includono, tanto più in presenza della AI generativa e del successo degli LLM all'interno del campo di studio. Il capitolo affronta quindi il lavoro suo e del suo gruppo negli ultimi 20 anni dedicato ai temi sulla comprensione del concetto di fiducia e del supporto di soluzioni distinte in diversi contesti di utilizzo. Quindi una interessante valutazione viene proposta, oggetto delle sue ricerche più recenti, sul tipo di futuro che si prospetta, sui possibili benefici sociali, ma anche sui rischi e sulle dovute preoccupazioni relative ai modi in cui questi sistemi potrebbero essere mal utilizzati. A Robin, stimata scienziata, con cui abbiamo condiviso una parte della ricca e intensa esperienza dei workshop internazionali su "Trust in Agent Societies", sento di dovere un grande ringraziamento e una speranza di poter trovare il modo di collaborare ancora.

Questo volume, l'ho già detto e forse lo ripeterò anche nel seguito, è una raccolta sorprendente, una fantastica carrellata di emozioni e inattesi (e non so quanto meritati) riconoscimenti. E poi c'è questo capitolo che ha per me un valore persino più speciale degli altri. Si tratta di quello scritto da **Cristiano Castelfranchi** (T^3: for a Multidimensional, Integrated and Dynamic Theory of Trust). E' una descrizione approfondita e ricca di spunti illuminanti sul nostro gruppo di ricerca, sul ruolo che ha svolto e sui contenuti che ci hanno coinvolto. Molte riflessioni sembrerebbero già acquisite, ma si aprono invece improvvisamente nuove prospettive e alcune delle conclusioni vengono capovolte e riproposte in chiavi nuove e originali. E' la cifra del genio e della cultura che Cristiano diffonde nel suo esercizio mentale. La forza dell'evoluzione ha permesso agli umani di costruire i loro percorsi sociali sulla base della collaborazione e sulla cultura del trasferimento delle conoscenze e dei metodi di lavoro. La mia grande fortuna è di aver incontrato nella vita professionale Cristiano e aver potuto raccogliere i suoi insegnamenti di maestro e padre scientifico. Non solo, Cristiano è anche un amico profondo, è quel fratello con cui poter condividere aspirazioni di vita individuale e sociale. E l'intensità e la durata del nostro rapporto mi permette di valutare oggi una straordinaria eredità che sento avermi formato e temprato e alla cui riconoscenza è difficile dare una misura.

3 I contributi socio-politici

E' questa sequela di contributi per certi versi ancora più sorprendente di quella relativa all'attività scientifica strettamente intesa. Non solo perché è di fatto un'anomalia nella tradizione canonica di questo genere di volumi, ma anche perché essere riusciti a scovare alcuni dei personaggi che si sono poi resi disponibili, è stata una abilità da parte di Alessandro, Filippo, Fabio e Luca da veri maestri del genere!
E' una testimonianza che gratifica e ricompensa il tempo e l'impegno che ho a costantemente profuso nelle cosiddette attività politico-sociali a favore della ricerca. Nel tentativo di promuovere e valorizzare questa attività umana (la ricerca e la scienza) così clamorosamente bistrattata nel nostro Paese, eppure così fondamentale per l'avanzamento dello sviluppo variamente inteso degli esseri umani e dei loro costrutti sociali.

La lettera di **Lucio Bianco** richiama tra le sue righe un periodo storico della battaglia per l'autonomia del CNR, intrapresa dalla comunità scientifica nazionale anche grazie ad una iniziativa, quella dell'Osservatorio sulla Ricerca da me ispirato e poi fondato assieme con scienziati della valenza di Carlo Bernardini, Tullio De Mauro, Margherita Hack, Francesco Lenci, Giorgio Parisi, Giulio Peruzzi, Roberto Battiston, Franco Pacini, Marcello Buiatti e molti altri che altrove verranno ancora richiamati in questo volume. La battaglia dell'Osservatorio e di Lucio per difendere l'Ente da una norma voluta dall'allora Ministra Moratti, si risolse positivamente.
Quella battaglia rappresentò una base per poi incontrarci e trovarci sistematicamente accordati in molteplici iniziative. Oggi ci legano un sentire e una condivisione di prospettive sulle politiche della ricerca e sul futuro del Paese. E un affetto e una stima di grande conforto.

Il breve ma intenso scritto di **Gaetano Manfredi** ripercorre un tema su cui decidemmo di avviare un'iniziativa nazionale di massimo livello. Era il breve periodo in cui fui suo Consigliere mentre egli era Ministro della Ricerca e dell'Università. Si trattava di approfondire seriamente e di individuare percorsi di ricucitura e di promozione del rapporto scienza-società. E' questo un rapporto controverso e complicatissimo da sempre ma, mai come nel XX secolo, le relazioni tra scienza ed effetti da essa derivati sulla società (il suo sviluppo, la sua organizzazione, la salute, l'ambiente, la distribuzione della ricchezza, l'etica, etc.) hanno avuto tanta rilevanza e impatto.

Una prima questione riguarda il fatto che l'avanzamento scientifico non sempre procede in modo lineare, a volte può intraprendere percorsi ambivalenti fino al maturare delle ipotesi prevalenti. Di fatto *dubbio* e *incertezza* sono parte integrante e motore attivo della ricerca scientifica e le *teorie* e i *modelli* di successo che la scienza è in grado di offrire, sono risultanza di processi lunghi, faticosi e irti di ostacoli. *Mai* comunque definitivi.

L'opposto delle pseudoscienze, che non si affidano al metodo scientifico ma piuttosto a categorie imperscrutabili e non definenti percorsi metodologici oggettivi. Solo basate su superstizioni o valutazioni del tutto inattendibili e fantasiose: figlie di speculazioni strumentali per finalità manipolatorie.

L'obiettivo di quell'impegno con Gaetano era di individuare percorsi per sensibilizzare la società rispetto al valore della scienza, di valorizzare il ruolo degli scienziati, di ricostruire un modello di politiche della ricerca capaci di riconnettere scienza e cittadini. La convinzione era che la forma di gestione e sviluppo che la scienza è stata in grado di darsi, possa rappresentare un modello anche per le forme della politica istituzionale e per avanzare il livello della consapevolezza e dello sviluppo democratico.

I pochi mesi di collaborazione con Gaetano mi hanno comunque convinto della straordinaria qualità dell'uomo e del politico. Ho potuto apprezzare la gentilezza d'animo, la profonda comprensione dei problemi (la forma mentis dello scienziato ne dà piena testimonianza) e la lucida capacità di risoluzione. L'augurio per il Paese, non so se per lui, è di un suo impiego futuro ai massimi livelli istituzionali.

Il capitolo di **Francesco Lenci** è la testimonianza di un fratello che ti vuole bene. Lo stesso bene che provo io per lui. Francesco è stato assieme a pochi altri, Carlo Bernardini e Giulio Peruzzi, nel nucleo storico e iniziale dell'Osservatorio sulla Ricerca. Nel suo commovente pezzo ne traccia parte del percorso. Ricostruisce in sintesi un periodo straordinario. Quello in cui un gruppetto di scienziati del CNR (un ente di ricerca di cui la stragrande parte della popolazione italiana ignora/va completamente l'esistenza) decide di coinvolgere le menti scientifiche e intellettuali più rilevanti del Paese per denunciare una riformina che, senza quei scalmanati idealisti, sarebbe passata del tutto indifferente. I rischi per l'autonomia della ricerca non li avrebbero presi sul serio neppure i più fini giuristi del settore. Ed invece, invocando principi costituzionali e rischi per lo sviluppo civile e sociale di una grande democrazia (rischi reali, non fintamente costruiti all'uopo), mobilitando su parole d'ordine incalzanti e proiettando scenari inquietanti sul controllo della scienza e della ricerca, ossia sugli indirizzi e sviluppi più avanzati di una società, quegli idealisti sollevarono una questione nazionale come mai era stata proposta prima d'allora.

Quel disegno fu fermato. Il CNR ha mantenuto la sua caratteristica di ente non strumentale e una diffusa consapevolezza del ruolo delle comunità scientifiche ebbe una sua significativa conferma. Il suo bel resoconto è lì a dimostrare che solo un'interpretazione adeguata della realtà e delle finalità profonde, ideali e condivise possono permettere, se perseguite con convinzione e determinazione, l'affermarsi di principi e valori che si ritengono corretti.

Ma ancora più commovente è il legame che quella stagione ha permesso di stabilire all'interno di quel gruppo. La sensibilità umana e affettiva che ci lega resterà un dono che conserveremo per sempre!

Daniele Archibugi nel suo bel pezzo introduce un problema rilevante, quello delle comunità scientifiche e della loro capacità di governance, riconoscibilità e accettabilità rispetto alla società. Questo problema di meta-livello ci offre la possibilità di analizzare pregi e difetti che una comunità di scienziati sviluppa e ci fa riflettere sulle reciproche influenze di questo meta-livello con sviluppo ed efficacia dell'attività scientifica vera e propria.

Questo ruolo ordinatore che le comunità scientifiche dovrebbero darsi è, come dice Daniele, spesso delegato e malcompreso. E, quelle volte che succede, è grazie alla buona volontà di qualcuno che si procede a questo ordinamento, più che ad un processo costitutivo e determinato. C'è una difficoltà reale nel costruire, da parte di questa comunità, un orizzonte comune e nell'individuare quegli obiettivi superiori, rispetto al puro lavoro di ricerca, che soli sono in grado di facilitare realmente le attività scientifiche di tutti. E' giusto che ci siano grammatiche differenti, molteplici, a livello di studio scientifico, in modo che si possano aggregare i vari ambiti di ricerca. Non va bene invece che analogamente non ci sia la capacità di comprendere il codice comune della missione complessiva della scienza. Qui si fa fatica.

Con Daniele abbiamo stabilito un rapporto di amicizia intenso proprio a valle di esperienze tra scienziati politici. Il contributo, piccolo ma testardo, che Daniele mi riconosce e di cui gli sono grato, credo di averlo dato per quello che resta un percorso fortemente in salita. Provare ad essere scienziati ma anche intellettuali della scienza.

L'articolo, intenso e profondo, di **Luigi Nicolais** è un vero e proprio monumento al pensiero critico, ossia al metodo di ragionare e di valutare l'oggetto di studio. Non importa quale sia questo oggetto, se si tratti di affrontare questioni umanistiche o scientifiche, di scienza dura o di scienza soft. Che poi, nella scienza moderna, sempre più interdisciplinare e interconnessa, la distinzione si assottiglia oltre il limite. Il pensiero critico invece è essenziale, ci assiste e ci guida nel mare magnum delle conoscenze e delle loro conseguenze. Ed è l'ancora che può sola salvarci dalla furia semplificatoria e riduttiva del tempo che affrontiamo. Per esempio, sostiene Gino: come farne a meno nei confronti dell'Intelligenza Artificiale? Che può "sia migliorare che distorcere la nostra comprensione del mondo. L'intelligenza artificiale può eseguire compiti con precisione e velocità, ma non ha la comprensione sfumata e la flessibilità del pensiero umano. Pertanto, è essenziale sviluppare capacità di pensiero critico per valutare le informazioni generate dai sistemi di intelligenza artificiale e garantire che l'intelligenza artificiale venga utilizzata in modo responsabile ed etico."

Pur avendo avuto nel corso della mia carriera scientifica, diversi Presidenti del Consiglio Nazionale delle Ricerche, considero Gino Nicolais il "mio Presidente", non solo perché ho trovato in lui la piena condivisione del mio progetto, come candidato alla direzione, per lo sviluppo dell'Istituto di Scienze e Tecnologie della Cognizione. Ma perché è grazie ad una piena e fattiva collaborazione con lui che ho sentito il CNR come la casa appropriata per sviluppare ricerca, per affrontare discussioni e confronti adeguati sul futuro della scienza e sugli impatti che essa può avere per la società. Le capacità intellettuali e al contempo pratiche di Gino ne fanno una delle più lucide menti della nostra cultura scientifica e prassi politica.

Una figura di scienziata di altissimo profilo che abbina grande sensibilità sull'orientamento politico per i temi della ricerca è **Lucia Votano**. Con Lucia ci conosciamo da lungo tempo grazie a Pietro Greco, amico comune e intellettuale visionario sulle questioni della scienza nei rapporti con la società, purtroppo prematuramente scomparso. Subito si è stabilito con lei un feeling profondo, una immediata coincidenza di punti di vista e una comune esigenza di porre rimedio alle inadeguatezze della governance della scienza e della ricerca nel Paese. Può capitare di sentirci di rado e immediatamente ricostruire analisi e valutazioni condivise. Questo suo omaggio mi onora.

C'è stato un periodo, non breve, in cui i direttori degli Istituti del CNR, la più diretta sensibilità della rete di ricerca dell'Ente, hanno trovato coesione e capacità di espressione comune: senza troppo rincorrere le richieste individuali dei propri istituti, pure legittime in situazioni critiche, ma soprattutto facendo sintesi di sistema. In quel periodo mi è capitato di condividere con vari di questi direttori un percorso irto di problemi ma anche pregno di potenziali ambizioni e speranze. **Corrado Bonifazi**, allora direttore dell'Istituto delle Popolazioni (IRPPS), è stato tra questi uno dei più vicini e attivi. Ricordo lunghe discussioni preliminari o successive ai vari incontri. Sempre una rassicurante lucidità e oggettiva visione connotava le sue opinioni. Credo che le amicizie possano avere origini e seguire percorsi tra i più vari, con Corrado ci siamo trovati in questo contesto e quando capita di risentirci, si riattiva immediatamente quel pensiero attivo che caratterizzava i nostri scambi.

L'esperienza di Consigliere del Ministro Fabio Mussi, durante il Governo Prodi del 2006-2008 ha segnato il mio percorso profession-ale come poche altre occasioni nella mia vita. Non è certo questa la sede per parlarne. L'omaggio che in questo volume mi ha dedicato **Nando Dalla Chiesa**, è però l'occasione per riferire mie impressioni su questo specifico rapporto. Devo ammettere che il racconto che fa Nando del nostro incontro mi ha commosso come pochi altri. Da una parte perché inatteso e dall'altra per il senso profondo che mi resti-tuisce quella esperienza. Negli incontri programmatici e di confronto che avemmo in quelle occasioni coglievo nei suoi atteggiamenti serietà e attenzione per i contenuti e le questioni di precipua valorialità. Una aperta curiosità, tipica dello studioso, verso gli argomenti degli inter-locutori. Mai un approccio strumentale o una posizione preconcetta di posizionamento. Un atteggiamento fortemente etico nell'approccio politico.

Sono state le ragioni per cui ho continuato, negli anni successivi, in cui pure non abbiamo avuto modo di incontrarci e confrontarci, a coinvolgerlo, su temi e questioni che mi parevano potessero stimolare un suo interesse. E' capitato, sfortunatamente, poche volte. Sarà un mio impegno riprovarci nel futuro.

La vicenda del mio incontro con **Pierpaolo Campostrini** è nar-rata da Pierpaolo nel bellissimo pezzo che mi dedica in questo volume. E' stata questa una delle occasioni per mettere alla prova le mie due anime più profonde: quella dell'uomo di scienza e quella dell'idealista politico. Si trattava di confrontare ipotesi nettamente alternative: una concretamente risolutiva e fondata, l'altra (in realtà erano varie e alternative) più aleatorie e vaghe nella loro efficacia. Entrambe però erano accompagnate da una suggestione ideologica (differente da ideale): la più concreta e razionale (il MOSE) denigrata come efficien-tista e tecnocratica, anti-ambientalista e nella sostanza di presumibile inefficacia; l'altra, alternativa e di tutela ambientale, evidentemente meno protettiva contro le maree ma considerata di salvaguardia eco-logica. Insomma politicamente dalla parte del MOSE erano schierati perlopiù i moderati, contro c'erano parte delle sinistre (in testa il Sindaco di Venezia, Cacciari). Delegato dal Ministro Mussi, incontrai i due schieramenti. Dopo approfondimento sui documenti che entrambe le parti mi proposero e dopo confronti di chiarimento e di analisi con loro, ebbi la chiara sensazione che la contrapposizione fosse tra una ricerca di soluzione molto concreta e fondata ed una posizione ideologica strumentale del tutto inappropriata al problema da affrontare. Quest'ultima era sostenuta da quanti erano per me un riferimento politico di valore. Non ebbi però dubbi su come dovessi relazionare al Ministro.

Il fatto che Pierpaolo oggi mi conforti con questo suo scritto è per me motivo di grande orgoglio e grande è la mia gratitudine per questo riconoscimento.

Massimo Cocco è un brillante e importante scienziato dell'Istituto Nazionale di Geofisica e Vulcanologia, anche lui come molti degli autori di questo volume e tanti altri non presenti in questa pubblicazione, rappresenta quella parte attiva della comunità scientifica che convintamente opera per soddisfare quanto richiamato da Massimo: "L'integrità della scienza e il ruolo degli scienziati si rafforzano attraverso la loro immagine nella società; più quell'immagine rappresenterà il reale beneficio che la scienza è in grado di produrre per la società, più la società sarà in grado di attribuirle quella funzione fondamentale".

Massimo è in questo senso anche un intellettuale finissimo. Spesso con lui ci siamo avviati su percorsi di analisi del ruolo della scienza e delle dinamiche di questa con la politica. Difficilmente ho trovato intuizioni e spunti illuminanti come quelli che lui è in grado di produrre su questi temi e, come con altri già citati in questo volume, ogni volta che mi capita di confrontarci provo la grande soddisfazione di mettermi alla prova con il pensiero profondo. Grazie Massimo!

Maria Chiara Carrozza è attualmente la Presidente del CNR. Con Maria Chiara le strade si sono intrecciate in diverse occasioni: ci siamo ritrovati a collaborare quando era responsabile per il PD dell'Università, in varie iniziative di mobilitazione per la ricerca e poi diventata Ministro nel Governo Letta mi ritrovai, senza che ne fossi preavvisato, a far parte, su sua nomina, del Search Committe per la selezione dei candidati alla Presidenza o come membri del CdA degli Enti di Ricerca vigilati dal MUR. Insomma la stima reciproca è stata intensa e prolungata. Da quando è Presidente del CNR in realtà, forse anche per una non completa convergenza sulle modalità con cui il Governo Draghi avviò una riforma del CNR, le nostre interazioni sono state più discrete. Trovare quindi in questo volume un suo scritto mi ha fatto un enorme piacere e mi ha anche parzialmente sorpreso. Anche perchè, le cose che Maria Chiara scrive sono di grande generosità nei confronti del mio percorso tanto strettamente scientifico, quanto di attenzione e impegno alle politiche della ricerca nei vari e articolati posizionamenti che ho assunto. Per esempio, sulla questione dell'autonomia della ricerca e sulla mia sensibilità al tema il suo apprezzamento lo trovo di grande gratificazione. E le fa onore l'onestà intellettuale che dimostra in tal senso. Al CNR ha operato con grandissimo impegno e serietà. Il suo percorso sono certo che lascerà una traccia importante.

Maurizio Franzini è un accademico e intellettuale di primissimo piano. Nella sua dedica sviluppa un'analisi, focalizzata con lo sguardo acuto e incisivo dell'economista, su un tema che sa avermi molto appassionato negli anni: la fiducia. Si sofferma, in particolare, su un aspetto di grande rilevanza sociale che la fiducia gioca, quello di stimolatore fondamentale della cooperatività. La capacità di trasformazione dei comportamenti (degli altri e dei propri) attraverso un'attitudine a considerare il proprio interlocutore capace e intenzionato a svolgere il compito che gli viene delegato. Affinchè questa trasformazione abbia effetto sembrerebbe necessario che l'interlocutore debba prendere coscienza della delega, ovvero che questa sia a lui manifesta. E' attraverso questo stimolo esplicito che si innesca quel meccanismo di riorganizzazione degli ingredienti cognitivi (credenze, intenzioni, scopi, etc.) che, come dimostrato nell'osservazione della socialità, figlia dell'evoluzione adattativa, permette di accrescere le potenzialità collaborative degli umani in interazione.

La costruzione della socialità organizzata ha di fatto, mano a mano, reso più automatici alcuni di questi meccanismi mentali, per cui è possibile che in molti casi, la disponibilità e l'impegno al miglioramento delle proprie prestazioni (e quindi la propria affidabilità) vengano promossi anche al di fuori della manifesta presenza di un delegante e di un interlocutore cui direttamente rendere conto. Esistono interlocutori terzi (al di sopra delle parti) o addirittura presidi cognitivi introiettati cui rendere conto (principi, valori, etica).

Tutto ciò è particolarmente interessante e magari capiterà un giorno di poterne discutere insieme, in modo scientificamente più approfondito con Maurizio che ringrazio per avermi offerto questo affettuoso e intenso pensiero.

Tra i compagni di strada con cui ci siamo ritrovati a ragionare e pensare su come poter contribuire per far uscire l'Italia da una stagione economica e da un clima civile e sociale assai difficile c'è **Daniela Palma**. Daniela è una brillante economista dell'ENEA. La sua analisi è molto bella e commovente per alcuni aspetti. Non solo perché richiama lucidamente tratti di quel percorso che riletti alla luce del presente dimostrano la lungimiranza e capacità anticipatoria delle nostre tesi, ma anche perché rievoca quella fusione di personalità e passioni che ebbe un ruolo decisivo per la maturazione che quelle analisi assunsero, mai faziose o estreme in chiave demagogica ma sempre sottoposte ad un vaglio costruttivo e collaborativo. E proprio da questo deriva la forza che riuscirono ad esprimere.

La perdita di alcune di quelle menti e passioni è un dolore che resta profondamente presente (penso a Carlo Bernardini, Pietro Greco, Tullio De Mauro, Franco Pacini, Marcello Buiatti, e molti altri) ma la strada costruita insieme -e questo suo intervento è un importante aiuto a definirne la traccia- un conforto che ci accompagna. A Daniela dedico un grande e affettuoso abbraccio.

Esistono colleganze politico-professionali che resistono al tempo pur non coltivandole in modo appropriato. Quella con **Alberto Silvani** è una di queste. Ci conosciamo da decenni, giovani ricercatori concentrati sulla scienza ma anche su come questa vada governata per renderla efficace, comprensibile e di valore. Studiosi di discipline differenti ma pronti a confrontarci nella sfera socio-politica di contesto. Il suo contributo mi ha meravigliato proprio perché, nonostante la premessa appena fatta, ricostruisce un dialogo tra di noi di intensa sostanza confermando la qualità di analisi che Alberto ha sempre dimostrato. La riflessione che sollecita è un'illuminazione sulla straordinaria necessità di individuare strumenti e forme capaci di permetterci il continuo dialogare e ricercare strade e percorsi di comprensione anche fuori dall'ordinaria consuetudine.

Questo suo regalo mi conforta sulla necessità di disseminare tracce riflessive e costruttive, nella convinzione che moltiplicando disseminazioni e disseminatori si continui almeno a tenere sveglia quella parte della cognizione che fa evolvere individui e società.

Voglio quindi ringraziare Alberto fuori da ogni retorica di forma per questo dono che considero straordinariamente importante.

Settimo Termini è un altro compagno di strada nel percorso svolto per migliorare il nostro mondo (quello della ricerca e dei saperi) che poi è uno dei fulcri fondamentali delle società moderne. Settimo è un intellettual e scienziato di primo piano. E' stato direttore dell'Istituto di Cibernetica del CNR e professore ordinario a Palermo. La sua capacità di ricostruire un periodo così importante non solo per noi ma per le sorti del nostro Paese mi ha trasmesso quel senso forte che assumono le vicende sociali quando superano la cronaca per puntare ai meccanismi profondi del cambiamento.

Non importa quale sia la risultante finale (o forse sì ma diamoci un alibi), conta molto la serietà e la qualità dei dialoghi, delle proposte, delle iniziative che assieme a compagni e amici appassionati quanto noi, siamo stati capaci di produrre. Una vicenda collettiva che trascende i pochi o tanti che la partoriscono. Diventa simbolicamente rilevante nelle intenzioni di fondo che l'hanno mossa. La differenza tra illocuzione e perlocuzione. L'intenzione del linguaggio e i suoi reali effetti. Ma senza l'intenzione non si raggiungeranno mai gli effetti voluti. E anche quando gli effetti sono divergenti dalle intenzioni, si avrà avuto modo di affermarle seppure con esiti distorti (e questo comporterà comunque una analisi del rapporto tra cause ed effetti e sperabilmente una maggiore comprensione del nesso tra le due). Dico tutto ciò per ringraziare Settimo. Per solidarizzare con un amico vero. In cui l'amicizia si colora di connotati. Un amico di scienza e di politica, un amico con cui, pur non sentendosi da tempo, è possibile immediatamente ristabilire dialoghi e nessi senza esitazione.

Come descritto da **Marco Pagani** nel suo omaggio in questo volume, la nostra amicizia attraversa ambiti istituzionali e privati. Con Marco, come il suo omaggio testimonia, abbiamo avuto esperienze di battaglie per il rinnovamento del CNR e al contempo per scoprire il miglior vino da abbinare con lo straordinario salmone svedese che portava da Stoccolma. La sua è una presenza incostante ma affettuosa e sodale. Unico difetto è la passione per una squadra gobba. L'alibi delle origini torinesi non tiene, data l'ampia compensazione della strabordante cultura di romanità.

Oliviero Stock ha rappresentato il mio primo interlocutore scientifico di grande valore. Avevo collaborato con vari colleghi, ma Oliviero era il salto di qualità e con lui (ed il chart-parsing ad isole) ho scoperto che la passione scientifica avrebbe avuto il sopravvento nella mia vita. I nostri rapporti sono rimasti rari seppure profondi ed è capitato anche di collaborare ancora, ma non è più successo di scrivere scienza insieme. E questo mi spiace moltissimo!

Però non ha ragione quando sostiene che il mio tempo scorre lento, l'arrivo del pensionamento è arrivato rapidissimo, spiazzandomi completamente.

Mi inorgoglisce enormemente il paragone che lui fa tra me e Raffaele Misiti: in fondo la mia anima sta proprio in quel collegamento e dialogo tra scienza e politica che Lello così bene rappresentava (come riportato dai ricordi dei suoi allievi).

E mi commuove la frase che Oliviero scrive: "Per quanto riguarda la comunicazione pubblica sulla ricerca, spesso ho avuto occasione di leggere un qualche articolo scritto da Rino, e il mio commento di solito è stato: speriamo i decisori politici lo leggano con attenzione".

E poi lo scambio di versi dopo che lo informai della mia traversata dello Stretto.

Io gli mandai i miei:

> *Perché attraversare lo Stretto?*
> *Perché inabissarsi in quei 3,14 km di nuoto?*
> *Non è certo per andare dalla Sicilia alla Calabria o viceversa,*
> *Non è un banale immergersi a Punta Faro per fuoriuscire sulla spiaggia di Cannitello*

Può servire forse a misurare il tempo che passa
A pesare nelle nostre braccia quelle vissute 66 rivoluzioni della Terra
intorno al Sole,
A sfidare le due impronte di titanio che ci affiancano permanenti
A voler sentire il battito del cuore messo alla prova della durezza delle
onde e delle correnti
A percepire cosa di "atletico" la giovinezza ci ha lasciato oltre il
rimpianto

E' vero: si tratta di sfidare il tratto di mare più impegnativo del
mediterraneo,
dove correnti e venti tempestosi sono frequenti e prorompenti
dove Ionio e Tirreno si incontrano/scontrano producendo fenomeni
idrodinamici unici
dove Scilla e Cariddi, ninfe furibonde permanentemente in lite, dis-
velano la leggenda
dove Polifemo accecato ancora scatena l'ira di Nettuno contro Ulisse
dove, come in nessun altro tratto di mare, ci si può imbattere in "vor-
tici" sì perigliosi.
dove terremoti e maremoti hanno distrutto nei secoli, più volte, le
città di Messina e Reggio,
dove la catastrofe del 1908 fu scatenata dalla faglia antica ma ancor
presente nel suo fondale
dove le placche geologiche di Africa ed Eurasia hanno deciso di convergere qui in movimenti tettonici
dove, di conseguenza, fratture profonde accumulano energia da libe-
rare in magnitudinali sconquassi
dove violentatori di ponti umanitari narcisisticamente intendono lu-
crare sul falso, inutile e dannoso mito del superamento di questo varco
naturale...

Ma l'attraversamento dello Stretto è anche e soprattutto un viaggio
per e nella propria anima,
Nel bisogno di conoscenza a cui essa continuamente attinge: ansia
della partenza, incognite del percorso, aspettative dei traguardi
Questo breve ma periglioso e sconosciuto ciclo, può apparirci metafo-
ra di vita esso stesso,
un viaggio nella nostra più profonda essenza,
bracciata dopo bracciata nelle acque gelide e sempre più oscure che
ci ricoprono, esso è un tentativo estremo per scoprire le vere ragioni
del nostro sentire
per affrontare le assenze dolorose che ci hanno colpito ed indagare a

fondo la memoria che ne resta
Sentirle in quel breve viaggio vicine
Coglierne una sorta di presenza nelle acque che scorrono su di noi,
Lasciarsi portare dalla loro spinta affettuosa: di madre, di padre, di
fratello, di parenti e amici cari.

Vuoti dell'anima che il mare colma nella sua intransigente pervasività
e coscienza "viva".

E c'è infine la presenza incombente di un'umanità dispersa
Di giustizia, solidarietà e diritti fatti strame
Di acque dove si è tragicamente concluso il bisogno di liberazione di
molti umani
Di impietose mani mai protese
Di urla soffocate nel silenzio assordante della nostra indifferenza.

Cercare quindi un'ultima disperata illusione di uscire da questi flutti
e trovare il mondo nuovo,
o almeno, in quell'ora, 8 minuti e 8 secondi di separazione dalla
routine del reale, immaginare che un mondo nuovo possa ancora es-
serci. . .

E lui rispose con quei versi brillanti ed ironici. Grazie Oliviero!

Il pezzo di **Paolo De Nardis**, sociologo di primissimo piano e grande intellettuale, presenta in modo sintetico, dotto ed efficace alcuni temi che abbiamo nel corso del tempo e in varie occasioni lungamente dibattuto. Per esempio l'annosa questione delle varie tipologie di ricerca e delle sedi più opportune in cui esercitarle. Seppure sempre più spesso la natura della ricerca mescola le differenti tipologie, spesso confondendole, concordavamo con Paolo che "quando si deve valutare la ricerca scientifica, ci si trova di fronte a un complesso universo di variabili di carattere strutturale, individuale ed epistemologico che rendono difficile il non considerare come intrecciati tra loro i fattori sociali, quelli cognitivi e quelli organizzativi." Il suo pezzo è un misto tra l'esercizio della riflessione colta e profonda sulle questioni di politica della ricerca e un omaggio affettuoso di un amico che stimo da tempo immemorabile e a cui voglio bene.

E' difficile immaginare chi meglio di **Salvatore Capasso**, attuale Direttore del Dipartimento Scienze Umane e Sociali del CNR, dipartimento a cui afferisce ISTC, l'Istituto che attualmente dirigo, potesse fornire un'analisi accurata e approfondita sulla relazione tra le scienze umane e sociali e le scienze dure. Capasso è professore di economia ed ha chiaro l'impatto che i fattori umani e sociali sono in grado di determinare, quando ben valutati e interpretati, sullo sviluppo tecnologico e produttivo, ma anche valoriale ed etico di una società. E d'altronde, nel dirigere il Dipartimento, Salvatore ha la necessità di tenere a mente queste potenzialità, di aderire alla straordinaria tradizione inter e multi disciplinare che il CNR ha coltivato fin dalle sue origini e che è parte costitutiva del proprio DNA. Tanto più che la scienza moderna ha esaltato questi fattori di mescolamento e reciproca contaminazione tra le scienze dure e quelle umane e sociali. Il convincente ed esplicativo capitolo di Salvatore è una bella dimostrazione della forza del più grande Ente di ricerca, quando invera la tradizione e seleziona gli scienziati adeguati per il proprio futuro. Sono davvero grato a Salvatore per il dono che mi ha fatto in questo volume.

E' molto difficile valutare il proprio lavoro, specie quando si tratta di tracce che indirizzano percorsi complessi e articolati. E il possibile riscontro è la risultante di varie componenti non necessariamente tutte valutabili e sotto controllo. In questo volume, **Francesco Profumo**, grande personalità scientifica e intellettuale del nostro Paese, ha voluto fornire un resoconto e una valutazione del lavoro che ho svolto da consigliere dell'allora Ministro Mussi. La sua generosità di giudizio è per me di grande conforto. Fu quel periodo una fase stravolgente della mia vita professionale: mi ritrovai improvvisamente –Mussi non mi diede molte possibilità di scelta– ad affrontare le problematiche della ricerca nel nostro Paese. E mi immersi in questa dimensione di "governance alta" con grande passione e con la forte determinazione di dover rappresentare le istanze che per anni avevamo – con altri scienziati- rappresentato ed invocato.

Il fatto che Francesco abbia l'esperienza e la capacità critica da Ministro oltre che da Rettore, Presidente del CNR e Professore di altissimo prestigio, mi inorgoglisce enormemente nel leggere i giudizi che propone. Dato che lo considero anche un amico sono portato a ritenere che la benevolenza affettuosa verso di me abbia avuto un ruolo importante per queste sue positive valutazioni. In ogni caso, lo ringrazio esprimendogli la mia più sentita riconoscenza.

4 Introduction

In this chapter I will report sensations, perceptions and considerations that the sequence of contributions in this volume, dedicated to me on the occasion of my retirement, has unleashed in me: in my mind and in my soul, that is, in the rational and affective forms of my thought. However, it is good that I first say it clearly: the forced and inevitable "retirement" has appeared to me as an irreconcilable and premature goal. Even with traits of injustice: why are university professors, doctors, magistrates, etc. retired at seventy and we at sixty-seven?
We have even tried to politically promote, with supportive colleagues like Daniele Archibugi, legislative initiatives for recovery. Nothing has been possible.
Obviously not everyone thinks this way.
Much depends on how much of one's being and feeling represents the work activity one exercises. What tool for exploring the world and oneself it determines. And how active one still feels in the construction of one's own work: in the particular case that concerns me, in reference to the knowledge and research that one investigates.
It is true that there are, also in order not to disperse a heritage that can be variously classified, "recovery networks" for this role (associating with the CNR, the possibility of teaching contracts with private universities, and so on), but the break that conceptually occurs with retirement underlines a limit to which one does not feel adequate. This happened to me.

In reality, my retirement came during my term as director of ISTC-CNR and, given that the procedure for selecting the new director was, at the time of retirement, in progress, the CNR proposed that I continue in this directorship, which at the time of publication of this volume is still ongoing, with a private law contract (unpaid). Moreover, in a period extraordinarily rich in deadlines and management changes of the CNR (change of method of accounting, VQR exercise, etc.): with a consequent request for commitment and activity that is anything but decreasing. I therefore find myself in the curious position of being retired but hyper-actively operational at the request of the Institution.

The generous and affectionate initiative of Alessandro, Filippo, Fabio and Luca for the drafting of this volume represents in fact an unexpected compensation compared to the profound discomfort of this change of perspective. An initiative that surprised me in some respects and moved me in others and for which I am grateful well beyond what I have directly expressed to them.

Just as I am grateful and empathetically grateful to all those who have offered their homage, often expressing judgments towards me that even if I do not know how much I deserve, I like to take note of and consider them sincere.

For this reason I will spend a few but, I hope, significant words for each of them.

5 Strictly scientific contributions

I am linked to **Munindar Singh** by a long and intense experience of collaboration in the workshops on Trust that were repeated at the AAMAS conference (Autonomous Agents and Multi-Agent Systems). This conference was the result of the new international scientific approach to Artificial Intelligence systems. An approach in which the vision of intelligence centered on a single agent was abandoned, replacing it with the interaction and relationship between multiple agents (distributed AI). The workshop on trust in particular has focused over the years on this property/attitude of intelligence (trust, precisely), so essential for the evolution of effectively collaborative systems and capable of identifying advanced solutions and balances in systems made up of autonomous and cognitive agents.

In his beautiful piece (*Social Artificial Intelligence and the Interplay of Norms, Values, and Trustworthiness*) Munindar, in the context of recognition for the work on trust that we have carried out with Cristiano Castelfranchi over the years, identifies a critical element, referring to an article written with Cristiano and Francesca Marzo. The objection that he himself reveals as contradictory to our broader theory, is to consider trustworthiness as an intrinsic property of the trusted. In reality, his criticism is absolutely correct: there are elements of substantial dynamism in the trustworthiness of any trustee who is capable of correlating with the context in which he operates and with the trustor with whom he interacts (as well as with his own history and contingent internal situation). Munindar is right! But in that paper we tried to approximate this dynamism and make it constant with those boundary conditions that we wanted to represent.

In any case, Munindar's piece is enjoyable and intelligent. The in-depth analysis of the close relationships between *norms, trust* and *values* will certainly represent one of the frontiers to correctly and profitably transfer the essential elements of social intelligence to mixed artificial systems.

A contribution of similar interest to that of Singh given the relationship between trust and social norms, was made by **Mario Paolucci** and **Giulia Andrighetto** (*Trust, Reputation, and Norms: A Simulative Approach to Cognitive Artifacts*). The cognitive artifacts that these researchers decide to relate to trust are precisely social norms and reputation. In fact, the relevance of this chapter also lies in underlining and reconstructing the fundamental role that ISTC has played in what they define as "Theoretical Cyber-Psychology" (TCP) where symbolic manipulation builds reasoning. Furthermore, the appeal to simulate trust more and better in relation to norms and reputation cannot remain unanswered. Not only to simulate but also to test the theoretical models developed in the real world. An appeal to leave the world of theory with greater determination and to transform the acquisitions into tools for practical and fruitful investigation of TCP. Although I have a greater aptitude for theoretical reasoning than for practical experimentation (on this I fully agree with Cristiano Castelfranchi), I believe that this assessment is not only correct but should be pursued with seriousness and depth. Thanks Mario and Giulia!

The chapter by **Anna Borghi** and **Luca Tummolini** (*Semantic Delegation: Understanding Concepts With A Little Help From Our Friends*) explores in original and in some ways intriguing forms a dimension of the socio-cognitive approach of trust and delegation when applied to the meaning of words, that is, to semantics. The development of the arguments is coherent and particularly elegant, as well as convincing. Delegation (and the trust that underlies it) is seen as a social mechanism for acquiring and mastering complex knowledge.

Particularly timely is how these considerations can represent a tool for understanding generative artificial intelligence (LLM) models. The conclusion reached by Anna and Luca seems to predict the need for an increasingly central role in the future for those concepts (*trust, delegation, autonomy*) on which, with Cristiano, we have been developing theories and hypotheses for some time. And an interesting space for collaboration with them opens up, which we consider a great gift and an extraordinary opportunity for the future.

The contribution by **Valeria Seidita** and **Antonio Chella** (*From Trust Theory to Explainable Agents: Enhancing Agent-Based Systems*) moves within the scope of the so-called Human-Robot Teaming Interaction (HRTI), that is, the discipline that aims to develop and study the collaboration between humans and robots in a team context. Valeria and Antonio focus in particular on the need to equip agents and robots with advanced cognitive capabilities, in particular the ability to self-model and justify actions. Referring to the trust model developed by us and integrating it into the BDI cycle, they show in this contribution how they have created artificial agents capable of understanding and explaining their own actions and of adapting to unknown and dynamic environments.

Furthermore, by equipping these agents with the ability to manage their own internal language (inner speech), they seem to be moving towards the generation of particularly evolved agents. With Valeria and Antonio, therefore, promising possibilities seem to emerge to resume the thread of a collaboration that has been very productive in the past years.

The chapter dedicated to me by **Elisa Colì** (*Trusting in the Online World: From Theoretical Studies to the Application of the Socio-Cognitive Model*) reflects on a very relevant question addressed together in various experiments and published works: what happens to trust, to this fundamental social attitude of ours, when our relationships move from the real world to the virtual world, for example, to online interactions?

The collaboration with Elisa began in recent years, immediately after the COVID pandemic, as she herself recalls in her contribution. But it has been a splendid forge of updated knowledge and intense collaboration. And it is still full of potential developments also thanks to her continuous solicitations and stimulating ideas. The tribute she dedicates to me in the introduction represents a splendid example of the relationship that can be established between an expert researcher and his younger collaborator. How the strength of research work can produce extraordinary scenarios of collaboration and friendship. Thank you Elisa!

Antonio Lieto, in his paper (*Trust in the Rational and Social Bands of Cognitive Architectures*) synthetic but effective, underlining the relevance of trust models, proposes some research directions in the field of cognitive architectures and models. In particular, starting from the bands of cognition proposed by Alain Newell and analyzing the architectural modalities that have seen their application, Antonio highlights the poor development of macro-cognitive models in these architectures and predicts the usefulness of looking at the knowledge acquired on trust models to update these architectures in the future in a more effective and adequate way. Reading these appreciations from a young but already well-known and established researcher, is a great pride for me.

A research topic that has attracted my interest and that of the ISTC T^3 research group and that deserves considerable attention in the perspective of the new generative AI, is the one related to the adaptation properties of the autonomous behavior of agents to different and changing situations and environments. This topic is simply defined as "*adjustable autonomy*" and **Amedeo Cesta, Gabriella Cortellessa** and **Angelo Oddi** have taken inspiration from it for their contribution (*Adjustable Autonomy in AI and Assistive Robotics: Experiences and Lessons Learned*). In this chapter they present some challenges faced in this perspective showing how the appropriate level of autonomy is a correct compromise between various dimensions and, when well exercised/designed, is crucial for the effectiveness and evolution of interactions between agents.

I would like to dedicate a special thanks to Amedeo with whom we have shared a long path of common experiences and dynamic and dialogic interactions and whose deep professionalism, strong sense of respect and institutional belonging and sincere loyalty that characterizes him.

One of the great problems that humanity faces every time a technological revolution impacts economic and social development concerns the transformation of the production system and the labor market. With Artificial Intelligence and the exponential development of its applications, this transformation is announced to be even more profound and rapid. **Pierluigi Zoccolotti** and **Fabio Lucidi** with their contribution in this volume (*How will artificial intelligence affect our lives? Some reflections in the areas of labor and health*) address this issue with the sagacity and authority of their analysis, considering some already assessable data of this phenomenon. The objective in the background that Pierluigi and Fabio set themselves is in any case to underline, in coherent (and recalled by them) analogy with what I have often indicated in my activity and practice as a researcher, how the role of science and scientists is to establish an intense dialogue with politics and society. Only thanks to this dialogue is it possible to evaluate some challenges in time and identify synergistic resolution paths. The comparison with Pierluigi and Fabio has often engaged us in reflections and opportunities for discussion that have transformed our professional relationship into a true and deep friendship. A friendship that honors and gratifies me.

An extraordinarily important theme in this phase of development of generative AI concerns the risks of replacing humans in a series of activities that are not only low-level but that concern some specific intellectual activities. **Fabio Paglieri** in his beautiful contribution to the volume (*Trust, delegation, and cognitive diminishment: the case of generative AI*) decides to address this problem to try to indicate whether we can decide to trust the new generative Artificial

Intelligence in this sense and how, to what extent it is useful or harmful. By addressing the theme (trusting generative AI) with the tools made available by the theories developed in the T^3 research group, Fabio shows a remarkable capacity and depth of analysis. But he demonstrates, at the same time, how coherent and productive the construction of scientific collaboration and the shared advancement of knowledge is. The piece is, as often happens with Fabio, enjoyable, intelligent and acutely argued. Fabio is also among the editors of this volume: my thanks are double.

The theme of Artificial Intelligence and its developments and impacts is also taken into consideration in the contribution of **Vieri Giuliano Santucci** and **Giovanni Pezzulo** (*Harmonizing Artificial Intelligence, Minds, and Society: A Cognitive Science Perspective*). The focus of their chapter is on the approaches that science can take to address the fundamental problem of the integration of AI with mind and society. It is necessary to involve the various research areas of cognitive science (psychology, neuroscience, cognitive robotics, linguistics, artificial intelligence, anthropology and philosophy) to fully understand and best develop how to harmonize AI with cognition and the collective subjects constituted in the human race. In this way it is possible to focus and clarify more clearly issues such as autonomy, adaptability and interaction. Vieri and Giovanni are scholars with different research experience: Vieri is younger but already brilliant; Giovanni with intuition and scientific value of the highest and recognized value. I have an intense and fruitful collaboration with Giovanni. With Vieri some attempts not yet completely defined (but I do not despair).

Jaime Simao Sichman can be considered a researcher of our research group in ISTC, even though he is a professor at the Brazilian University of São Paulo. In fact, Jaime has not only carried out research as a doctoral student at our institute, but has also taken on the philosophy of the socio-cognitive approach to artificial intelligence, expressing our tradition in the international community of Multi-Agent Systems and Autonomous Agents. In his contribution (**Reputation Interoperability in Multi-Agent Systems**) he summarizes the results of a certain number of research projects developed over the last 20 years with the theme of reputation at its center, one of the central themes of the socio-cognitive approach to Distributed Artificial Intelligence.

Jaime is therefore not only a talented international scientist but a true extension of ISTC research in the world. His only flaw is that he has sympathies for a football team from the capital, unmentionable, second-rate, as well as of disqualified reputation. I have tried several times to redeem him towards the great and "maggical" Rome but all attempts have failed. Nevertheless I love him!

Stefania Costantini and **Andrea Formisano** in their contribution (*Epistemic Logics Programs: a survey of motivations, recent contributions and new developments*) intend to offer both conceptual and practical tools suitable for investigating the existing and new semantics of epistemic logic programs (ELP). According to the authors these tools should be useful to both researchers and users, facilitating a deeper exploration and understanding in this area. The semantics of ELPs is provided through methods that characterize worldviews, which can be different, where each is a set of answer sets, as opposed to a single set of answer sets as in the more traditional Answer Set Programming (ASP). Stefania and Andrea propose a method for the rapid prototyping of a solver (execution engines) that provides the worldviews of a given program for the semantic class under consideration.

Stefania and I share a recent but profound sharing of scientific and science policy issues. My thanks for this contribution are sincere.

The two main basic ingredients of the mental attitude of trust are motivational and epistemic (cognitive) aspects. **Emiliano Lorini** in his contribution (*Should I Trust? On the Epistemic Foundation of Arguments*) focuses his investigation on the latter, analyzing the concept of argument that constitutes the basis for building the theoretical foundations supporting trust. Emiliano is an expert in formal models of interaction between cognitive agents, using logic and game theory. In this case, for the formalization of the concept of argument he uses a semantics of epistemic logic that is based on basic beliefs. In this way he can represent the reasons for believing or not believing something. Emiliano, a bit like Jaime, was born in the ISTC school and it is legitimate to consider him as an extension of the cognitive model that ISTC has exported to the international scientific community.

A controversial issue, also addressed by Cristiano and me in various studies, is how important transparent behavior is to increase interpersonal and social trust. **Aron Szekely** and **Luca Tummolini** (*When Transparency Shapes Trust: A Signalling Approach*) in their paper address this issue with wit and originality, also referring to important trends: from Open Governance initiatives to the Open Science movement. Obviously, the authors rightly argue, the problem shifts to the signals that define transparency and determine trustworthiness. Their suggestion is that integrating Signalling Theory and Trust Theory could be the appropriate way to seriously address the issue. The stimulus is attractive and the arguments brought forward are very convincing. It would be nice to soon start a collaboration with which to investigate this fascinating problem.

The relationship between trust and coordination and the essential role that this relationship can assume for the engineering of intelligent systems and in particular of the so-called socio-technical ones is explored by **Andrea Omicini** and **Alessandro Ricci** (*Coordination and Trust in MAS Towards Intelligent Socio-technical Systems*). The two authors investigate some intersections of these two concepts in particular in multi-agent systems and socio-technical systems. In particular, they focus on the management of interaction in complex intelligent systems and on common issues of dependence, as well as on the substantial differentiation between objective and subjective points of view. Andrea and Alessandro's approach is certainly convincing. Their prediction on a relevant role of the trust/coordination relationship in future AI is fully shareable.

The commitment to research can sometimes be a harbinger of human relationships resulting from joint challenges of knowledge. With Andrea and Alessandro this happened.

Robert Demolombe (*Reasoning about trust in information transmitted by a sequence of agents*) develops an interesting exercise to investigate how an agent can trust a certain piece of information that is transmitted to him and that is the result of a passage between various agents, before reaching him. The conceptual acuity of the analysis is based on some interesting properties that Robert has introduced in the past in the context of information communication and that represent for me an extraordinary heritage of formal and conceptual cleanliness: validity, sincerity, competence, completeness, cooperativeness and vigilance.

The piece should be savored for the formal clarity never detached from reality and indeed explanatory of the conceptual articulation. I have known Robert since I began to organize the Workshop on "Trust in Agent Societies" at the International Conference of Autonomous Agents and Multi-Agent Systems (AAMAS Conference). This contribution of his is a great gift.

The chapter by **Filippo Cantucci** and **Alessandro Sapienza** (*Trust and Cognition in Human and Hybrid Societies*) retraces common works of recent times. With Filippo and Alessandro we have recently faced new research challenges. In particular, important questions have emerged with the new season of Artificial Intelligence, not only generative, which have revealed the substantiality of previous acquisitions but at the same time the need to develop new approaches and directions. The acquisition of new project funding and consequent scientific professionalism (I am thinking of the arrival of **Francesco Stella**), have provided new life on topics whose relevance seems to grow and take on renewed vigor in international trends.

The relationship with Filippo and Alessandro is now long-standing and of proven effectiveness. My gratitude for this gift, made together with Fabio Paglieri and Luca Tummolini, in organizing this volume, is enormous.

Trust is a theme that cuts across different fields of research. The contribution of **Francesca Camilli**, **Tarek el Sehity** and **Raffaella Pocobello** (*Trust the Process: Exploring Trust as a Foundational Element in Open Dialogue*) addresses it by highlighting the essential role it plays in mental health crises. The chapter shows how there is evidence that individuals with psychosis or at risk of it show lower basic trust than healthy people. With consequent inhibition or impairment of social relationships. In this sense, the authors argue that a substantial part of mental health intervention should aim to restore trust in one's own perceptions, feelings and decisions.

The challenge that the authors have been pursuing for some years is to emphasize collaborative therapeutic environments, such as open dialogue, where the restoration of trust is actively promoted through its fundamental principles. In this challenge, their investigations therefore meet with the work we have done on the theory of trust.

Finally, referring to the concluding words of Raffaella, with whom I share an intense esteem and a relationship of friendship and sympathy, I am happy to be able to say that it was a great opportunity for me to bet both on the potential of this line of research and on the ability of the researchers involved in pursuing the objectives they had set themselves. These kinds of successes are the most important reason for dedicating time and effort to practices that might otherwise appear as empty, purely bureaucratic activities.

In her beautiful and generous contribution, **Robin Cohen** (*From Trust Modeling to Trusted AI: Enlightened by the work of Falcone*) addresses a central theme of the reception of Artificial Intelligence systems in our daily lives. In particular, she highlights the need for responsible AI and the ways to achieve it. Her reflection addresses the role of users' mental models, the effort to make the behavior of systems that include it "explainable", especially in the presence of generative AI and the success of LLMs within the field of study. The chapter then addresses her and her group's work over the last 20 years dedicated to the themes of understanding the concept of trust and the support of distinct solutions in different contexts of use. Then an interesting assessment is proposed, the subject of her most recent research, on the type of future that lies ahead, on the possible social benefits, but also on the risks and due concerns regarding the ways in which these systems could be misused. To Robin, an esteemed scientist, with whom we shared part of the rich and intense experience of the international workshops on "Trust in Agent Societies", I feel I owe a great deal of thanks and a hope that we can find a way to collaborate again.

This volume, as I have already said and perhaps I will repeat in the following, is a surprising collection, a fantastic array of emotions and unexpected (and I don't know how well-deserved) recognitions. And then there is this chapter that for me has an even more special value than the others. It is the one written by **Cristiano Castelfranchi** (T^3: for a Multidimensional, Integrated and Dynamic Theory of Trust). It is an in-depth description full of illuminating ideas about our research group, the role it played and the contents that involved us. Many reflections would seem already acquired, but instead new perspectives suddenly open up and some of the conclusions are overturned and re-proposed in new and original ways. It is the figure of genius and culture that Cristiano spreads in his mental exercise.

The power of evolution has allowed humans to build their social paths based on collaboration and the culture of transferring knowledge and working methods. My great fortune is to have met Cristiano in my professional life and to have been able to gather his teachings as a maestro and scientific father. Not only that, Cristiano is also a deep friend, he is that brother with whom I can share individual and social life aspirations. And the intensity and duration of our relationship allows me to evaluate today an extraordinary legacy that I feel has formed and tempered me and whose gratitude is difficult to measure.

6 Socio-political contributions

This series of contributions is in some ways even more surprising than that relating to scientific activity strictly understood. Not only because it is in fact an anomaly in the canonical tradition of this type of volume, but also because having managed to find some of the characters who then made themselves available, was a skill on the part of Alessandro, Filippo, Fabio and Luca as true masters of the genre!

It is a testimony that gratifies and rewards the time and effort that I have constantly spent in the so-called political-social activities in favor of research. In an attempt to promote and enhance this human activity (research and science) so clamorously mistreated in our country, yet so fundamental for the advancement of the variously understood development of human beings and their social constructs.

Lucio Bianco's letter recalls between its lines a historical period of the battle for the autonomy of the CNR, undertaken by the national scientific community also thanks to an initiative, that of the Research Observatory inspired by me and then founded together with scientists of the caliber of Carlo Bernardini, Tullio De Mauro, Margherita Hack, Francesco Lenci, Giorgio Parisi, Giulio Peruzzi, Roberto Battiston, Franco Pacini, Marcello Buiatti and many others who will be recalled elsewhere in this volume. The battle of the Observatory and Lucio to defend the Institution from a rule wanted by the then Minister Moratti, was resolved positively.

That battle was a basis for us to meet and find ourselves systematically in agreement in multiple initiatives. Today we are bound by a feeling and a sharing of perspectives on research policies and the future of the country. And a great comforting affection and esteem.

Gaetano Manfredi's short but intense writing retraces a theme on which we decided to launch a top-level national initiative. It was the short period in which I was his Advisor while he was Minister of Research and University. It was a matter of seriously examining and identifying ways to mend and promote the relationship between science and society. This has always been a controversial and extremely complicated relationship but, never as in the twentieth century, have the relationships between science and the effects it derives from on society (its development, its organization, health, the environment, the distribution of wealth, ethics, etc.) had such relevance and impact.

A first issue concerns the fact that scientific progress does not always proceed in a linear way, at times it can take ambivalent paths until the prevailing hypotheses mature. In fact, doubt and uncertainty are an integral part and active engine of scientific research and the successful theories and models that science is able to offer are the result of long, tiring processes fraught with obstacles. However, they are never definitive.

The opposite of pseudo-sciences, which do not rely on the scientific method but rather on inscrutable categories and do not define objective methodological paths. Only based on superstitions or completely unreliable and imaginative evaluations: children of instrumental speculations for manipulative purposes.

The objective of that commitment with Gaetano was to identify ways to sensitize society with respect to the value of science, to enhance the role of scientists, to reconstruct a model of research policies capable of reconnecting science and citizens. The belief was that the form of management and development that science has been able to give itself, can also represent a model for the forms of institutional politics and to advance the level of awareness and democratic development.

The few months of collaboration with Gaetano have nevertheless convinced me of the extraordinary quality of the man and the politician. I was able to appreciate the kindness of soul, the deep understanding of the problems (the scientist's mindset bears full witness to this) and the lucid ability to resolve. My wish for Italy, I don't know if for him, is that he will be employed at the highest institutional levels in the future.

Francesco Lenci's chapter is the testimony of a brother who loves you. The same love that I feel for him. Francesco was together with a few others, Carlo Bernardini and Giulio Peruzzi, in the historical and initial nucleus of the Observatory on Research. In his moving piece he traces part of its path. He reconstructs in summary an extraordinary period. The one in which a small group of scientists from the

CNR (a research body whose existence the vast majority of the Italian population is completely unaware of) decides to involve the most important scientific and intellectual minds in the country to denounce a small reform that, without those hotheaded idealists, would have passed completely indifferent. The risks to the autonomy of research would not have been taken seriously even by the finest jurists in the sector. And instead, by invoking constitutional principles and risks for the civil and social development of a great democracy (real risks, not faked ones constructed for this purpose), mobilizing on pressing slogans and projecting disturbing scenarios on the control of science and research, that is, on the most advanced directions and developments of a society, those idealists raised a national question as had never been proposed before.

That plan was stopped. The CNR maintained its characteristic of a non-instrumental body and a widespread awareness of the role of scientific communities had its significant confirmation. His beautiful account is there to demonstrate that only an adequate interpretation of reality and of the profound, ideal and shared purposes can allow, if pursued with conviction and determination, the affirmation of principles and values that are considered correct.

But even more moving is the bond that that season allowed to be established within that group. The human and emotional sensitivity that binds us will remain a gift that we will preserve forever!

Daniele Archibugi in his beautiful piece introduces a relevant problem, that of scientific communities and their capacity for governance, recognizability and acceptability with respect to society. This meta-level problem offers us the opportunity to analyze the strengths and weaknesses that a community of scientists develops and makes us reflect on the reciprocal influences of this meta-level with the development and effectiveness of actual scientific activity.

This ordering role that scientific communities should give themselves is, as Daniele says, often delegated and misunderstood. And, when it happens, it is thanks to the good will of someone that this ordering is carried out, rather than a constitutive and determined process.

There is a real difficulty in building, on the part of this community,

581

a common horizon and in identifying those superior objectives, with respect to pure research work, which alone are able to truly facilitate the scientific activities of all. It is right that there are different, multiple grammars at the level of scientific study, so that the various areas of research can be aggregated. It is not good, however, that there is similarly no ability to understand the common code of the overall mission of science. Here it is difficult.

With Daniele we have established an intense friendship right downstream of experiences between political scientists. The contribution, small but stubborn, that Daniele recognizes in me and for which I am grateful, I believe I have given for what remains a very uphill path. Trying to be scientists but also intellectuals of science.

The article, intense and profound, by **Luigi Nicolais** is a true monument to critical thinking, that is, to the method of reasoning and evaluating the object of study. It does not matter what this object is, whether it is a question of dealing with humanistic or scientific questions, of hard science or soft science. And then, in modern science, increasingly interdisciplinary and interconnected, the distinction becomes thinner beyond the limit. Critical thinking, on the other hand, is essential, it assists us and guides us in the sea magnum of knowledge and its consequences. And it is the anchor that can only save us from the simplifying and reductive fury of the time we face. For example, Gino argues: how can we do without it in relation to Artificial Intelligence? Which can "both improve and distort our understanding of the world. Artificial intelligence can perform tasks with precision and speed, but it does not have the nuanced understanding and flexibility of human thought. Therefore, it is essential to develop critical thinking skills to evaluate the information generated by artificial intelligence systems and ensure that artificial intelligence is used responsibly and ethically."

Although I have had several Presidents of the National Research Council during my scientific career, I consider Gino Nicolais "my President", not only because I found in him the full sharing of my project, as a candidate for the direction, for the development of the Institute of Cognitive Sciences and Technologies. But because it is thanks to a full and active collaboration with him that I felt the CNR as the appropriate home to develop research, to face adequate discussions and comparisons on the future of science and the impacts that it can have on society. Gino's intellectual and at the same time practical abilities make him one of the most lucid minds of our scientific culture and political practice.

A figure of a scientist of the highest profile who combines great sensitivity on the political orientation for research themes is **Lucia Votano**. Lucia and I have known each other for a long time thanks to Pietro Greco, a mutual friend and a visionary intellectual on the issues of science in relations with society, who unfortunately passed away prematurely. A deep feeling was immediately established with her, an immediate coincidence of points of view and a common need to remedy the inadequacies of the governance of science and research in the country. It can happen that we hear from each other rarely and immediately reconstruct shared analyses and evaluations. This tribute of hers honors me.

There was a period, not a short one, in which the directors of the CNR Institutes, the most direct sensitivity of the research network of the Institution, found cohesion and the ability to express themselves together: without chasing too much the individual requests of their institutes, even legitimate in critical situations, but above all by making a system synthesis. In that period, I happened to share with several of these directors a path full of problems but also full of potential ambitions and hopes. **Corrado Bonifazi**, then director of the Institute of Populations (IRPPS), was one of the closest and most active of these. I remember long discussions before or after the various meetings. A reassuring lucidity and objective vision always characterized his opinions. I believe that friendships can have the most varied origins and paths, with Corrado we found ourselves in this context and when we happen to hear from each other again, that active thought that characterized our exchanges is immediately reactivated.

My experience as Advisor to Minister Fabio Mussi during the Prodi Government of 2006-2008 has marked my professional path like

few other occasions in my life. This is certainly not the place to talk
about it. The homage that **Nando Dalla Chiesa** has dedicated to
me in this volume is, however, an opportunity to report my impressions on this specific relationship. I must admit that Nando's account
of our meeting moved me like few others. On the one hand because it
was unexpected and on the other for the profound meaning that that
experience gives me. In the programmatic and comparison meetings
that we had on those occasions I perceived in his attitudes seriousness and attention to the contents and questions of primary value.
An open curiosity, typical of the scholar, towards the arguments of
the interlocutors. Never an instrumental approach or a preconceived
position of positioning. A strongly ethical attitude in the political
approach. These were the reasons why I continued, in the following
years, even though we did not have the opportunity to meet and discuss, to involve him on themes and questions that I thought could
stimulate his interest. Unfortunately, it happened only a few times.
It will be my commitment to try again in the future.

The story of my meeting with **Pierpaolo Campostrini** is narrated by Pierpaolo in the beautiful piece he dedicates to me in this
volume. This was one of the occasions to test my two deepest souls:
that of the man of science and that of the political idealist. It was
a matter of comparing clearly alternative hypotheses: one concretely
decisive and well-founded, the other (in reality they were various and
alternative) more random and vague in their effectiveness. Both, however, were accompanied by an ideological suggestion (different from
the ideal): the more concrete and rational (MOSE) denigrated as
efficiency-oriented and technocratic, anti-environmentalist and essentially presumably ineffective; the other, alternative and environmental
protection, evidently less protective against the tides but considered
ecological safeguard. In short, politically, the moderates were mostly
on the side of MOSE, while part of the left was against it (led by the
Mayor of Venice, Cacciari). Delegated by Minister Mussi, I met the
two sides. After studying the documents that both sides proposed to
me and after clarifying and analyzing them with them, I had the clear
feeling that the opposition was between a search for a very concrete

and well-founded solution and an instrumental ideological position that was completely inappropriate for the problem to be addressed. The latter was supported by those who were a valuable political reference for me. However, I had no doubts about how I should relate to the Minister.

The fact that Pierpaolo comforts me today with this writing of his is a source of great pride for me and I am very grateful for this recognition.

Massimo Cocco is a brilliant and important scientist of the National Institute of Geophysics and Volcanology, he too, like many of the authors of this volume and many others not present in this publication, represents that active part of the scientific community that works with conviction to satisfy what Massimo recalls: "The integrity of science and the role of scientists are strengthened through their image in society; the more that image represents the real benefit that science is able to produce for society, the more society will be able to attribute that fundamental function to it". In this sense, Massimo is also a very fine intellectual. Often with him we have started on paths of analysis of the role of science and its dynamics with politics. I have rarely found insights and enlightening ideas like those that he is able to produce on these issues and, as with others already mentioned in this volume, every time I happen to compare ourselves I feel the great satisfaction of testing myself with deep thought. Thank you Massimo!

Maria Chiara Carrozza is currently the President of the CNR. With Maria Chiara, our paths have crossed on several occasions: we found ourselves collaborating when she was responsible for the PD of the University, in various initiatives to mobilize for research and then when she became Minister in the Letta Government, I found myself, without being warned, to be part, upon her nomination, of the Search Committee for the selection of candidates for the Presidency or as members of the Board of Directors of the Research Institutes supervised by the MUR. In short, the mutual esteem has been intense and prolonged. Since she has been President of the CNR in reality, perhaps also due to a not complete convergence on the ways in which the Draghi Government started a reform of the CNR, our interactions have been more discreet. Therefore, finding one of her writings in this volume gave me enormous pleasure and also partially surprised me. Also because, the things that Maria Chiara writes are of great generosity towards my path, both strictly scientific, as well as attention and commitment to research policies in the various and complex positions that I have assumed. For example, on the issue of the autonomy of research and my sensitivity to the topic, I find your appreciation very gratifying. And the intellectual honesty you demonstrate in this regard does you credit. At the CNR you worked with great commitment and seriousness. I am sure that your path will leave an important mark.

Maurizio Franzini is a leading academic and intellectual. In his dedication he develops an analysis, focused with the sharp and incisive gaze of the economist, on a topic that he knows has fascinated me a lot over the years: trust. He focuses, in particular, on an aspect of great social relevance that trust plays, that of a fundamental stimulator of cooperation. The ability to transform behaviors (of others and of one's own) through an attitude of considering one's interlocutor capable and willing to carry out the task that is delegated to him. In order for this transformation to have an effect, it would seem necessary that the interlocutor must become aware of the delegation, or that it be manifest to him. It is through this explicit stimulus that the mechanism of reorganization of cognitive ingredients (beliefs, intentions, goals, etc.) is triggered, which, as demonstrated in the observation of sociality, the daughter of adaptive evolution, allows for the increase of the collaborative potential of humans in interaction.

The construction of organized sociality has in fact, gradually, made some of these mental mechanisms more automatic, so it is possible that in many cases, the availability and commitment to improving one's performance (and therefore one's reliability) are promoted even outside the obvious presence of a delegating person and an interlocutor to whom one is directly accountable. There are third-party interlocutors (above the parties) or even introjected cognitive devices to whom one is accountable (principles, values, ethics).

All this is particularly interesting and perhaps one day we will be able to discuss it together, in a more scientifically in-depth way with Maurizio, whom I thank for having offered me this affectionate and intense thought.

Among the traveling companions with whom we found ourselves reasoning and thinking about how we could contribute to get Italy out of a very difficult economic season and civil and social climate is **Daniela Palma**. Daniela is a brilliant economist at ENEA. Her analysis is very beautiful and moving in some respects. Not only because it clearly recalls traits of that path that, reread in the light of the present, demonstrate the foresight and anticipatory capacity of our theses, but also because it evokes that fusion of personalities and passions that had a decisive role in the maturation that those analyses assumed, never biased or extreme in a demagogic way but always subjected to a constructive and collaborative scrutiny. And it is precisely from this that the strength that they managed to express comes. The loss of some of those minds and passions is a pain that remains deeply present (I think of Carlo Bernardini, Pietro Greco, Tullio De Mauro, Franco Pacini, Marcello Buiatti, and many others) but the road built together - and this intervention of his is an important help in defining the path - a comfort that accompanies us. To Daniela I dedicate a big and affectionate hug.

There are political-professional connections that resist time even if they are not cultivated in an appropriate way. The one with **Alberto Silvani** is one of these. We have known each other for decades, young researchers focused on science but also on how it should be governed to make it effective, understandable and valuable. Scholars of different disciplines but ready to confront each other in the socio-political sphere of context. His contribution amazed me precisely because, despite the premise just made, it reconstructs a dialogue between us of intense substance confirming the quality of analysis that Alberto has always demonstrated. The reflection that it prompts is an illumination on the extraordinary need to identify tools and forms capable of allowing us to continuously dialogue and seek ways and paths of understanding even outside the ordinary custom.

This gift of his comforts me on the need to disseminate reflective and constructive traces, in the belief that by multiplying disseminations and disseminators we continue to at least keep awake that part of cognition that makes individuals and society evolve. I would therefore like to thank Alberto beyond any formal rhetoric for this gift that I consider extraordinarily important.

Settimo Termini is another traveling companion in the journey undertaken to improve our world (that of research and knowledge) which is one of the fundamental fulcrums of modern societies. Settimo is a leading intellectual and scientist. He was director of the Institute of Cybernetics of the CNR and full professor in Palermo. His ability to reconstruct a period so important not only for us but for the fate of our country has given me that strong sense that social events take on when they go beyond the news to focus on the profound mechanisms of change.

It doesn't matter what the final result is (or maybe it does but let's give ourselves an alibi), what matters a lot is the seriousness and quality of the dialogues, proposals, initiatives that together with comrades and friends as passionate as we are, we have been able to produce. A collective event that transcends the few or many who give birth to it. It becomes symbolically relevant in the underlying intentions that moved it. The difference between illocution and perlocution. The intention of the language and its real effects. But without the intention, the desired effects will never be achieved. And even when the effects are divergent from the intentions, they will have been affirmed even with distorted results (and this will still involve an analysis of the relationship between causes and effects and hopefully a greater understanding of the connection between the two). I say all this to thank Settimo. To show solidarity with a true friend. Where friendship is colored with connotations. A friend of science and politics, a friend with whom, even if you haven't heard from each other for a while, it is possible to immediately re-establish dialogues and connections without hesitation.

As described by **Marco Pagani** in his tribute in this volume, our friendship spans institutional and private spheres. With Marco, as his tribute testifies, we have had experiences of battles for the renewal of the CNR and at the same time to discover the best wine to pair with the extraordinary Swedish salmon that he brought from Stockholm. His presence is inconstant but affectionate and sociable. The only flaw is his passion for a team from the Hunchback. The excuse of his Turin origins does not hold, given the ample compensation of the overflowing culture of Romanity.

Oliviero Stock was my first scientific interlocutor of great value. I had collaborated with various colleagues, but Oliviero was the leap in quality and with him (and the island chart-parsing) I discovered that scientific passion would have prevailed in my life. Our relationships remained rare although deep and it also happened that we collaborated again, but it no longer happened that we wrote science together. And I am very sorry about that!

But he is not right when he claims that my time passes slowly, the arrival of retirement came very quickly, completely disconcerting me. I am enormously proud of the comparison he makes between me and Raffaele Misiti: after all, my soul is precisely in that connection and dialogue between science and politics that Lello represented so well (as reported by the memories of his students).

And I am moved by the sentence that Oliviero writes: "As regards public communication on research, I have often had the opportunity to read some article written by Rino, and my comment has usually been: let's hope the political decision makers read it carefully".

And then the exchange of verses after I informed him of my crossing of the Strait.

I sent him mine:

> *Why cross the Strait?*
> *Why sink in those 3.14 km of swimming?*
> *It is certainly not to go from Sicily to Calabria or vice versa,*
> *It is not a trivial dive at Punta Faro to emerge on the beach of Cannitello*
>
> *It can perhaps serve to measure the passing of time*
> *To weigh in our arms those 66 revolutions of the Earth around the Sun,*
> *To challenge the two titanium imprints that are permanently alongside us*

589

To want to feel the heartbeat put to the test by the hardness of the waves and currents
To perceive what "athletic" youth has left us beyond regret

It's true: it is about challenging the most challenging stretch of sea in the Mediterranean,
where stormy currents and winds are frequent and bursting
where the Ionian and Tyrrhenian meet/clash producing unique hydrodynamic phenomena
where Scylla and Charybdis, furious nymphs permanently in dispute, reveal the legend
where blinded Polyphemus still unleashes Neptune's wrath against Ulysses
where, like in no other stretch of sea, you can to encounter such dangerous "vortices".
where earthquakes and tsunamis have destroyed the cities of Messina and Reggio over the centuries, several times,
where the catastrophe of 1908 was triggered by the ancient fault that is still present in its seabed
where the geological plates of Africa and Eurasia have decided to converge here in tectonic movements
where, consequently, deep fractures accumulate energy to be released in magnitudinal upheavals
where rapists of humanitarian bridges narcissistically intend to profit from the false, useless and harmful myth of overcoming this natural gap...

But crossing the Strait is also and above all a journey for and into one's own soul,
In the need for knowledge from which it continually draws: anxiety of departure, unknowns of the journey, expectations of the goals
This short but perilous and unknown cycle, can appear to us as a metaphor for life itself,
a journey into our deepest essence,
stroke after stroke in the icy and increasingly dark waters that cover us, it is an extreme attempt to discover the true reasons for our feelings
to face the painful absences that have struck us and to investigate in depth the memory that remains
To feel them close in that brief journey
To grasp a sort of presence in the waters that flow over us,
To let ourselves be carried by their affectionate push: of a mother, of a father, of a brother, of relatives and dear friends.

Voids of the soul that the sea fills in its uncompromising pervasiveness and "living" conscience.

And finally there is the looming presence of a dispersed humanity
Of justice, solidarity and rights turned to straw
Of waters where the need for liberation of many humans has tragically
ended
Of merciless hands never stretched out
Of screams stifled in the deafening silence of our indifference.

So, to seek a last desperate illusion of getting out of these waves and
finding the new world,
or at least, in that 1 hour, 8 minutes and 8 seconds of separation
from the routine of reality, to imagine that a new world could still
exist...

And he responded with those brilliant and ironic verses. Thanks Oliviero!

The piece by **Paolo De Nardis**, a leading sociologist and great intellectual, presents in a synthetic, erudite and effective way some themes that we have long debated over time and on various occasions. For example, the age-old question of the various types of research and the most appropriate places in which to carry them out. Although the nature of research increasingly mixes the different types, often confusing them, we agreed with Paolo that "when you have to evaluate scientific research, you are faced with a complex universe of structural, individual and epistemological variables that make it difficult not to consider social, cognitive and organizational factors as intertwined." His piece is a mix between the exercise of cultured and profound reflection on research policy issues and an affectionate tribute from a friend whom I have esteemed for time immemorial and whom I love.

It is difficult to imagine who better than **Salvatore Capasso**, current Director of the Department of Human and Social Sciences of the CNR, the department to which ISTC, the Institute I currently direct, belongs, could provide an accurate and in-depth analysis of the relationship between the human and social sciences and the hard sciences. Capasso is a professor of economics and is clear about the impact that human and social factors are able to determine, when well assessed and interpreted, on the technological and productive development, but also on the values and ethics of a society. And moreover,

591

in directing the Department, Salvatore has the need to keep these potentialities in mind, to adhere to the extraordinary inter and multidisciplinary tradition that the CNR has cultivated since its origins and that is an integral part of its DNA. Even more so since modern science has exalted these factors of mixing and mutual contamination between the hard sciences and the human and social sciences. Salvatore's convincing and explanatory chapter is a beautiful demonstration of the strength of the largest research institution, when it verifies tradition and selects the right scientists for its future. I am truly grateful to Salvatore for the gift he has given me in this volume.

It is very difficult to evaluate one's own work, especially when it involves tracks that direct complex and articulated paths. And the possible feedback is the result of various components that are not necessarily all assessable and under control. In this volume, **Francesco Profumo**, a great scientific and intellectual personality of our country, wanted to provide an account and an evaluation of the work I carried out as an advisor to the then Minister Mussi. His generosity of judgment is a great comfort to me. That period was an overwhelming phase in my professional life: I suddenly found myself -Mussi did not give me many possibilities of choice- facing the problems of research in our country. And I immersed myself in this dimension of "high governance" with great passion and with the strong determination of having to represent the instances that for years we had - with other scientists - represented and invoked.

The fact that Francesco has the experience and critical capacity of a Minister as well as a Rector, President of the CNR and Professor of the highest prestige, makes me extremely proud when reading the judgments he offers. Since I also consider him a friend, I am led to believe that his affectionate benevolence towards me has played an important role in these positive evaluations. In any case, I thank him by expressing my most heartfelt gratitude.